METAL IONS IN BIOLOGICAL SYSTEMS

VOLUME 43

Biogeochemical Cycles of Elements

METAL IONS IN BIOLOGICAL SYSTEMS

Edited by

Astrid Sigel
Helmut Sigel

Department of Chemistry
Inorganic Chemistry
University of Basel
CH-4056 Basel, Switzerland

and **Roland K. O. Sigel**

Institute of Inorganic Chemistry
University of Zürich
CH-8057 Zürich, Switzerland

VOLUME 43

Biogeochemical Cycles of Elements

CRC Press
Taylor & Francis Group
Boca Raton London New York

CRC Press is an imprint of the
Taylor & Francis Group, an **informa** business
A TAYLOR & FRANCIS BOOK

The figure on the cover is Figure 2 of Chapter 9 by N.J. O'Driscoll, A. Rencz, and D.R.S. Lean.

CRC Press
Taylor & Francis Group
6000 Broken Sound Parkway NW, Suite 300
Boca Raton, FL 33487-2742

First issued in paperback 2019

© 2005 by Taylor Francis Group, LLC
CRC Press is an imprint of Taylor & Francis Group, an Informa business

No claim to original U.S. Government works

ISBN-13: 978-0-8493-3807-6 (hbk)
ISBN-13: 978-0-367-39327-4 (pbk)

Library of Congress Card Number 2004061452

Library of Congress Cataloging-in-Publication Data

Biogeochemical cycles of elements / editors, Astrid Sigel, Helmut Sigel, Roland K.O. Sigel.
    p. cm. -- (Metal ions in biological systems ; v. 43)
  Includes bibliographical references.
  ISBN 0-8493-3807-7 (alk. paper)
  1. Biogeochemical cycles. 2. Nonmetals. I. Sigel, Astrid. II. Sigel, Helmut. III. Sigel
Roland K. O. IV. Series.

QP532.M47 vol. 43
[QH344]
572'.51--dc22
[577'.14]   2004061452

Visit the Taylor & Francis Web site at
http://www.taylorandfrancis.com

and the CRC Press Web site at
http://www.crcpress.com

Preface to the Series

Recently, the importance of metal ions to the vital functions of living organisms, hence their health and well-being, has become increasingly apparent. As a result, the long-neglected field of "bioinorganic chemistry" is now developing at a rapid pace. The research centers on the synthesis, stability, formation, structure, and reactivity of biological metal ion-containing compounds of low and high molecular weight. The metabolism and transport of metal ions and their complexes are being studied, and new models for complicated natural structures and processes are being devised and tested. The focal point of our attention is the connection between the chemistry of metal ions and their role for life.

No doubt, we are only at the brink of this process. Thus, it is with the intention of linking coordination chemistry and biochemistry in their widest sense that the *Metal Ions in Biological Systems* series reflects the growing field of "bioinorganic chemistry". We hope, also, that this series will help to break down the barriers between the historically separate spheres of chemistry, biochemistry, biology, medicine, and physics, with the expectation that a good deal of future outstanding discoveries will be made in the interdisciplinary areas of science.

Should this series prove a stimulus for new activities in this fascinating "field", it would serve its purpose and would be a satisfactory result for the efforts spent by the authors.

Fall 1973 *Helmut Sigel*

Preface to Volume 43

This volume, devoted to *Biogeochemical Cycles of Elements*, opens with an introductory chapter indicating the role of transition elements for the biogeochemical cycles of the bulk or main group non-metal elements, the interactions between the cycles of different elements, and the change that the cycles experienced as a result of the evolution of life. All these subjects are further substantiated in the following five chapters in which the biogeochemistry of hydrogen, oxygen, nitrogen, sulfur, and phosphorus are individually dealt with. For example, it is revealing to learn about the important role that hydrogen has played in Earth's geochemistry and biology since the earliest stages of this planet's history. In fact, this role still persists today because the reversible oxidation of molecular hydrogen, $H_2 \rightleftharpoons 2H^+ + 2e^-$, is among the most widely utilized reactions in the microbial world. Interestingly, hydrogenases, the enzymes which catalyze this transformation, are present in all three phylogenetic domains of life.

Iron, phytoplankton growth, and the carbon cycle are in the focus of Chapter 7. Iron is the most abundant transition metal ion in most terrestrial organisms being involved via iron-bearing molecules in photosynthesis, respiratory electron transport, nitrate and nitrite reduction, N_2 fixation, sulfate reduction, and so forth. However, iron is vanishingly scarce in seawater and therefore, it is a limiting nutrient for marine organisms, which need this metal ion to support their growth. Low iron concentrations thus limit primary oceanic production, that is, the transfer of carbon dioxide into organic carbon by photosynthetic plankton (phytoplankton), and this leaves its mark on the global carbon cycle, reflecting a very complex relationship.

Cadmium (Chapter 8) is one of the all-star pollutants, along with a few other elements such as mercury (Chapter 9), lead (Chapter 10), and arsenic (see Volume 44, below), and is of great toxicity to most living systems, including man. However, recently it has become apparent that cadmium is also a nutrient for marine phytoplankton, playing a role similar to other micronutrients such as zinc and cobalt, and this has dramatically spurred the interest inherent in the biogeochemistry of this element.

Mercury and lead serve no known biological function, but both are useful metals and employed by mankind since ancient times. Both are also toxic environmental contaminants and thus of prime concern for several governmental agencies. All this has led to intense research on the biogeochemistry of mercury and lead, including their speciation in the environment, their atmospheric transport, and the processes affecting their fate; these topics and more are summarized in Chapters 9 and 10.

Closely related to the present book is the forthcoming Volume 44, *Biogeochemistry, Availability, and Transport of Metals in the Environment.*

Astrid Sigel
Helmut Sigel
Roland K. O. Sigel

Contents of Volume 43

Preface to the Series . . . . *iii*

Preface to Volume 43 . . . . *v*

Contributors . . . . *xi*

Contents of Previous Volumes . . . . *xiii*

   Handbook on Toxicity of Inorganic Compounds . . . . *xlv*

   Handbook on Metals in Clinical and Analytical Chemistry . . . . *xlv*

   Handbook on Metalloproteins . . . . *xlv*

Chapter 1
The Biogeochemical Cycles of the Elements and the
Evolution of Life *1*
Peter M. H. Kroneck
1. Introduction *1*
2. Modern Earth: Cycling of the Biological Elements *2*
3. Evolution of Life *3*
4. Outlook *6*
References *6*

Chapter 2
Biogeochemistry of Dihydrogen (H_2) *9*
Tori M. Hoehler
1. Introduction *10*
2. H_2 from the Planetary Matrix *13*
3. H_2 Cycling in Anaerobic Ecosystems *17*
4. H_2 Cycling in Phototrophic Ecosystems *32*
5. Summary *39*

Acknowledgments *41*
References *41*

Chapter 3
Dioxygen over Geological Time **49**
Norman H. Sleep
1. Introduction *50*
2. Long-Term Crustal Reservoirs and Cycles *51*
3. Modern and Ancient Biological Redox Cycles *53*
4. Mantle Cycle *59*
5. Carbon Burial *66*
6. Conclusion: Oxygen Build-Up Over Geological Time *68*
Acknowledgments *71*
Abbreviations *71*
References *71*

Chapter 4
The Nitrogen Cycle: Its Biology **75**
Marc Rudolf and Peter M. H. Kroneck
1. Introduction *76*
2. Nitrogen Fixation *78*
3. Respiratory Processes: Energy Conservation with
   Inorganic Nitrogen Compounds *85*
4. Assimilatory Processes: Building the Essential
   Molecules of Life *95*
5. Outlook *96*
Acknowledgments *97*
Abbreviations *97*
References *98*

Chapter 5
The Biological Cycle of Sulfur **105**
Oliver Klimmek
1. Sulfur in Biology *106*
2. Chemistry of Elemental Sulfur *107*
3. Polysulfide Sulfur as an Intermediate in
   Sulfur Respiration *108*
4. Polysulfide Sulfur Respiration of Bacteria *108*
5. Sulfur Respiration of Archaea *123*
6. Polysulfide Sulfur Transferases *124*
7. Conclusions and Outlook *127*
Acknowledgment *127*
Abbreviations *127*
References *128*

Chapter 6

Biological Cycling of Phosphorus *131*
Bernhard Schink
1. Introduction: Chemistry of Phosphorus Minerals *131*
2. Phosphates in Biology *133*
3. Metabolism of Phosphorus Compounds with
   C–P Linkages *138*
4. Metabolism of Reduced Inorganic
   Phosphorus Compounds *143*
5. Formation of Phosphine *145*
6. General Conclusions *147*
Acknowledgments *148*
Abbreviations *148*
References *148*

Chapter 7

Iron, Phytoplankton Growth, and the Carbon Cycle *153*
Joseph H. Street and Adina Paytan
1. Iron, an Essential Nutrient for Marine Organisms *154*
2. Iron Chemistry in Seawater *158*
3. Iron Distribution and Cycling in the Ocean *160*
4. Iron Limitation of Marine Primary Productivity and Control
   on Ecosystem Structure *170*
5. The Role of Iron in Regulating Atmospheric CO_2 .... *180*
6. Summary and Conclusions *183*
Acknowledgments *184*
Abbreviations *184*
References *185*

Chapter 8

The Biogeochemistry of Cadmium *195*
François M. M. Morel and Elizabeth G. Malcolm
1. Introduction *196*
2. Concentrations: Sources and Sinks *197*
3. Chemical Speciation *201*
4. Biological Effects *204*
5. The Biogeochemistry of Cadmium as an Algal Nutrient
   in the Sea *209*
6. Cadmium as a Paleotracer *215*
7. Envoi *216*
Acknowledgments *217*
Abbreviations *217*
References *217*

Chapter 9
The Biogeochemistry and Fate of Mercury
in the Environment *221*
Nelson J. O'Driscoll, Andrew Rencz, and David R. S. Lean
1. Introduction 222
2. Mercury Speciation in the Environment 222
3. Processes Affecting Atmospheric Transport and Fate 223
4. Processes Affecting Aquatic Transport and Fate 226
5. Effects of a Changing Landscape on Mercury Fate 229
6. Big Dam West Lake Mercury Mass Balance 230
7. General Conclusions 233
Acknowledgments 234
Abbreviations 234
References 235

Chapter 10
Biogeochemistry and Cycling of Lead *239*
William Shotyk and Gaël Le Roux
1. Introduction 240
2. Chemistry of Lead and Behavior in the Environment 242
3. Lead Isotopes and Their Measurement 245
4. Ancient and Modern Uses of Lead 247
5. Emissions of Lead to the Environment 250
6. Inputs and Fate of Anthropogenic Lead in the Biosphere 252
7. Temporal Trends in Atmospheric Lead Deposition 255
8. Environmental Lead Exposure and Human Health 263
9. Summary and Conclusions 267
Acknowledgments 267
Abbreviations 268
References 268

Subject Index *277*

Contributors

Numbers in parentheses indicate the pages on which the authors' contributions begin.

Tori M. Hoehler *Exobiology Branch, NASA Ames Research Center, Mail Stop 239-4, Moffett Field, CA 94035-1000, USA* (Fax: +1-650-604-1088, tori.m.hoehler@nasa.gov) (9)

Oliver Klimmek *Institut für Mikrobiologie, Johann Wolfgang Goethe-Universität, Marie-Curie-Strasse 9, D-60439 Frankfurt am Main, Germany* (Fax: +49-69-79829527, klimmek@em.uni-frankfurt.de) (105)

Peter M. H. Kroneck *Mathematisch-Naturwissenschaftliche Sektion, Fachbereich Biologie, Universität Konstanz, Postfach M665, D-78457 Konstanz, Germany* (peter.kroneck@uni-konstanz.de) (1, 75)

David R. S. Lean *Biology Department, Faculty of Science, University of Ottawa, P. O. Box 450, Stn. A., Ottawa, Ontario, K1N 6N5, Canada* (221)

Gaël Le Roux *Institute of Environmental Geochemistry, University of Heidelberg, Im Neuenheimer Feld 236, D-69120 Heidelberg, Germany* (gleroux@ugc.uni-heidelberg.de) (239)

Elizabeth G. Malcolm *Department of Geosciences, Princeton University, 153 Guyot Hall, Princeton, NJ 08544, USA* (emalcolm@princeton.edu) (195)

François M. M. Morel *Department of Geosciences, Princeton University, 153 Guyot Hall, Princeton, NJ 08544, USA* (morel@princeton.edu) (195)

Nelson J. O'Driscoll *Biology Department, Faculty of Science, University of Ottawa, P. O. Box 450, Stn. A., Ottawa, Ontario, K1N 6N5, Canada* (nodrisco@science.uottawa.ca) (221)

Adina Paytan *Department of Geological and Environmental Sciences, Braun Hall (Bldg 320), Room 207, Stanford University, Stanford, CA 94305-2115, USA* (Fax: +1-650-725-0979, apaytan@pangea.stanford.edu) (153)

Andrew Rencz *Geological Survey of Canada, 601 Booth Street, Ottawa, Ontario, K1A 0E8, Canada* (rencz@nrcan.gc.ca) (221)

Marc Rudolf *Fachbereich Biologie, Universität Konstanz, Postfach M665, D-78457 Konstanz, Germany* (75)

Bernhard Schink *Fachbereich Biologie, Universität Konstanz, Postfach 5560 <M654>, Universitätsstrasse 10, D-78457 Konstanz, Germany* (Fax: +49-7531-884047, bernhard.schink@uni-konstanz.de) (131)

William Shotyk *Institute of Environmental Geochemistry, University of Heidelberg, Im Neuenheimer Feld 236, D-69120 Heidelberg, Germany* (Fax: +49-6221-545228, shotyk@ugc.uni-heidelberg.de) (239)

Norman H. Sleep *Department of Geophysics, Stanford University, Stanford, CA 94305, USA* (Fax: +1-650-7257344, norm@pangea.stanford.edu) (49)

Joseph H. Street *Department of Geological and Environmental Sciences, Braun Hall (Bldg 320), Room 207, Stanford University, Stanford, CA 94305-2115, USA* (Fax: +1-650-725-0979, jstreet@stanford.edu) (153)

Contents of Previous Volumes

Volume 1. **Simple Complexes***

Volume 2. **Mixed-Ligand Complexes***

Volume 3. **High Molecular Complexes***

Volume 4. **Metal Ions as Probes***

Volume 5. **Reactivity of Coordination Compounds***

Volume 6. **Biological Action of Metal Ions***

Volume 7. **Iron in Model and Natural Compounds***

Volume 8. **Nucleotides and Derivatives: Their Ligating
Ambivalency***

Volume 9. **Amino Acids and Derivatives as Ambivalent
Ligands**

1. Complexes of α-Amino Acids with Chelatable Side Chain
   Donor Atoms
   R. Bruce Martin

2. Metal Complexes of Aspartic Acid and Glutamic Acid
   *Christopher A. Evans, Roger Guevremont, and
   Dallas L. Rabenstein*

3. The Coordination Chemistry of *L*-Cysteine and *D*-Penicillamine
   Arthur Gergely and Imre Sóvágó

4. Glutathione and Its Metal Complexes
   *Dallas L. Rabenstein, Roger Guevremont, and
   Christopher A. Evans*

*Out of print

5. Coordination Chemistry of *L*-Dopa and Related Ligands
   Arthur Gergely and Tamás Kiss

6. Stereoselectivity in the Metal Complexes of Amino Acids
   and Dipeptides
   Leslie D. Pettit and Robert J. W. Hefford

7. Protonation and Complexation of Macromolecular Polypeptides:
   Corticotropin Fragments and Basic Trypsin Inhibitor (Kunitz Base)
   Kálmán Burger

Author Index–Subject Index

Volume 10.   Carcinogenicity and Metal Ions

1. The Function of Metal Ions in Genetic Regulation
   Gunther L. Eichhorn

2. A Comparison of Carcinogenic Metals
   C. Peter Flessel, Arthur Furst, and Shirley B. Radding

3. The Role of Metals in Tumor Development and Inhibition
   Haleem J. Issaq

4. Paramagnetic Metal Ions in Tissue During Malignant Development
   Nicholas J. F. Dodd

5. Ceruloplasmin and Iron Transferrin in Human Malignant Disease
   Margaret A. Foster, Trevor Pocklington, and Audrey A. Dawson

6. Human Leukemia and Trace Elements
   E. L. Andronikashvili and L. M. Mosulishvili

7. Zinc and Tumor Growth
   Andre M. van Rij and Walter J. Pories

8. Cyanocobalamin and Tumor Growth
   Sofija Kanopkaitė and Gediminas Bražėnas

9. The Role of Selenium as a Cancer-Protecting
   Trace Element
   Birger Jansson

10. Tumor Diagnosis Using Radioactive Metal Ions and
    Their Complexes
    Akira Yokoyama and Hideo Saji

Author Index–Subject Index

Volume 11.   Metal Complexes as Anticancer Agents*

Volume 12.   Properties of Copper*

Volume 13.   Copper Proteins*

---

*Out of print

Volume 14. Inorganic Drugs in Deficiency and Disease

1. Drug-Metal Ion Interaction in the Gut
   P. F. D'Arcy and J. C. McElnay

2. Zinc Deficiency and Its Therapy
   Ananda S. Prasad

3. The Pharmacological Use of Zinc
   George J. Brewer

4. The Anti-inflammatory Activities of Copper Complexes
   John R. J. Sorenson

5. Iron-Containing Drugs
   David A. Brown and M. V. Chidambaram

6. Gold Complexes as Metallo-Drugs
   Kailash C. Dash and Hubert Schmidbaur

7. Metal Ions and Chelating Agents in Antiviral Chemotherapy
   D. D. Perrin and Hans Stünzi

8. Complexes of Hallucinogenic Drugs
   Wolfram Hänsel

9. Lithium in Psychiatry
   Nicholas J. Birch

Author Index–Subject Index

Volume 15. Zinc and Its Role in Biology and Nutrition

1. Categories of Zinc Metalloenzymes
   Alphonse Galdes and Bert L. Vallee

2. Models for Zn(II) Binding Sites in Enzymes
   Robert S. Brown, Joan Huguet, and Neville J. Curtis

3. An Insight on the Active Site of Zinc Enzymes Through
   Metal Substitution
   Ivano Bertini and Claudio Luchinat

4. The Role of Zinc in DNA and RNA Polymerases
   Felicia Ying-Hsiueh Wu and Cheng-Wen Wu

5. The Role of Zinc in Snake Toxins
   Anthony T. Tu

6. Spectroscopic Properties of Metallothionein
   Milan Vašák and Jeremias H. R. Kägi

7. Interaction of Zinc with Erythrocytes
   Joseph M. Rifkind

8. Zinc Absorption and Excretion in Relation to Nutrition
   Manfred Kirchgessner and Edgar Weigand

9. Nutritional Influence of Zinc on the Activity of Enzymes
   and Hormones
   Manfred Kirchgessner and Hans-Peter Roth

10. Zinc Deficiency Syndrome During Parenteral Nutrition
    Karin Ladefoged and Stig Jarnum

Author Index – Subject Index

**Volume 16.   Methods Involving Metal Ions and Complexes
                in Clinical Chemistry**

1. Some Aspects of Nutritional Trace Element Research
   Clare E. Casey and Marion F. Robinson

2. Metals and Immunity
   Lucy Treagan

3. Therapeutic Chelating Agents
   Mark M. Jones

4. Computer-Directed Chelate Therapy of Renal Stone Disease
   Martin Rubin and Arthur E. Martell

5. Determination of Trace Metals in Biological Materials by Stable
   Isotope Dilution
   Claude Veillon and Robert Alvarez

6. Trace Elements in Clinical Chemistry Determined by Neutron
   Activation Analysis
   Kaj Heydorn

7. Determination of Lithium, Sodium, and Potassium in
   Clinical Chemistry
   Adam Uldall and Arne Jensen

8. Determination of Magnesium and Calcium in Serum
   Arne Jensen and Erik Riber

9. Determination of Manganese, Iron, Cobalt, Nickel, Copper, and
   Zinc in Clinical Chemistry
   Arne Jensen, Erik Riber, Poul Persson, and Kaj Heydorn

10. Determination of Lead, Cadmium, and Mercury in
    Clinical Chemistry
    Arne Jensen, Jytte Molin Christensen, and Poul Persson

11. Determination of Chromium in Urine and Blood
    Ole Jøns, Arne Jensen, and Poul Persson

12. Determination of Aluminum in Clinical Chemistry
    Arne Jensen, Erik Riber, and Poul Persson

13. Determination of Gold in Clinical Chemistry
    Arne Jensen, Erik Riber, Poul Persson, and Kaj Heydorn

14. Determination of Phosphates in Clinical Chemistry
   Arne Jensen and Adam Uldall

15. Identification and Quantification of Some Drugs in Body
   Fluids by Metal Chelate Formation
   R. Bourdon, M. Galliot, and J. Hoffelt

16. Metal Complexes of Sulfanilamides in Pharmaceutical Analysis
   and Therapy
   Auke Bult

17. Basis for the Clinical Use of Gallium and
   Indium Radionuclides
   Raymond L. Hayes and Karl F. Hübner

18. Aspects of Technetium Chemistry as Related to
   Nuclear Medicine
   Hans G. Seiler

Author Index–Subject Index

Volume 17.   Calcium and Its Role in Biology

1. Bioinorganic Chemistry of Calcium
   R. Bruce Martin

2. Crystal Structure Studies of Calcium Complexes and Implications
   for Biological Systems
   H. Einspahr and C. E. Bugg

3. Intestinal and Renal Absorption of Calcium
   Piotr Gmaj and Heini Murer

4. Calcium Transport Across Biological Membranes
   Ernesto Carafoli, Giuseppe Inesi, and Barry Rosen

5. Physiological Aspects of Mitochondrial Calcium Transport
   Gary Fiskum

6. Mode of Action of the Regulatory Protein Calmodulin
   *Jos A. Cox, Michelle Comte, Armand Malnoë, Danielle Burger,
   and Eric A. Stein*

7. Calcium and Brain Proteins
   S. Alamà

8. The Role of Ca^{2+} in the Regulation and Mechanism of Exocytosis
   Carl E. Creutz

9. Calcium Function in Blood Coagulation
   Gary L. Nelsestuen

10. The Role of Calcium in the Regulation of the Skeletal Muscle
   Contraction-Relaxation Cycle
   Henry G. Zot and James D. Potter

11. Calcification of Vertebrate Hard Tissues
    Roy E. Wuthier

Author Index–Subject Index

Volume 18.   Circulation of Metals in the Environment

 1. Introduction to "Circulation of Metals in the Environment"
    Peter Baccini

 2. Analytical Chemistry Applied to Metal Ions in the Environment
    Arne Jensen and Sven Erik Jørgensen

 3. Processes of Metal Ions in the Environment
    Sven Erik Jørgensen and Arne Jensen

 4. Surface Complexation
    Paul W. Schindler

 5. Relationships Between Biological Availability and
    Chemical Measurements
    David R. Turner

 6. Natural Organic Matter and Metal-Organic Interactions in
    Aquatic Systems
    Jacques Buffle

 7. Evolutionary Aspects of Metal Ion Transport Through
    Cell Membranes
    John M. Wood

 8. Regulation of Trace Metal Concentrations in Fresh Water Systems
    Peter Baccini

 9. Cycling of Metal Ions in the Soil Environment
    Garrison Sposito and Albert L. Page

10. Microbiological Strategies in Resistance to Metal Ion Toxicity
    John M. Wood

11. Conclusions and Outlook
    Peter Baccini

Author Index–Subject Index

Volume 19.   Antibiotics and Their Complexes

 1. The Discovery of Ionophores: An Historical Account
    Berton C. Pressman

 2. Tetracyclines and Daunorubicin
    R. Bruce Martin

 3. Interaction of Metal Ions with Streptonigrin and Biological
    Properties of the Complexes
    Joseph Hajdu

4. Bleomycin Antibiotics: Metal Complexes and Their
   Biological Action
   Yukio Sugiura, Tomohisa Takita, and Hamao Umezawa

5. Interaction Between Valinomycin and Metal Ions
   K. R. K. Easwaran

6. Beauvericin and the Other Enniatins
   Larry K. Steinrauf

7. Complexing Properties of Gramicidins
   James F. Hinton and Roger E. Koeppe II

8. Nactins: Their Complexes and Biological Properties
   Yoshiharu Nawata, Kunio Ando, and Yoichi Iitaka

9. Cation Complexes of the Monovalent and Polyvalent Carboxylic
   Ionophores: Lasalocid (X-537A), Monensin, A-23187 (Calcimycin)
   and Related Antibiotics
   George R. Painter and Berton C. Pressman

10. Complexes of D-Cycloserine and Related Amino Acids with
    Antibiotic Properties
    Paul O'Brien

11. Iron-Containing Antibiotics
    J. B. Neilands and J. R. Valenta

12. Cation-Ionophore Interactions: Quantification of the Factors
    Underlying Selective Complexation by Means of Theoretical
    Computations
    Nohad Gresh and Alberte Pullman

Author Index–Subject Index

Volume 20. Concepts on Metal Ion Toxicity

1. Distribution of Potentially Hazardous Trace Metals
   Garrison Sposito

2. Bioinorganic Chemistry of Metal Ion Toxicity
   R. Bruce Martin

3. The Interrelation Between Essentiality and Toxicity of Metals in the
   Aquatic Ecosystem
   Elie Eichenberger

4. Metal Ion Speciation and Toxicity in Aquatic Systems
   Gordon K. Pagenkopf

5. Metal Toxicity to Agricultural Crops
   Frank T. Bingham, Frank J. Peryea, and Wesley M. Jarrell

6. Metal Ion Toxicity in Man and Animals
   Paul B. Hammond and Ernest C. Foulkes

7. Human Nutrition and Metal Ion Toxicity
   M. R. Spivey Fox and Richard M. Jacobs

8. Chromosome Damage in Individuals Exposed to Heavy Metals
   Alain Léonard

9. Metal Ion Carcinogenesis: Mechanistic Aspects
   Max Costa and J. Daniel Heck

10. Methods for the In Vitro Assessment of Metal Ion Toxicity
    J. Daniel Heck and Max Costa

11. Some Problems Encountered in the Analysis of Biological
    Materials for Toxic Trace Elements
    Hans G. Seiler

Author Index–Subject Index

**Volume 21.   Applications of Nuclear Magnetic Resonance to
               Paramagnetic Species**

1. Nuclear Relaxation Times as a Source of Structural Information
   Gil Navon and Gianni Valensin

2. Nuclear Relaxation in NMR of Paramagnetic Systems
   Ivano Bertini, Claudio Luchinat, and Luigi Messori

3. NMR Studies of Magnetically Coupled Metalloproteins
   Lawrence Que, Jr., and Michael J. Maroney

4. Proton NMR Studies of Biological Problems Involving
   Paramagnetic Heme Proteins
   James D. Satterlee

5. Metal-Porphyrin Induced NMR Dipolar Shifts and Their Use in
   Conformational Analysis
   Nigel J. Clayden, Geoffrey R. Moore, and Glyn Williams

6. Relaxometry of Paramagnetic Ions in Tissue
   Seymour H. Koenig and Rodney D. Brown III

Author Index–Subject Index

**Volume 22.   ENDOR, EPR, and Electron Spin Echo for Probing
               Coordination Spheres**

1. ENDOR: Probing the Coordination Environment
   in Metalloproteins
   Jürgen Hüttermann and Reinhard Kappl

2. Identification of Oxygen Ligands in Metal-Nucleotide-Protein
   Complexes by Observation of the Mn(II)-^{17}O Superhyperfine
   Coupling
   Hans R. Kalbitzer

3. Use of EPR Spectroscopy for Studying Solution Equilibria
   Harald Gampp

4. Applications of EPR Saturation Methods to Paramagnetic Metal
   Ions in Proteins
   Marvin W. Makinen and Gregg B. Wells

5. Electron Spin Echo: Applications to Biological Systems
   Yuri D. Tsvetkov and Sergei A. Dikanov

Author Index–Subject Index

Volume 23. Nickel and Its Role in Biology

1. Nickel in the Natural Environment
   Robert W. Boyle and Heather W. Robinson

2. Nickel in Aquatic Systems
   Pamela Stokes

3. Nickel and Plants
   Margaret E. Farago and Monica M. Cole

4. Nickel Metabolism in Man and Animals
   Evert Nieboer, Rickey T. Tom, and W. (Bill) E. Sanford

5. Nickel Ion Binding to Amino Acids and Peptides
   R. Bruce Martin

6. Nickel in Proteins and Enzymes
   Robert K. Andrews, Robert L. Blakeley, and Burt Zerner

7. Nickel-Containing Hydrogenases
   *José J. G. Moura, Isabel Moura, Miguel Teixeira,
   António V. Xavier, Guy D. Fauque, and Jean LeGall*

8. Nickel Ion Binding to Nucleosides and Nucleotides
   R. Bruce Martin

9. Interactions Between Nickel and DNA: Considerations About the
   Role of Nickel in Carcinogensis
   E. L. Andronikashvili, V. G. Bregadze, and J. R. Monaselidze

10. Toxicology of Nickel Compounds
    Evert Nieboer, Franco E. Rossetto, and C. Rajeshwari Menon

11. Analysis of Nickel in Biological Materials
    Hans G. Seiler

Author Index–Subject Index

Volume 24. Aluminum and Its Role in Biology

1. Bioinorganic Chemistry of Aluminum
   R. Bruce Martin

2. Aluminum in the Environment
   Charles T. Driscoll and Williams D. Schecher

3. The Physiology of Aluminum Phytotoxicity
   Gregory J. Taylor

4. The Physiology of Aluminum Tolerance
   Gregory J. Taylor

5. Aluminum in the Diet and Mineral Metabolism
   Janet L. Greger

6. Aluminum Intoxication: History of Its Clinical Recognition
   and Management
   David N. S. Kerr and M. K. Ward

7. Aluminum and Alzheimer's Disease, Methodologic Approaches
   Daniel P. Perl

8. Mechanisms of Aluminum Neurotoxicity—Relevance to
   Human Disease
   Theo P. A. Kruck and Donald R. McLachlan

9. Aluminum Toxicity and Chronic Renal Failure
   Michael R. Wills and John Savory

10. Analysis of Aluminum in Biological Materials
    John Savory and Michael R. Wills

Author Index–Subject Index

**Volume 25. Interrelations Among Metal Ions, Enzymes, and
Gene Expression**

1. Metal Ion-Induced Mutagensis In Vitro:
   Molecular Mechanisms
   Kathleen M. Downey and Antero G. So

2. Metallonucleases: Real and Artificial
   Lena A. Basile and Jacqueline K. Barton

3. Metalloregulatory Proteins: Metal-Responsive Molecular Switches
   Governing Gene Expression
   Thomas V. O'Halloran

4. Yeast Metallothionein: Gene Function and Regulation by
   Metal Ions
   David J. Ecker, Tauseef R. Butt, and Stanley T. Crooke

5. Zinc-Binding Proteins Involved in Nucleic Acid Replication
   Joseph E. Coleman and David P. Giedroc

6. "Zinc Fingers": The Role of Zinc(II) in Transcription Factor IIIA
   and Related Proteins
   Jeremy M. Berg

7. Site-Directed Mutagensis and Structure-Function Relations in EF-Hand Ca^{2+}-Binding Proteins
   Sture Forsén, Torbjörn Drakenberg, Sara Linse, Peter Brodin, Peter Sellers, Charlotta Johansson, Eva Thulin, and Thomas Grundström

8. Genetic Alteration of Active Site Residues of Staphylococcal Nuclease: Insights into the Enzyme Mechanism
   Albert S. Mildvan and Engin H. Serpersu

9. Alcohol Dehydrogenase: Structure, Catalysis, and Site-Directed Mutagensis
   Y. Pocker

10. Probing the Mechanism of Action of Carboyxypeptide A by Inorganic, Organic, and Mutagenic Modifications
    David S. Auld, James F. Riordan, and Bert L. Vallee

11. Site-Directed Mutagensis of *E. coli* Alkaline Phosphatase: Probing the Active-Site Mechanism and the Signal Sequence-Mediated Transport of the Enzyme
    John E. Butler-Ransohoff, Debra A. Kendall, and Emil Thomas Kaiser

12. Site-Directed Mutagensis of Heme Proteins
    Patrick R. Stayton, William M. Atkins, Barry A. Springer, and Stephen G. Sligar

13. Exploring Structure-Function Relationships in Yeast Cytochrome c Peroxidase Using Mutagensis and Crystallography
    J. Matthew Mauro, Mark A. Miller, Stephen L. Edwards, Jimin Wang, Laurence A. Fishel, and Joseph Kraut

Author Index–Subject Index

Volume 26. Compendium on Magnesium and Its Role in Biology, Nutrition, and Physiology

1. Bioinorganic Chemistry of Magnesium
   R. Bruce Martin

2. Magnesium in the Environment
   Raili Jokinen

3. Magnesium in Plants: Uptake, Distribution, Function, and Utilization by Man and Animals
   Stanley R. Wilkinson, Ross M. Welch, Henry F. Mayland, and David L. Grunes

4. Magnesium in Animal Nutrition
   H. Meyer and J. Zentak

5. Dietary Magnesium and Drinking Water: Effects on Human Health Status
   John R. Marier

6. Magnesium in Biology and Medicine: An Overview
   Nicholas J. Birch

7. Role of Magnesium in Enzyme Systems
   Frank W. Heaton

8. Magnesium: A Regulated and Regulatory Cation
   Michael E. Maguire

9. Magnesium Transport in Prokaryotic Cells
   Marshall D. Snavely

10. Hormonal Regulation of Magnesium Homeostasis in Cultured Mammalian Cells
    Robert D. Grubbs

11. Functional Compartmentation of Intracellular Magnesium
    T. Günther

12. Membrane Transport of Magnesium
    T. Günther and H. Ebel

13. Intestinal Magnesium Absorption
    H. Ebel

14. The Renal Handling of Magnesium
    Michael P. Ryan

15. Magnesium and Liver Cirrhosis
    Leon Cohen

16. Hypomagnesemia and Hypermagnesemia
    Nachman Brautbar, Atul T. Roy, Philip Hom, and David B. N. Lee

17. Systemic Stress and the Role of Magnesium
    H. G. Classen

18. Magnesium and Lipid Metabolism
    Yves Rayssiguier

19. Magnesium and the Cardiovascular System: Experimental and Clinical Aspects Updated
    Burton M. Altura and Bella T. Altura

20. Magnesium and the Peripheral (Extradural) Nervous System: Metabolism, Neurophysiological Functions, and Clinical Disorders
    Jerry G. Chutkow

21. Magnesium and the Central (Intradural) Nervous System: Metabolism, Neurophysiological Functions, and Clinical Disorders
    Jerry G. Chutkow

22. The Role of Magnesium in the Regulation of Muscle Function
    Christopher H. Fry and Sarah K. Hall

23. Magnesium in Bone and Tooth
Colin Robinson and John A. Weatherell

24. Magnesium and Osteoporosis
Leon Cohen

25. The Role of Magnesium in Pregnancy, for the Newborn, and in Children's Diseases
Ludwig Spätling

26. Magnesium and Its Role in Allergy
Nicole Hunziker

27. Magnesium and Its Relationship to Oncology
Jean Durlach, Michel Bara, and Andrée Guiet-Bara

28. The Assessment of Magnesium Status in Humans
Ronald J. Elin

29. Magnesium and Placebo Effects in Human Medicine
H. G. Classen

30. Determination of Magnesium in Biological Materials
Hans G. Seiler

Author Index–Subject Index

Volume 27. Electron Transfer Reactions in Metalloproteins

1. Mediation of Electron Transfer by Peptides and Proteins: Current Status
Stephan S. Isied

2. Electron and Atom Group Transfer Properties of Protein Systems
Hans E. M. Christensen, Lars S. Conrad, Jens Ulstrup, and Kurt V. Mikkelsen

3. Electron Tunneling Pathways in Proteins
David N. Beratan, José Nelson Onuchic, and Harry B. Gray

4. Diprotein Complexes and Their Electron Transfer Reactions
Nenad M. Kostić

5. Electron Transfer Between Bound Proteins
George McLendon

6. Long-Range Electron Transfer in Cytochrome c Derivatives with Covalently Attached Probe Complexes
Robert A. Scott, David W. Conrad, Marly K. Eidsness, Antonius C. F. Gorren, and Sten A. Wallin

7. Photoinduced Electron Transfer Reactions in Metalloprotein Complexes Labeled with Ruthenium Polypyridine
Francis Millett and Bill Durham

8. Stereoselective Effects in Electron Transfer Reactions Involving Synthetic Metal Complexes and Metalloproteins
Klaus Bernauer

9. Properties and Electron Transfer Reactivity of [2Fe-2S] Ferredoxins
A. Geoffrey Sykes

10. Electron Transfer in Photosynthetic Reaction Centers
Sethulakshmi Kartha, Ranjan Das, and James A. Norris

11. Modeling the Primary Electron Transfer Events of Photosynthesis
Michael R. Wasielewski

12. Electrochemical Investigation of Metalloproteins and Enzymes Using a Model Incorporating Microscopic Aspects of the Electrode-Solution Interface
Alan M. Bond and H. Allen O. Hill

Author Index–Subject Index

Volume 28. Degradation of Environmental Pollutants by Microorganisms and Their Metalloenzymes

1. General Strategies in the Biodegradation of Pollutants
Thomas W. Egli

2. Oxidation of Aromatic Pollutants by Lignin-Degrading Fungi and Their Extracellular Peroxidases
Kenneth E. Hammel

3. Biodegradation of Tannins
James A. Field and G. Lettinga

4. Aerobic Biodegradation of Aromatic Hydrocarbons by Bacteria
Shigeaki Harayama and Kenneth N. Timmis

5. Degradation of Halogenated Aromatics by Actinomycetes
Bruno Winter and Wolfgang Zimmermann

6. Enzymes Catalyzing Oxidative Coupling Reactions of Pollutants
Jean-Marc Bollag

7. Mechanism of Action of Peroxidases
Helen Anni and Takashi Yonetani

8. Mechanistic Aspects of Dihydroxybenzoate Dioxygenases
John D. Lipscomb and Allen M. Orville

9. Aerobic and Anaerobic Degradation of Halogenated Aliphatics
Dick B. Janssen and Bernard Witholt

10. Mechanisms of Reductive Dehalogenation by Transition Metal Cofactors Found in Anaerobic Bacteria
Lawrence P. Wackett and Craig A. Schanke

11. Bacterial Degradation of Hemicelluloses
    Wolfgang Zimmermann

12. Degradation of Cellulose and Effects of Metal Ions on Cellulases
    Anil Goyal and Douglas E. Eveleigh

13. Metalloproteases and Their Role in Biotechnology
    Guido Grandi and Giuliano Galli

14. Metal-Dependent Conversion of Inorganic Nitrogen and
    Sulfur Compounds
    Peter M. H. Kroneck, Joachim Beuerle, and Wolfram Schumacher

Author Index – Subject Index

**Volume 29.   Biological Properties of Metal Alkyl
                Derivatives**

1. Global Bioalkylation of the Heavy Elements
   John S. Thayer

2. Analysis of Organometallic Compounds in the Environment
   Darren Mennie and Peter J. Craig

3. Biogeochemistry of Methylgermanium Species in
   Natural Waters
   Brent L. Lewis and H. Peter Mayer

4. Biological Properties of Alkyltin Compounds
   Yasuaki Arakawa and Osamu Wada

5. Biological Properties of Alkyl Derivatives of Lead
   Yukio Yamamura and Fumio Arai

6. Metabolism of Alkyl Arsenic and Antimony Compounds
   Marie Vahter and Erminio Marafante

7. Biological Alkylation of Selenium and Tellurium
   Ulrich Karlson and William T. Frankenberger, Jr.

8. Making and Breaking the Co-Alkyl Bond in B_{12} Derivatives
   John M. Pratt

9. Methane Formation by Methanogenic Bacteria: Redox Chemistry
   of Coenzyme F430
   Bernhard Jaun

10. Synthesis and Degradation of Organomercurials by Bacteria—A
    Comment by the Editors
    Helmut Sigel and Astrid Sigel

11. Biogenesis and Metabolic Role of Halomethanes in Fungi
    David B. Harper

Author Index – Subject Index

Volume 30.   Metalloenzymes Involving Amino Acid–Residue and Related Radicals

1. Free Radicals and Metalloenzymes: General Considerations
   Ei-Ichiro Ochiai

2. Peroxidases: Structure, Function, and Engineering
   Thomas L. Poulos and Roger E. Fenna

3. Photosystem II
   Curtis W. Hoganson and Gerald T. Babcock

4. Ribonucleotide Reductase in Mammalian Systems
   Lars Thelander and Astrid Gräslund

5. Manganese-Dependent Ribonucleotide Reduction and Overproduction of Nucleotides in Coryneform Bacteria
   George Auling and Hartmut Follmann

6. Prostaglandin Endoperoxide Synthases
   William L. Smith and Lawrence J. Marnett

7. Diol Dehydrase from *Clostridium glycolicum:* The Non-B_{12}-Dependent Enzyme
   Maris G. N. Hartmanis

8. Diol Dehydrase and Glycerol Dehydrase, Coenzyme B_{12}-Dependent Isozymes
   Tetsuo Toraya

9. Adenosylcobalamin (Vitamin B_{12} Coenzyme)-Dependent Enzymes
   Ei-Ichiro Ochiai

10. *S*-Adenosylmethionine-Dependent Radical Formation in Anaerobic Systems
    Kenny K. Wong and John W. Kozarich

11. The Free Radical-Coupled Copper Active Site of Galactose Oxidase
    James W. Whittaker

12. Amine Oxidases
    Peter F. Knowles and David M. Dooley

13. Bacterial Transport of and Resistance to Copper
    Nigel L. Brown, Barry T. O. Lee, and Simon Silver

Author Index–Subject Index

Volume 31.   Vanadium and Its Role for Life

1. Inorganic Considerations on the Function of Vanadium in Biological Systems
   Dieter Rehder

2. Solution Properties of Vandium(III) with Regard to
   Biological Systems
   *Roland Meier, Martin Boddin, Steffi Mitzenheim, and
   Kan Kanamori*

3. The Vanadyl Ion: Molecular Structure of Coordinating Ligands by
   Electron Paramagnetic Resonance and Electron Nuclear Double
   Resonance Spectroscopy
   Marvin W. Makinen and Devkumar Mustafi

4. Vanadyl(IV) Complexes of Nucleotides
   Enrique J. Baran

5. Interactions of Vanadates with Biogenic Ligands
   Debbie C. Crans

6. Use of Vanadate-Induced Photocleavage for Detecting Phosphate
   Binding Sites in Proteins
   Andras Muhlrad and Israel Ringel

7. Vanadium-Protein Interactions
   N. Dennis Chasteen

8. Stimulation of Enzyme Activity by Oxovanadium Complexes
   Paul J. Stankiewicz and Alan S. Tracy

9. Inhibition of Phosphate-Metabolizing Enzymes by
   Oxovanadium(V) Complexes
   Paul J. Stankiewicz, Alan S Tracy, and Debbie C. Crans

10. Vanadium-Dependent Haloperoxidases
    Hans Vilter

11. Vanadium Nitrogenases of *Azotobacter*
    Robert R. Eady

12. Amavadin, the Vanadium Compound of Amanitae
    Ernst Bayer

13. Vanadium in Ascidians and the Chemistry of Tunichromes
    *Mitchell J. Smith, Daniel E. Ryan, Koji Nakanishi,
    Patrick Frank, and Keith O. Hodgson*

14. Biochemical Significance of Vanadium in a Polychaete Worm
    Toshiaki Ishii, Izumi Nakai, and Kenji Okoshi

15. Vanadium Transport in Animal Systems
    Kenneth Kustin and William E. Robinson

16. Vanadium in Mammalian Physiology and Nutrition
    Forrest H. Nielsen

17. Vanadium Compounds as Insulin Mimics
    *Chris Orvig, Katherine H. Thompson, Mary Battell, and
    John H. McNeill*

18. Antitumor Activity of Vanadium Compounds
    Cirila Djordjevic

19. Methods for the Spectroscopic Characterization of Vanadium
    Centers in Biological and Related Chemical Systems
    C. David Garner, David Collison, and Frank E. Mabbs

20. Analytical Procedures for the Determination of Vanadium in
    Biological Materials
    Hans G. Seiler

Author Index–Subject Index

**Volume 32.   Interactions of Metal Ions with Nucleotides,
               Nucleic Acids, and Their Constituents**

1. Phosphate-Metal Ion Interactions of Nucleotides and
   Polynucleotides
   Cindy Klevickis and Charles M. Grisham

2. Sugar-Metal Ion Interactions
   Shigenobu Yano and Masami Otsuka

3. Dichotomy of Metal Ion Binding to N1 and N7 of Purines
   R. Bruce Martin

4. General Conclusions from Solid State Studies of Nucleotide-Metal
   Ion Complexes
   Katsuyuki Aoki

5. Solution Structures of Nucleotide Metal Ion
   Complexes. Isomeric Equilibria
   Helmut Sigel and Bin Song

6. Stacking Interactions Involving Nucleotides and Metal
   Ion Complexes
   Osamu Yamauchi, Akira Odani, Hideki Masuda, and Helmut Sigel

7. Effect of Metal Ions on the Hydrolytic Reactions of Nucleosides
   and Their Phosphoesters
   Satu Kuusela and Harri Lönnberg

8. Metal Complexes of Sulfur-Containing Purine Derivatives
   Erich Dubler

9. Mechanistic Insight from Kinetic Studies on the Interaction of
   Model Palladium(II) Complexes with Nucleic Acid Components
   Tobias Rau and Rudi van Eldik

10. Platinum(II)-Nucleobase Interactions. A Kinetic Approach
    Jorma Arpalahti

11. NMR Studies of Oligonucleotide-Metal Ion Interactions
    Einar Sletten and Nils Åge Frøystein

12. Metal Ion Interactions with DNA: Considerations on Structure, Stability, and Effects from Metal Ion Binding
    Vasil G. Bregadze

13. Electron Transfer Reactions Through the DNA Double Helix
    Thomas J. Meade

14. The Role of Metal Ions in Ribozymes
    Anna Marie Pyle

15. Ternary Metal Ion-Nucleic Acid Base-Protein Complexes
    Michal Sabat

16. Metal-Responsive Gene Expression and the Zinc-Metalloregulatory Model
    David A. Suhy and Thomas V. O'Halloran

17. The Role of Iron-Sulfur Proteins Involved in Gene Regulation
    M. Claire Kennedy

18. Current Status of Structure-Activity Relationships of Platinum Anticancer Drugs: Activation of the *trans*-Geometry
    Nicholas Farrell

19. Cisplatin and Derived Anticancer Drugs: Mechanism and Current Status of DNA Binding
    Marieke J. Bloemink and Jan Reedijk

20. Proteins that Bind to and Mediate the Biological Activity of Platinum Anticancer Drug-DNA Adducts
    Joyce P. Whitehead and Stephen J. Lippard

21. Interactions of Metallopharmaceuticals with DNA
    Michael J. Clarke and Michael Stubbs

Subject Index

Volume 33. Probing of Nucleic Acids by Metal Ion Complexes of Small Molecules

1. Molecular Modeling of Transition Metal Complexes with Nucleic Acids and Their Consituents
   Jiří Kozelka

2. Zinc Complexes as Targeting Agents for Nucleic Acids
   Eiichi Kimura and Mitsuhiko Shionoya

3. Metallocene Interactions with DNA and DNA Processing Enzymes
   Louis Y. Kuo, Andrew H. Liu, and Tobin J. Marks

4. Evidences for a Catalytic Activity of the DNA Double Helix in the Reaction between DNA, Platinum(II), and Intercalators
   Marc Boudvillain, Rozenn Dalbiès, and Marc Leng

5. *Trans*-Diammineplatinum(II)—What Makes it Different from *cis*-DDP? Coordination Chemistry of a Neglected Relative of Cisplatin and Its Interaction with Nucleic Acids
Bernhard Lippert

6. Metal Ions in Multiple-Stranded DNA
Michal Sabat and Bernhard Lippert

7. DNA Interactions with Substitution-Inert Transition Metal Ion Complexes
Bengt Nordén, Per Lincoln, Björn Åkerman, and Eimer Tuite

8. Effect of Metal Ions on the Fluorescence of Dyes Bound to DNA
Vasil G. Bregadze, Jemal G. Chkhaberidze, and Irine G. Khutsishvili

9. Photolytic Covalent Binding of Metal Complexes to DNA
Mark A. Billadeau and Harry Morrison

10. Electrochemically Activated Nucleic Acid Oxidation
Dean H. Johnston, Thomas W. Welch, and H. Holden Thorp

11. Electron Transfer between Metal Complexes Bound to DNA: Is DNA a Wire?
Eric D. A. Stemp and Jacqueline K. Barton

12. Porphyrin and Metalloporphyrin Interactions with Nucleic Acids
Robert F. Pasternack and Esther J. Gibbs

13. Selective DNA Cleavage by Metalloporphyrin Derivatives
Geneviève Pratviel, Jean Bernadou, and Bernard Meunier

14. Synthetic Metallopeptides as Probes of Protein-DNA Interactions
Eric C. Long, Paula Denney Eason, and Qi Liang

15. Targeting of Nucleic Acids by Iron Complexes
Alexandra Draganescu and Thomas D. Tullius

16. Nucleic Acid Chemistry of the Cuprous Complexes of 1,10-Phenanthroline and Derivatives
David S. Sigman, Ralf Landgraf, David M. Perrin, and Lori Pearson

17. Specific DNA Cleavage by Manganese(III) Complexes
Dennis J. Gravert and John H. Griffin

18. Nickel Complexes as Probes of Guanine Sites in Nucleic Acid Folding
Cynthia J. Burrows and Steven E. Rokita

19. Hydrolytic Cleavage of RNA Catalyzed by Metal Ion Complexes
Janet R. Morrow

20. RNA Recognition and Cleavage by Iron(II)-Bleomycin
Jean-Marc Battigello, Mei Cui, and Barbara J. Carter

21. Metallobleomycin-DNA Interactions: Structures and Reactions
    Related to Bleomycin-Induced DNA Damage
    David H. Petering, Qunkai Mao, Wenbao Li,
    Eugene DeRose, and William E. Antholine

Subject Index

Volume 34. Mercury and Its Effects on Environment and Biology

 1. Analytical Methods for the Determination of Mercury(II) and
    Methylmercury Compounds: The Problem of Speciation
    James H. Weber

 2. Mercury in Lakes and Rivers
    Markus Meili

 3. Biogeochemical Cycling of Mercury in the Marine Environment
    William F. Fitzgerald and Robert P. Mason

 4. Catchments as a Source of Mercury and Methylmercury in Boreal
    Surface Waters
    Kevin H. Bishop and Ying-Hua Lee

 5. Gold Rushes and Mercury Pollution
    Jerome O. Nriagu and Henry K. T. Wong

 6. Accumulation of Mercury in Soil and Effects on the Soil Biota
    Lage Bringmark

 7. Biogeochemistry of Mercury in the Air-Soil-Plant System
    Ki-Hyun Kim, Paul J Hanson, Mark O. Barnett, and
    Steven E. Lindberg

 8. Microbial Transformation of Mercury Species and Their
    Importance in the Biogeochemical Cycle of Mercury
    Franco Baldi

 9. Bioaccumulation of Mercury in the Aquatic Food Chain in Newly
    Flooded Areas
    R. A. (Drew) Bodaly, Vincent L. St. Louis, Michael J. Paterson,
    Robert J. P. Fudge, Britt D. Hall, David M. Rosenberg, and
    John W. M. Rudd

10. Mercury in the Food Web: Accumulation and Transfer Mechanisms
    Alain Boudou and Francis Ribeyre

11. Physiology and Toxicology of Mercury
    László Magos

12. Metabolism of Methylmercury in the Brain and Its
    Toxicological Significance
    N. Karle Mottet, Marie E. Vahter, Jay S. Charleston, and
    Lars T. Friberg

13. Maternal-Fetal Mercury Transport and Fetal
    Methylmercury Poisoning
    Rikuzo Hamada, Kimiyoshi Arimura, and Mitsuhiro Osame

14. Effects of Mercury on the Immune System
    K. Michael Pollard and Per Hultman

15. The Impact of Mercury Released from Dental 'Silver' Fillings
    on Antibiotic Resistances in the Primate Oral and Intestinal
    Bacterial Flora
    *Cynthia A. Liebert, Joy Wireman, Tracy Smith, and
    Anne O. Summers*

16. Inhibition of Brain Tubulin-Guanosine 5'-Triphosphate Interactions
    by Mercury: Similarity to Observations in Alzheimer's
    Diseased Brain
    James C. Pendergrass and Boyd E. Haley

17. Interaction of Mercury with Nucleic Acids and Their Components
    Einar Sletten and Willy Nerdal

18. Mercury Responsive Gene Regulation and Mercury-199 as a Probe
    of Protein Structure
    David L. Huffman, Lisa M. Utschig, and Thomas V. O'Halloran

19. Bacterial Mercury Resistance Genes
    John L. Hobman and Nigel L. Brown

Subject Index

**Volume 35.   Iron Transport and Storage in Microorganisms,
                Plants, and Animals**

1. Biological Cycling of Iron in the Ocean
   Neil M. Price and François M. M. Morel

2. Microbial Iron Transport: Iron Acquisition by
   Pathogenic Microorganisms
   B. Rowe Byers and Jean E. L. Arceneaux

3. Bacterial Iron Transport: Mechanisms, Genetics, and Regulation
   Volkmar Braun, Klaus Hantke, and Wolfgang Köster

4. Molecular Biology of Iron Transport in Fungi
   Sally A. Leong and Günther Winkelmann

5. Soil Microorganisms and Iron Uptake by Higher Plants
   Janette Palma Fett, Kristin LeVier, and Mary Lou Guerinot

6. Iron Transport in Graminaceous Plants
   Satoshi Mori

7. Coordination Chemistry of Siderophores: Thermodynamics and
   Kinetics of Iron Chelation and Release
   Anne-Marie Albrecht-Gary and Alvin L. Crumbliss

8. Biomimetic Siderophores: From Structural Probes to
   Diagnostic Tools
   Abraham Shanzer and Jacqueline Libman[†]

9. The Physical Chemistry of Bacterial Outermembrane Siderophore
   Receptor Proteins
   Dick van der Helm

10. The Iron Responsive Element (IRE) Family of mRNA Regulators
    Elizabeth C. Theil

11. Structure-Function Relationships in the Ferritins
    *Pauline M. Harrison, Paul C. Hempstead, Peter J. Artymiuk, and
    Simon C. Andrews*

12. Ferritin. Uptake, Storage and Release of Iron
    N. Dennis Chasteen

13. Ferritin. Its Mineralization
    Annie K. Powell

14. Iron Storage and Ferritin in Plants
    Jean-François Briat and Stéphane Lobréaux

15. Transferrin, the Transferrin Receptor, and the Uptake of
    Iron by Cells
    Philip Aisen

16. Iron Homeostasis
    Robert R. Crichton and Roberta J. Ward

17. The Role of Other Metal Ions in Iron Transport
    Barry Chiswell, Kelvin O'Hallaran, and Jarrod Wall

18. Iron Chelators for Clinical Use
    Gary S. Tilbrook and Robert C. Hider

Subject Index

**Volume 36. Interrelations Between Free Radicals and
Metal Ions in Life Processes**

1. The Mechanism of "Fenton-like" Reactions and Their Importance
   for Biological Systems. A Biologist's View
   Stefan I. Liochev

2. Reactions of Aliphatic Carbon-Centered and Aliphatic-Peroxyl
   Radicals with Transition-Metal Complexes as a Plausible Source
   for Biological Damage Induced by Radical Processes
   Dan Meyerstein

3. Free Radicals as a Result of Dioxygen Metabolism
   Bruce P. Branchaud

[†]Deceased

4. Free Radicals as a Source of Uncommon Oxidation States of
   Transition Metals
   George V. Buxton and Quinto G. Mulazzani

5. Biological Chemistry of Copper-Zinc Superoxide Dismutase and Its
   Link to Amyotrophic Lateral Sclerosis
   *Thomas J. Lyons, Edith Butler Gralla, and
   Joan Selverstone Valentine*

6. DNA Damage Mediated by Metal Ions with Special Reference to
   Copper and Iron
   José-Luis Sagripanti

7. Radical Migration Through the DNA Helix: Chemistry
   at a Distance
   Shana O. Kelley and Jacqueline K. Barton

8. Involvement of Metal Ions in Lipid Peroxidation:
   Biological Implications
   Odile Sergent, Isabelle Morel, and Josiane Cillard

9. Formation of Methemoglobin and Free Radicals in Erythrocytes
   Hans Nohl and Klaus Stolze

10. Role of Free Radicals and Metal Ions in the Pathogenesis of
    Alzheimer's Disease
    *Craig S. Atwood, Xudong Huang, Robert D. Moir,
    Rudolph E. Tanzi, and Ashley I. Bush*

11. Metal Binding and Radical Generation of Proteins in Human
    Neurological Disease and Aging
    Gerd Multhaup and Colin L. Masters

12. Thiyl Radicals in Biochemically Important Thiols in the Presence
    of Metal Ions
    Hans-Jürgen Hartmann, Christian Sievers, and Ulrich Weser

13. Methylmercury-Induced Generation of Free Radicals:
    Biological Implications
    Theodore A. Sarafian

14. Role of Free Radicals in Metal-Induced Carcinogenesis
    Joseph R. Landolph

15. pH-Dependent Organocobalt Sources for Active Radical Species:
    A New Type of Anticancer Agents
    Mark E. Vol'pin,[†] Ilia Ya. Levitin, and Sergei P. Osinsky

16. Detection of Chromatin-Associated Hydroxyl Radicals Generated
    by DNA-Bound Metal Compounds and Antitumor Antibiotics
    G. Mike Makrigiorgos

[†]Deceased

17. Nitric Oxide (NO): Formation and Biological Roles in Mammalian Systems
    Jon M. Fukuto and David A. Wink

18. Chemistry of Peroxynitrate and Its Relevance to Biological Systems
    Willem H. Koppenol

19. Novel Nitric Oxide-Liberating Heme Proteins from the Saliva of Bloodsucking Insects
    F. Ann Walker, José M. C. Ribeiro, and William R. Montfort

20. Nitrogen Monoxide-Related Diseases and Nitrogen Monoxide Scavengers as Potential Drugs
    Simon P. Fricker

21. Therapeutics of Nitric Oxide Modulation
    Ho-Leung Fung, Brian P. Booth, and Mohammad Tabrizi-Fard

Subject Index

Volume 37.   Manganese and Its Role in Biological Processes

1. Manganese in Natural Waters and Earth's Crust. Its Availability to Organisms
   James J. Morgan

2. Manganese Transport in Microorganisms
   Valeria Cizewski Culotta

3. Manganese Uptake and Transport in Plants
   Zdenko Rengel

4. Manganese Metabolism in Animals and Humans Including the Toxicity of Manganese
   Carl L. Keen, Jodi L. Ensunsa, and Michael S. Clegg

5. Interrelations Between Manganese and Other Metal Ions in Health and Disease
   James C. K. Lai, Margaret J. Minski, Alex W. K. Chan, and Louis Lim

6. The Use of Manganese as a Probe for Elucidating the Role of Magnesium Ions in Ribozymes
   Andrew L. Feig

7. Mn^{2+} as a Probe of Divalent Metal Ion Binding and Function in Enzymes and Other Proteins
   George H. Reed and Russell R. Poyner

8. Enzymes and Proteins Containing Manganese: An Overview
   James D. Crowley, Deborah A. Traynor, and David C. Weatherburn

9. Manganese(II) in Concanavalin A and Other Lectin Proteins
   A. Joseph Kalb(Gilboa), Jarjis Habash, Nicola S. Hunter,
   Helen J. Price, James Raferty, and John R. Helliwell

10. Manganese-Activated Phosphatases
    Frank Rusnak

11. Manganese(II) as a Probe for the Mechanism and Specificity of
    Restriction Endonucleases
    Geoffrey S. Baldwin, Niall A. Gormley, and Stephen E. Halford

12. Role of the Binuclear Manganese(II) Site in Xylose Isomerase
    Ralf Bogumil, Reinhard Kappl, and Jürgen Hüttermann

13. Arginase: A Binuclear Manganese Metalloenzyme
    David E. Ash, J. David Cox, and David W. Christianson

14. The Use of Model Complexes to Elucidate the Structure and
    Function of Manganese Redox Enzymes
    Vincent L. Pecoraro and Wen-Yuan Hsieh

15. Manganese(II)-Dependent Extradiol-Cleaving
    Catechol Dioxygenases
    Lawerence Que, Jr. and Mark F. Reynolds

16. Manganese Catalases
    Derek W. Yoder, Jungwon Hwang, and James E. Penner-Hahn

17. Manganese Peroxidase
    Michael H. Gold, Heather L. Youngs, and
    Maarten D. Sollewijn Gelpke

18. Manganese Superoxide Dismutase
    James W. Whittaker

19. Mechanistic Aspects of the Tyrosyl Radical-Manganese Complex
    in Photosynthetic Water Oxidation
    Curtis W. Hoganson and Gerald T. Babcock

20. The Polypeptides of Photosystem II and Their Influence on
    Manganotyrosyl-Based Oxygen Evolution
    Richard J. Debus

Subject Index

Volume 38.   Probing of Proteins by Metal Ions and Their
Low-Molecular-Weight Complexes

1. Peptide Bond Characteristics
   R. Bruce Martin

2. Lanthanide Ion-Mediated Peptide Hydrolysis
   Makoto Komiyama

3. Co(III)-Promoted Hydrolysis of Amides and Small Peptides
   David A. Buckingham and Charles R. Clark

4. Synthetic Cu(II) and Ni(II) Peptidases
   Gregory M. Polzin and Judith N. Burstyn

5. Palladium(II) and Platinum(II) Complexes as Synthetic Peptidases
   Nebojša Milović and Nenad M. Kostić

6. Protease Activity of 1,10-Phenanthroline-Copper Systems
   Makoto Kito and Reiko Urade

7. Specific Protein Degradation by Copper(II) Ions
   Geoffrey Allen

8. Artificial Iron-Dependent Proteases
   Saul A. Datwyler and Claude F. Meares

9. Hydroxyl Radical Footprinting of Proteins Using Metal
   Ion Complexes
   Tomasz Heyduk, Noel Baichoo, and Ewa Heyduk

10. Nickel- and Cobalt-Dependent Oxidation and Cross-Linking
    of Proteins
    Steven E. Rokita and Cynthia J. Burrows

11. Effects of Metal Ions on the Oxidation and Nitrosation of Cysteine
    Residues in Proteins and Enzymes
    Ann M. English and Dean A. Wilcox

12. Protein Cross-Linking Mediated by Metal Ion Complexes
    Kathlynn C. Brown and Thomas Kodadek

13. Ferrocenoyl Amino Acids and Peptides: Probing Peptide Structure
    Heinz-Bernhard Kraatz and Marek Galka

14. Synthetic Analogs of Zinc Enzymes
    Gerard Parkin

15. Mimicking Biological Electron Transfer and Oxygen Activation
    Involving Iron and Copper Proteins: A Bio(in)organic
    Supramolecular Approach
    Martinus C. Feiters

Subject Index

Volume 39.   Molybdenum and Tungsten.
Their Roles in Biological Processes

1. The Biogeochemistry of Molybdenum and Tungsten
   Edward I. Stiefel

2. Transport, Homeostasis, Regulation, and the Binding of Molybdate
   and Tungstate to Proteins
   Richard N. Pau and David M. Lawson

3. Molybdenum Nitrogenases. A Crystallographic and
   Mechanistic View
   David M. Lawson and Barry E. Smith

4. Chemical Dinitrogen Fixation by Molybdenum and Tungsten
   Complexes: Insights from Coordination Chemistry
   Masanobu Hidai and Yasushi Mizobe

5. Biosynthesis of the Nitrogenase Iron-Molybdenum-Cofactor from
   Azotobacter vinelandii
   Jeverson Frazzon and Dennis R. Dean

6. Molybdenum Enzymes Containing the Pyranopterin
   Cofactor. An Overview
   Russ Hille

7. The Molybdenum and Tungsten Cofactors:
   A Crystallographic View
   Holger Dobbek and Robert Huber

8. Models for the Pyranopterin-Containing Molybdenum and
   Tungsten Cofactors
   Berthold Fischer and Sharon J. Nieter Burgmayer

9. Biosynthesis and Molecular Biology of the Molybdenum
   Cofactor (Moco)
   Ralf R. Mendel and Günter Schwarz

10. Molybdenum in Nitrate Reductase and Nitrite Oxidoreductase
    Peter M. H. Kroneck and Dietmar J. Abt

11. The Molybdenum-Containing Hydroxylases of Nicotinate,
    Isonicotinate, and Nicotine
    Jan R. Andreesen and Susanne Fetzner

12. The Molybdenum-Containing Xanthine Oxidoreductases and
    Picolinate Dehydrogenases
    Emil F. Pai and Takeshi Nishino

13. Enzymes of the Xanthine Oxidase Family: The Role
    of Molybdenum
    David J. Lowe

14. The Molybdenum-Containing Hydroxylases of Quinoline,
    Isoquinoline, and Quinaldine
    Reinhard Kappl, Jürgen Hüttermann, and Susanne Fetzner

15. Molybdenum Enzymes in Reactions Involving
    Aldehydes and Acids
    *Maria João Romão, Carlos A. Cunha, Carlos D. Brondino,
    and José J. G. Moura*

16. Molybdenum and Tungsten Enzymes in C1 Metabolism
    Julia A. Vorholt and Rudolf K. Thauer

17. Molybdenum Enzymes and Sulfur Metabolism
    John H. Enemark and Michele Mader Cosper

18. Comparison of Selenium-Containing Molybdoenzymes
Vadim N. Gladyshev

19. Tungsten-Dependent Aldehyde Oxidoreductase: A New Family of Enzymes Containing the Pterin Cofactor
Roopali Roy and Michael W. W. Adams

20. Tungsten-Substituted Molybdenum Enzymes
C. David Garner and Lisa J. Stewart

21. Molybdenum Metabolism and Requirements in Humans
Judith R. Turnlund

22. Metabolism and Toxicity of Tungsten in Humans and Animals
Florence Lagarde and Maurice Leroy

Subject Index

Volume 40. The Lanthanides and Their Interrelations with Biosystems

1. Distribution of the Lanthanides in the Earth's Crust
Stuart Ross Taylor and Scott McLennan

2. Mobilization of Lanthanides through the Terrestrial Biosphere
Robert A. Bulman

3. Complexes of Lanthanide Ions with Amino Acids, Nucleotides, and Other Ligands of Biological Interest in Solution
Herbert B. Silber and Sarah J. Paquette

4. Biologically Relevant Structural Coordination Chemistry of Simple Lanthanide Ion Complexes
Jack M. Harrowfield

5. Lanthanide Ions as Probes in Studies of Metal Ion-Dependent Enzymes
Etsuro Yoshimura and Tokuko Watanabe

6. Lanthanide Chelates as Fluorescent Labels for Diagnostics and Biotechnology
Kazuko Matsumoto and Jingli Yuan

7. Responsive Luminescent Lanthanide Complexes
David Parker and J. A. Gareth Williams

8. Lanthanide Ions as Probes of Electron Transfer in Proteins
Ronald M. Supkowski and William DeW. Horrocks, Jr.

9. Lanthanide Ions as Luminescent Probes of Proteins and Nucleic Acids
Claudia Turro, Patty K.-L. Fu, and Patricia M. Bradley

10. Lanthanide-Promoted Peptide Bond Hydrolysis
Makoto Komiyama and Tohru Takarada

11. Lanthanide-Catalyzed Hydrolysis of Phosphate Esters and
    Nucleic Acids
    Hans-Jörg Schneider and Anatoly K. Yatsimirsky

12. Sequence-Selective Scission of DNA and RNA by
    Lanthanide Ions and Their Complexes
    Makoto Komiyama

13. Lanthanide Ions as Probes for Metal Ions in the Structure and
    Catalytic Mechanism of Ribozymes
    Roland K. O. Sigel and Anna Marie Pyle

14. Lanthanides as Shift and Relaxation Agents in Elucidating the
    Structure of Proteins and Nucleic Acids
    Carlos F. G. C. Geraldes and Claudio Luchinat

15. Lanthanide Ions as Magnetic Resonance Imaging Agents. Nuclear
    and Electronic Relaxation Properties. Applications
    Lothar Helm, Éva Tóth, and André E. Merbach

16. Interactions of Lanthanides and Their Complexes with Proteins.
    Conclusions Regarding Magnetic Resonance Imaging
    Silvio Aime, Alessandro Barge, Mauro Botta, and Enzo Terreno

17. Metabolism and Toxicity of the Lanthanides
    Robert A. Bulman

18. Cell Responses to Lanthanides and Potential Pharmacological
    Actions of Lanthanides
    Kui Wang, Yi Cheng, Xiaoda Yang, and Rongchang Li
Subject Index

Volume 41.   Metal Ions and Their Complexes in Medication

1. Speciation Dependent Intake and Uptake of Essential Elements
    Janette Davidge and David R. Williams

2. Magnesium in Human Therapy
    *Hans-Georg Classen, Heimo Franz Schimatschek, and
    Konrad Wink*

3. Calcium Status and Supplementation
    Lasse Larsson and Per Magnusson

4. Zinc Deficiency: Its Characterization and Treatment
    Ananda S. Prasad

5. The Use and Role of Zinc and Its Compounds in Wound Healing
    Paul W. Jones and David R. Williams

6. Iron Chelators and Their Therapeutic Potential
    Robert R. Crichton and Roberta J. Ward

7. Vanadium Compounds in the Treatment of Diabetes
    Katherine H. Thompson and Chris Orvig

8. Copper and Zinc Complexes as Antiinflammatory Drugs
   Carolyn T. Dillon, Trevor W. Hambley, Brendan J. Kennedy,
   Peter A. Lay, Jane E. Weder, and Qingdi Zhou

9. Gold Complexes in the Treatment of Rheumatoid Arthritis
   Luigi Messori and Giordana Marcon

10. The Medical Use of Lithium
    Nicholas J. Birch

11. Bismuth in Medicine
    Hongzhe Sun, Li Zhang, and Ka-Yee Szeto

12. Metal Complexes as Chemotherapeutic Agents Against Tropical
    Diseases: Malaria, Trypanosomiasis, and Leishmaniasis
    Roberto A. Sánchez-Delgado, Attilio Anzellotti, and Liliana Suárez

13. Metal Complexes as Therapeutic Agents in Nitrogen
    Monoxide Modulation
    Simon P. Fricker

Subject Index

Volume 42.   Metal Complexes in Tumor Diagnosis and as
Anticancer Agents

1. Magnetic Resonance Contrast Agents for Medical and
   Molecular Imaging
   Matthew J. Allen and Thomas J. Meade

2. Luminescent Lanthanide Probes as Diagnostic and
   Therapeutic Tools
   Jean-Claude Bünzli

3. Radiolanthanides in Nuclear Medicine
   Frank Rösch and Eva Forssell-Aronsson

4. Radiometallo-Labeled Peptides in Tumor Diagnosis and Therapy
   Mihaela Ginj and Helmut R. Maecke

5. Cisplatin and Related Anticancer Drugs: Recent
   Advances and Insights
   Katie R. Barnes and Stephen J. Lippard

6. The Effect of Cytoprotective Agents in Platinum
   Anticancer Therapy
   Michael A. Jakupec, Markus Galanski, and Bernhard K. Keppler

7. Antitumor Active *Trans*-Platinum Compounds
   Giovanni Natile and Mauro Coluccia

8. Polynuclear Platinum Drugs
   Nicholas P. Farrell

9. Platinum(IV) Anticancer Complexes
   Matthew D. Hall, Rachael Dolman, and Trevor W. Hambley

10. Ruthenium Anticancer Drugs
    Enzo Alessio, Giovanni Mestroni, Alberta Bergamo, and
    Gianni Sava

11. Antitumor Titanium Compounds and Related Metallocenes
    Francesco Caruso and Miriam Rossi

12. Gold Complexes as Antitumor Agents
    Luigi Messori and Giordana Marcon

13. Gallium and Other Main Group Metal Compounds as
    Antitumor Agents
    Michael A. Jakupec and Bernhard K. Keppler

14. Metal Ion Dependent Antibiotics in Chemotherapy
    David H. Petering, Chuanwu Xia, and William E. Antholine

Volume 44.   Biogeochemistry, Availability, and Transport
            of Metals in the Environment

 1. Atmospheric Transport of Metals
    Torunn Berg and Eiliv Steinnes

 2. The Marine Biogeochemistry of Iron
    Alison Butler and Jessica D. Martin

 3. Speciation and Bioavailability of Trace Metals in
    Freshwater Environments
    Laura Sigg and Renata Behra

 4. Bioavailability and Biogeochemistry of Metals in the
    Terrestrial Environment
    Kerstin Michel and Bernard Ludwig

 5. Heavy Metal Uptake by Plants and Cyanobacteria
    Hendrik Küpper and Peter M. H. Kroneck

 6. Arsenic: Its Biogeochemistry and Transport in Groundwater
    Charles F. Harvey and Roger D. Beckie

 7. Anthropogenic Impacts on the Biogeochemistry and Cycling
    of Antimony
    William Shotyk, Michael Krachler, and Bin Chen

 8. Microbial Transformations of Radionuclides: Fundamental
    Mechanisms and Biogeochemical Implications
    Jon R. Lloyd and Joanna C. Renshaw

 9. Biogeochemistry of Carbonates: Recorders of Past Oceans
    and Climate
    Rosalind E. M. Rickaby and Daniel P. Schrag

Subject Index

Comments and suggestions with regard to contents, topics, and the like for future volumes of the series are welcome.

The following Marcel Dekker, Inc., books are also of interest for any reader involved with bioinorganic chemistry or who is dealing with metals or other inorganic compounds:

HANDBOOK ON TOXICITY OF INORGANIC COMPOUNDS

Edited by Hans G. Seiler and Helmut Sigel, with Astrid Sigel

In 74 chapters, written by 84 international authorities, this book covers the physiology, toxicity, and levels of tolerance, including prescriptions for detoxification, for all elements of the Periodic Table (up to atomic number 103). The book also contains short summary sections for each element, dealing with the distribution of the elements, their chemistry, technological uses, and ecotoxicity as well as their analytical chemistry.

HANDBOOK ON METALS IN CLINICAL AND ANALYTICAL CHEMISTRY

Edited by Hans G. Seiler, Astrid Sigel, and Helmut Sigel

This book is written by 80 international authorities and covers over 3500 references. The first part (15 chapters) focuses on sample treatment, quality control, etc., and on the detailed description of the analytical procedures relevant for clinical chemistry. The second part (43 chapters) is devoted to a total of 61 metals and metalloids; all these contributions are identically organized covering the clinical relevance and analytical determination of each element as well as, in short summary sections, its chemistry, distribution, and technical uses.

HANDBOOK ON METALLOPROTEINS

Edited by Ivano Bertini, Astrid Sigel, and Helmut Sigel

This book consists of 23 chapters written by 43 international authorities. It summarizes a large part of today's knowledge on metalloproteins, emphasizing their structure–function relationships, and it encompasses the metal ions of life: sodium, potassium, magnesium, calcium, vanadium, chromium, manganese, iron, cobalt, nickel, copper, zinc, molybdenum, and tungsten.

The Biogeochemical Cycles of the Elements and the Evolution of Life

Peter M. H. Kroneck

Universität Konstanz, Mathematisch-Naturwissenschaftliche Sektion, Fachbereich Biologie, D-78457 Konstanz, Germany

1. Introduction   1
2. Modern Earth: Cycling of the Biological Elements   2
3. Evolution of Life   3
4. Outlook   6
References   6

1. INTRODUCTION

The biogeochemical cycling of the basic elements for life including carbon, nitrogen, oxygen, and sulfur, has attracted the interest of many researchers over the past decades in view of its importance for the earth with a rapidly growing population, and in view of its impact on our environment and climate [1]. Of similar importance, the biogeochemistry of the essential transition metals has been extensively studied because of their function as cofactors, or as part of cofactors in enzymes, and as structural elements in proteins. Note that metalloproteins comprise one third to half of all known proteins. Many essential life processes, including photosynthesis, respiration, nitrogen fixation, strictly depend on transition metal ions and their ability to catalyze multi-electron transformations. Other essential life processes, such as proteolysis and the equilibration of carbon dioxide and bicarbonate are hydrolytic transformations that are also catalyzed by metalloenzymes [2]. The essential transition metal ions for terrestrial organisms include vanadium to zinc of the first-row transition metal series and molybdenum and tungsten in the second- and third-row series.

Iron is the most abundant transition metal ion in most terrestrial organisms. Iron levels are usually high in most lakes, streams, and rivers whereas levels of other transition metals vary widely [2]. With regard to these essential micronutrients, the ocean, particularly far from land, is the most extreme environment for life on earth [3]. Iron is a limiting nutrient for primary production in large areas of the oceans. Dissolved iron(III) in the upper oceans occurs almost entirely in the form of complexes with strong ligands presumed to be of biological origin. Recently, it has been shown that photolysis of Fe(III)−siderophore complexes led to formation of lower-affinity Fe(III) ligands and the reduction of Fe(III), increasing the availability of siderophore-bound iron for uptake by planktonic assemblages [4].

The biogeochemical cycles of the elements have changed through time as a result of the evolution of life and their interaction with the environment. The interactions between cycles of different elements and between life and its environment raises fundamental evolutionary questions [5]. Is the selection of a given transition metal ion through evolution a consequence of availability or of unique functionality? Did changing metal availability through time change biogeochemical cycling of major elements on a global scale? Answers to these important questions will help us to understand the Earth's environmental history and evolutionary relationships.

This book provides an overview of the modern biogeochemical cycles of the essential elements for life. This chapter intends to introduce the reader to several aspects of the cycles of the major biological elements and the key transition metal ions on modern Earth as discussed in detail in the following chapters.

2. MODERN EARTH: CYCLING OF THE BIOLOGICAL ELEMENTS

The basic elements for life, such as carbon, nitrogen, and sulfur, can exist in the biosphere in several oxidation states, ranging from $C(+4)$ in carbon dioxide to $C(-4)$ in methane, $N(5+)$ in nitrate to $N(-3)$ in ammonia, or $S(+6)$ in sulfate to $S(-2)$ in hydrogen sulfide. Interconversions of these various species constitute their global biogeochemical cycles which are sustained by complex biological processes, with bacteria playing a prominent role [3,6–10].

The evolution of atmospheric oxygen is intimately intertwined with the evolution of the Earth, both from a biological and geological perspective. Atmospheric O_2 is controlled principally by the long-term (multi-million-year) geochemical cycles of carbon and sulfur. The effect on O_2 of the cycles of other elements that exhibit variable oxidation states has been shown to be far less significant quantitatively [11]. The first documented description how the carbon and sulfur cycles affect O_2 was given in 1845 [12]. The following global reactions were deduced [Eqs. (1) and (2)].

$$CO_2 + H_2O \rightleftharpoons CH_2O + O_2 \tag{1}$$

$$2Fe_2O_3 + 8SO_4^{2-} + 16H^+ \rightleftharpoons 15O_2 + 4FeS_2 + 8H_2O \tag{2}$$

Reaction (1), going from left to right, represents net photosynthesis (photosynthesis minus respiration) as represented by the burial of organic matter (CH_2O) in sediments. Going from right to left, Reaction (1) represents two processes: (i) the oxidation of old sedimentary organic matter, and (ii) the sum of several reactions involving the thermal breakdown of organic matter via diagenesis, metamorphism, and magmatism with the resulting reduced carbon-containing compounds released to the Earth's surface where they are oxidized to CO_2 by atmospheric or oceanic O_2 [11].

Reaction (2), going from right to left, represents the oxidation of pyrite during weathering of the continents (organic sulfur is lumpered here with pyrite for simplification) and the sum of thermal pyrite decomposition and the oxidation of resulting reduced sulfur-containing gases produced by metamorphism and magmatism. Going from left to right, Reaction (2) is the sum of several reactions. These are (i) photosynthesis and initial burial of organic matter [Reaction (1)], (ii) early diagenetic bacterial sulfate reduction and the production of H_2S, with organic matter serving as the reducing agent, and (iii) the precipitation of pyrite via the reaction of H_2S with Fe_2O_3. The geochemical sulfur cycle also involves hydrothermal reactions between oceanic crust and seawater sulfate at mid-ocean rises, and these reactions have been advanced as major controls on atmospheric O_2 [11].

3. EVOLUTION OF LIFE

It is thought that the Earth's early atmosphere had very likely CO_2, CO, and N_2 as main components by the time of the origin of life. NO, H_2, H_2O, and sulfuric gases could also have been present, and probably trace amounts of O_2 [13]. Small amounts of dioxygen could have been generated from photolysis of water [14]. The level of atmospheric dioxygen began to rise $\sim$2.4–2.6 gigayears ago and the present concentration of free oxygen was only reached 500 million years ago. The exact history of dioxygen accumulation is still a matter of debate, but the generally accepted theory is that the emergence of organisms capable of performing oxygenic photosynthesis—probably ancestors of the present-day cyanobacteria—was responsible for this [13]. It has been argued that cyanobacteria altered the solubility of metals on a global scale as a by-product of dioxygen-evolving photosynthesis. It is reasonable to propose that the greater autonomy provided by photosynthesis allowed early cyanobacteria to colonize vast vacant habitats, life previously being restricted to niches with exploitable sources of chemical energy.

Molecular oxygen did not immediately accumulate in the atmosphere and waters, due largely to the presence of abundant ferrous iron, which reacted with the oxygen to precipitate massive banded ferric iron formations. As larger concentrations of atmospheric O_2 began to build up, redox actions of metal sulfides with dixoygen could occur, such as the transformation of insoluble molybdenum disulfide to soluble molybdate [Eq. (3)]. This reaction may have allowed life to proliferate by making the highly soluble molybdate ion available for incorporation into critical metalloenzymes [15].

$$2MoS_2 + 7O_2 + 2H_2O \rightleftharpoons 2MoO_4^{2-} + 4SO_2 + 4H^+ \tag{3}$$

Concerning the evolution of the first metabolic cycles, an autotrophic origin of life has been discussed most recently [16]. The central problem within the theory is the first process of carbon fixation. An autocatalytic cycle has been proposed that can be reconstructed retrodictively from the extant reductive citric acid cycle by replacing thioesters by thioacids and assuming that the reducing power is obtained from the oxidative formation of pyrite, FeS_2, from ferrous sulfide (FeS), and hydrogen sulfide. This cycle would be strictly chemoautotrophic, and photoautotrophy is not required. The discovery of the occurrence of several forms of iron sulfide in sulfur bacteria leads one to speculate as to whether sulfur bacteria have a homeostatic device based on Fe/S rather than on Fe/O solids, and whether this was the earlier form of iron store. An Fe/S homeostatic economy for primitive life could have had many other ramifications [17].

Almost all forms of life conserve energy for growth and development by the process of oxidative phosphorylation [18]. This involves the spontaneous, thermodynamically favorable transfer of electrons from a suitable donor through a complex chain of membrane-bound metal-dependent redox proteins, enzymes, and electron carriers to a suitable acceptor. The concomitant pumping of protons across the membrane enables energy to be conserved by chemiosmotic coupling in the form of a proton-motive force. The enzyme ATP synthase uses this force to couple proton transfer back across the membrane to the synthesis of ATP from ADP and inorganic phosphate, P_i [19]. Higher forms of life typically use NADH (from glucose oxidation) as the electron donor, and molecular oxygen serves as the acceptor, a process accomplished by the aerobic respiratory chain of mitochondria. Microorganisms, on the other hand, vary by the electron donors and acceptors that they use for respiratory purposes. Some microbes use the NADH/oxygen/donor/acceptor pair, but others may use a variety of compounds, such as methane, ammonia, dihydrogen, or carbon monoxide as donors, and carbon dioxide, nitrate, sulfate, or metal ions such as Fe(3+) as acceptors [19].

Specialized metalloenzymes donate electrons directly or indirectly into complex respiratory systems that ultimately, by means of both organic (quinones) and protein redox carriers, reduce the added electron acceptor by the appropriate enzyme complex. Most recently, a simple energy-conserving system was discovered in *Pyrococcus furiosus*, an archeon which grows optimally near

100°C. This microorganism has an anaerobic respiratory system that consists of a single metalloenzyme, a membrane-bound hydrogenase. It does not require an added electron acceptor as the hydrogenase reduces protons, the simplest of acceptors, to dihydrogen gas by using electrons from the cytoplasmic redox protein ferredoxin. The ability of microorganisms such as *P. furiosus* to couple hydrogen production to energy conservation has important ramifications not only in the evolution of respiratory systems but also in the origin of life [20].

It is also not surprising that several molecular links do exist between the evolution of enzymes used in the cycles of the basic elements of life. A prime example is given by denitrification enzymes and cytochrome *c* oxidase,

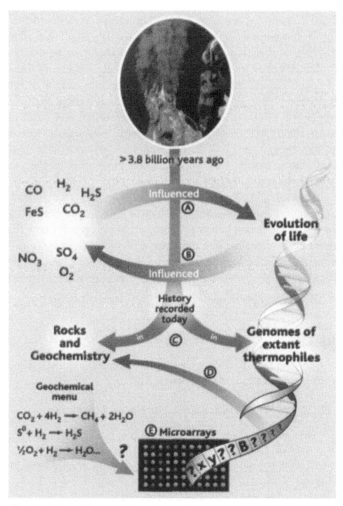

Figure 1 Geochemistry of hydrothermal ecosystems and the evolution of life (taken from Ref. [23], with permission).

the terminal oxidase of aerobic respiration [21]. Denitrification is a respiratory process in which nitrate (NO_3^-) is reduced stepwise to nitrite (NO_2^-), nitric oxide (NO), and nitrous oxide (N_2O) to finally yield dinitrogen gas (N_2). The most important is the homology between nitric oxide reductase and heme/copper cytochrome oxidases, and the presence of the mixed-valence $[Cu^{1.5+}(S_{Cys})_2Cu^{1.5+}]$ CuA electron-transfer center in nitrous oxide reductases, the quinol-dependent NO reductase from the gram-positive *Bacillus azotoformans*, and heme/copper cytochrome oxidases [21,22].

4. OUTLOOK

Environmental conditions are some of the strongest forces that drive evolution. Because of hydrothermal systems which have prevailed throughout Earth's history, the extant organisms in these systems and their genomes are living records of changes over geological time. The deeply branching lineages of the universal tree of life are all occupied by thermophiles, although this is still a matter of debate [23] (Fig. 1). Both genomic and geochemical information must be regarded as records of evolutionary changes that have occurred in these hydrothermal systems. Genomes of several thermophilic microorganisms have been sequenced, and geochemical evidence of ecosystem evolution has become available in nearly 4 billion years of history of hydrothermally altered rocks, as well as in active hydrothermal ecosystems.

Once decoded and integrated, these genomic and geochemical clues can reveal the geological and evolutionary history of how biogeochemical interactions turn hot water and rocks into habitats. Clues to the strategies for how life thrives in these dynamic ecosystems are beginning to be elucidated through a confluence of biogeochemistry, microbiology, ecology, molecular biology, and genomics. These efforts have the potential to reveal how ecosystems originate, the extent of the subsurface biosphere, and the driving forces of evolution [23–26].

REFERENCES

1. Smil V. Cycles of Life: Civilization and the Biosphere, Scientific American Library. New York: W. H. Freeman and Company, 1997.
2. Butler A. Science 1998; 281:207–210.
3. Morel FMM, Price NM. Science 2003; 300:944–947.
4. Barbeau K, Rue EL. Bruland KW, Butler A. Nature 2001; 413:409–413.
5. Williams RJP, Frausto da Silva JJR. The Natural Selection of the Chemical Elements. Oxford: Clarendon Press, 1996.
6. Brune A, Frenzel P, Cypionka H. FEMS Microbiol Rev 2000; 24:691–710.
7. Conrad R. Microbiol Rev 1996; 60:609–640.
8. Zumft WG. Microbiol Mol Rev 1997; 61:533–616.
9. Simon J. FEMS Microbiol Rev 2002; 26:285–309.
10. Anbar AD, Knoll AH. Science 2002; 297:1137–1142.

11. Berner RA, Berling DJ, Dudley R, Robinson JM, Wildman RA Jr. Ann Rev Earth Planet Sci 2003; 31:105–134.
12. Ebelmen JJ. Ann Rev Mines 1845; 12:627–654.
13. Kasting FJ. Science 1993; 259:902–926.
14. Bekker A, Holland HD, Wang P-L, Rumble D III, Hannah JL, Coetzee LL, Beukes NN. Nature 2004; 427:117–120.
15. Stiefel E. In: Sigel A, Sigel H, eds. Metal Ions in Biological Systems. Vol. 39, New York, Basel: M. Dekker Inc, 2002:1–29.
16. Wächtershäuser G. Proc Natl Acad Sci USA 1990; 87:200–204.
17. Williams RJP. Nature 1990; 343:213–214.
18. Mitchell P. Nature 1961; 191:144–148.
19. Thauer RK, Jungermann K, Decker K. Bacteriol Rev 1977; 41:100–180.
20. Sapra R, Bagramyan K, Adams MWW. Proc Natl Acad Sci USA 2003; 7545–7550.
21. Wasser IM, deVries S, Moënne-Loccoz P, Schröder I, Karlin KD. Chem Rev 2002; 102:1201–1234.
22. Hendriks J, Oubrie A, Castresana J, Urbani A, Gemeinhardt S, Saraste M. Biochim Biophys Acta 2000; 1459:266–273.
23. Reysenbach A-L, Shock E. Science, 2002; 296:1077–1082.
24. Newman DK, Banfield JF. Science 2002; 296:1071–1077.
25. Williams RJP, Frausto da Silva JJR. Bringing Chemistry to Life. From Matter to Man, Oxford: Oxford University Press, 1999.
26. Sigel A, Sigel H, Sigel RKO, eds. Biogeochemistry, Availability, and Transport of Metals in the Environment, Metal Ions in Biological Systems. Vol. 44, New York: Marcel Dekker, Inc, 2005.

2

Biogeochemistry of Dihydrogen (H$_2$)

Tori M. Hoehler

*Exobiology Branch, NASA Ames Research Center,
Moffett Field, California 94035-1000, USA*

1. Introduction 10
   1.1. Antiquity and Ubiquity of H$_2$ in the Microbial World 10
   1.2. The Evolving Role of H$_2$ in Biogeochemistry 11
2. H$_2$ from the Planetary Matrix 13
   2.1. Abiotic Mechanisms of H$_2$ Production 13
   2.2. Atmospheric Chemistry Involving H$_2$ 14
   2.3. Abiotic H$_2$ as an Energy Source for Photosynthesis-
      Independent Ecosystems 16
3. H$_2$ Cycling in Anaerobic Ecosystems 17
   3.1. Interspecies H$_2$ Transfer 17
      3.1.1. Organic Decomposition in Anaerobic
         Ecosystems 17
      3.1.2. Thermodynamics of Microbial H$_2$ Metabolism 20
      3.1.3. Obligate Interspecies H$_2$ Transfer 21
      3.1.4. Facultative Interspecies H$_2$ Transfer 22
      3.1.5. Enzymatic and Metabolic Reversal 23
      3.1.6. H$_2$ Leakage 25
   3.2. Factors Controlling H$_2$ Concentrations 26
      3.2.1. The Production–Consumption Steady State 26
      3.2.2. Thermodynamic Controls 27
      3.2.3. Kinetic and Other Controls 30
   3.3. Implications for Biogeochemistry 32
4. H$_2$ Cycling in Phototrophic Ecosystems 32

    4.1. H_2 in the Metabolism of Phototrophic Microorganisms   32
        4.1.1.   Photosynthetic Metabolism   32
        4.1.2. H_2 Consumption   34
        4.1.3. H_2 Production   34
    4.2. H_2 Cycling in Photosynthetic Microbial Mats   36
    4.3.   Implications for Biogeochemistry   39
5. Summary   39
Acknowledgments   41
References   41

1. INTRODUCTION

Hydrogen has played an important role in Earth's geochemistry and biology since the earliest stages of the planet's history. H_2 is postulated to have been a significant component of the early, prebiotic atmosphere, where it had direct or indirect effects on atmospheric redox chemistry and radiation budget, and contributed to planetary oxidation by escape to space. With the origins and evolution of life came a broad variety of metabolisms involving H_2, and an expanded role for H_2 in planetary biogeochemistry. H_2 levels in the modern atmosphere are perhaps three orders of magnitude or more lower than during Earth's earliest history, but the significance of H_2 cycling for planetary chemistry persists through its central role in the microbial world.

1.1. Antiquity and Ubiquity of H_2 in the Microbial World

The reversible oxidation of molecular hydrogen, $H_2 \rightleftharpoons H^+ + 2e^-$, is simultaneously among the simplest, most ancient, and most widely utilized reactions in microbial biochemistry. Hydrogenases, the enzymes which catalyze this transformation, are present in all three phylogenetic domains of life [1–3]. They are particularly widely represented in the bacteria and Archaea, including the most deeply branching and least derived lineages [2,3], suggesting that H_2 metabolism may have been a very early biochemical innovation. Similarly, the use of H^+ gradients to store the energy generated by metabolic chemistry is extremely widespread in these domains [4]. Because H_2 oxidation liberates electrons, energy and H^+ in a single step, it could have filled a dual metabolic and energy-conserving role in very early organisms. The widespread availability of H_2 in geothermal emanations (e.g., [5–8]) further suggests that it would have been available from the very beginning as an abundant energy source on which to base the earliest forms of biochemistry. Indeed, some origin of life theories suggest that H_2 cycling may have been a component of the first metabolic redox chemistry [9,10].

Since its origins in biochemistry, H_2-based redox chemistry has come to be associated with an extremely broad range of microbial metabolic strategies, including oxygenic and anoxygenic phototrophy, anaerobic chemolitho- and chemoheterotrophy, and aerobic chemolithotrophy [3,11,12]. H_2/H^+ can serve as electron donor, acceptor, or both, in these processes. This broad metabolic utility and versatility casts H_2 in a highly central role in the ecology of microbial communities, and in the geochemistry they mediate.

1.2. The Evolving Role of H_2 in Biogeochemistry

The utilization of H_2 in metabolic chemistry has remained a common thread throughout the long history of microbial evolution. The increasing complexity of individual organisms and microbial ecosystems has been mirrored by an increasing complexity in the role played by H_2. Successive biochemical advances have often retained and expanded on earlier aspects of H_2 metabolism. This continuity and expansion in the microbial utilization of H_2 corresponds to a continually evolving role for H_2 in the biogeochemistry of the Earth.

Many origin of life theories hold that the first metabolisms were chemoautotrophic—systems that relied on chemical energy sources and readily available reducing power to synthesize organic matter from CO_2 [10,13]. In the modern, oxygenated oceans and atmosphere, H_2 is present at low levels that would scarcely be useful for anaerobic metabolisms of this sort. But the early Earth may have offered an abundant supply of H_2 for microbial activity, by virtue of an atmosphere 3–4 orders of magnitude richer in H_2 [14–16], and elevated rates of geothermal emanation [17] that likely contributed H_2 to the environment at several times the modern rate. Many of the organisms most deeply rooted in the phylogenetic tree are capable of H_2 metabolism [2,3], so that Earth's early biosphere could potentially have been supported by direct transfer of H_2 from abiotic (geochemical) sources to biological consumers. Possible early and modern representatives of such systems are addressed in Section 2.

The accumulation of complex organic matter, either from abiotic sources (terrestrial or extraterrestrial), or from the productivity of a chemosynthetic biosphere, presented an additional energy source that became accessible with the increasing complexity of microbial metabolism and communities. Under the anaerobic conditions that prevailed prior to ~2.4 billion years ago [18], the complete degradation of complex organics cannot be accomplished by a single organism, but requires the combined and sequential activities of several types of microbes [19]. In modern anoxic ecosystems, such as aquatic sediments or animal digestive systems, microbes accomplish the overall transformation of organic matter by breaking the process up into individual steps [19]. These individual steps are linked through the transfer of electrons from one group of microbes to another, using an electron carrier such as H_2. In these systems, H_2 cycling lies at the organism–organism interface, and is an important mediator of electron flow, energy regulation, and community interactions. In the past and present, such systems represent a major control on the redox chemistry of the oceans and atmosphere [20], and the ultimate biological filter on material passing into the rock record. The importance of H_2 in the microbial ecology and biogeochemistry of anoxic systems is detailed in Section 3.

The microbial biosphere achieved its full expression of metabolic and ecological complexity when organisms evolved the ability to harness light energy—first, using electron donors such as H_2S, Fe^{2+}, and H_2 to conduct anoxygenic photosynthesis, and later, using H_2O to conduct oxygenic photosynthesis (the type of photosynthesis found in green plants). Each advance harbored the potential to significantly increase the productivity of the biosphere, and therefore, to increase the biological impact on Earth's geochemistry [21]. Yet each retained a capacity to metabolize H_2 by a variety of means [22–25]. As such, these organisms had the potential to interact very directly with the preexisting (H_2-sensitive) biosphere. The fossil record indicates that much of this new productivity was concentrated in microbial mats and stromatolites, which remained globally widespread for a significant fraction of Earth's history [26,27]. These systems place the great productivity of photosynthetic organisms in close physical proximity with the biocatalytic potential of anaerobic microbes. H_2 serves at the interface of the two groups, providing a functional link that affects the ecology, chemistry, and, ultimately, the contribution of these systems to global geochemistry. The role of H_2 in these photosynthetic microbial ecosystems is discussed in Section 4.

H_2 retained a metabolic role even with the advent of eukaryotic, and ultimately multi-cellular organisms. In a variety of deep branching anaerobic eukaryotes, metabolism is tied to an H_2-based symbiosis with bacterial or Archaeal partners (see review in Ref. [1]). The driver of most of these symbioses is an H_2-generating organelle called the hydrogenosome, which appears in some cases to be derived from a mitochondrion [28]. These hydrogenosomes carry out fermentation, rather than oxidative phosphorylation, of pyruvate—and thereby produce H_2. Microbial symbionts [which may include methanogens, sulfate reducers, and (in one case) photosynthetic purple non-sulfur bacteria] increase the energetic yield of this fermentation by consuming the end product H_2 (Section 3). The symbiosis significantly enhances the growth of the host, and is in some cases so evolved that methanogen symbionts have lost their cell walls, and exhibit synchronization of cell division with the host. Green algae also exhibit an active H_2 metabolism, and are considered a promising potential source of "biohydrogen" for alternative energy systems [29]. Even animals play host to a hydrogen cycle, by virtue of the anaerobic microbes they host in their digestive tracts. This cycling can be of global biogeochemical significance: for example, microbial methanogenesis in animal digestive systems comprises $\sim 20\%$ of the net global emission of methane to the atmosphere [30].

The role of H_2 as a continuous thread through the evolution of metabolic- and ecosystem-level complexity, and in the associated biogeochemical cycling, is explored in more detail in the following sections.

2. H_2 FROM THE PLANETARY MATRIX

2.1. Abiotic Mechanisms of H_2 Production

H_2 is produced by the interaction of water and rocks through a variety of mechanisms, and the frequent presence of H_2 in geothermal emanations (e.g., [5,7,8]) suggests that such processes occur widely in the crust. Perhaps the best known mechanism involves the reduction of water to H_2 by the ferrous iron component in rocks, as shown, for example, by the oxidation of fayalite (F) to quartz (Q) and magnetite (M):

$$3Fe_2SiO_4(F) + 2H_2O \longrightarrow 3SiO_2(Q) + 2Fe_3O_4(M) + 2H_2 \qquad (1)$$

The capacity for H_2 generation is greatest in minerals with high ferrous iron contents, such as the ultramafic minerals olivine and pyroxene, where the process is called serpentinization [31]. Basalts, with a lower ferrous iron content, have a lower capacity to produce H_2, and granites lower still (at least with respect to this mechanism). At present, the Earth's crust consists primarily of basaltic to granitic rocks, but ultramafics were likely abundant in the early, predifferentiated crust [32]. Exposure of such rocks continues in the present day as a result of tectonic activity—e.g., faulting at divergent plate boundaries, or obduction of ophiolite sequences onto continental crust. Fluid emanations from these sites are frequently characterized by high H_2 contents [6,33–35], consistent with active serpentinization. The process of serpentinization occurs much more rapidly at temperatures of several hundred degrees than at temperatures characteristic of Earth's surface environments [31], and requires constant exposure of fresh mineral surfaces to water.

Water can also be reduced to H_2 by metallic iron or nickel. Although not particularly relevant to Earth's modern crustal geochemistry, this process may have been an important contributor of H_2 to the environment during Earth's earliest history, when impact events may have transported large quantities of metallic iron and nickel to our planet's surface [36]. Indeed, it is hypothesized that the moon-forming impact could have generated a quantity of H_2 equivalent to a 50 bar atmosphere (N. Sleep, personal communication). Subsequent large impacts may also have resulted in episodes of significant H_2 production [36].

Ferrous sulfides, which are most concentrated in hydrothermal vent structures, can also be a source of H_2. In this case, electrons originate with the sulfide, which is oxidized to pyrite [37]:

$$FeS + H_2S \longrightarrow FeS_2 + H_2 \qquad (2)$$

A prominent origin of life theory suggests that such a process may have provided energy and reducing equivalents for the earliest biochemistry [10,38].

H_2 is generated in a solid-state reaction within silicate minerals that have incorporated trace water impurities, typically as $-OH$, into their lattice during formation. As the rock cools, formation of peroxy linkages from $-OH$ impurities liberates H_2 [39,40]:

$$O_3Si-O-SiO_3 + H_2O \longrightarrow O_3Si-OH + HO-SiO_3 \tag{3}$$
$$O_3Si-OH + HO-SiO_3 \longrightarrow O_3Si-O-O-SiO_3 + H_2 \tag{4}$$

Release of H_2 to the environment depends on (very slow) diffusion through the mineral matrix, or fracturing/crushing of individual mineral grains. Crushing experiments employing a range of silicate mineral types have shown a release of H_2, with an estimated potential H_2 reservoir of up to $5 \, L \, m^{-3}$ of rock [40]. Crushing of silicate rocks may also liberate H_2 when the broken Si–O bonds reduce water from the surrounding environment to H_2 [41]. This process depends directly on the exposure of fresh surfaces, since the reaction occurs only in a monolayer where the crystal lattice is disrupted.

A final mechanism of H_2 generation depends on the radiolytic cleavage of water by α, β, or γ radiation produced during decay of isotopes in the host rocks [42]. This process is believed responsible for elevated H_2 concentrations (up to millimolar levels) in fluid emanations from rocks of the East European platform [43], Canadian and Fennoscandian shield [44], and Witwatersrand Basin, South Africa [45]. This mechanism is generally most important for granites, in which the content of the principal radiogenic elements U, Th, and, K is generally highest among crustal rock types [46].

Among the processes that could prevail in widespread crustal rocks (i.e., excluding pyritization and impact deposition of reduced iron and nickel), serpentinization offers the greatest potential capacity for H_2 generation (per unit volume of rock), by orders of magnitude. However, multiple factors—including rock composition, fluid flow and chemistry, temperature regime, and mechanical processing—ultimately determine which mode of lithogenic H_2 production may prevail in a given environment.

2.2. Atmospheric Chemistry Involving H_2

The principal significance of H_2 in modern biogeochemical cycling is its contribution to microbial metabolism [47]. Nonetheless, atmospheric chemical processes involving H_2 have contributed to the long-term chemical evolution of our planet, and may have played a role in establishing the conditions that fostered the origin and early evolution of life. H_2 is a trace component of the modern atmosphere, with an abundance of $\sim$0.5 ppm [48] and a residence time of <1 year [49]. Estimates of its abundance in the early, prebiotic atmosphere have varied substantially, and depend principally on theories regarding the oxidation state of the early mantle—which, in turn, buffers the oxidation state of outgassing material [50]. The mantle has had an approximately constant oxidation state for at least the last 3.9 billion years [51], so the H_2 content of volcanic gases during this period is thus expected to approximate modern levels of up to a few percent [52]. The abundance of H_2 in volcanic gases and in the atmosphere prior to this time is not clear. Early models that assume a significant elemental iron and nickel component to the mantle predict a strongly reducing atmosphere, with H_2 (along with CH_4 and NH_3) in abundance (e.g., [53]). As discussed earlier, this could also have occurred at least transiently with the delivery of iron and nickel by large meteorite impacts [36]. Theories favoring an early segregation of metallic nickel and iron into the core (out of the mantle) predict a mildly reducing to neutral atmosphere of N_2, CO_2, and H_2O, with an H_2 abundance of hundreds to thousands of parts per million [14–16]—still 3–4 orders of magnitude higher than in the present day. Exclusive of mantle redox buffering, fluid cycling through a primarily ultramafic early crust could have contributed significant H_2 to the atmosphere via enhanced serpentinization.

H_2 itself is considered photochemically inert in the lower atmosphere [48], but it participates in a variety of reactions involving the radical products of photolysis. By serving as a sink for the oxidizing OH radical [49], and influencing the abundance of the atmospheric reductants H, HCO, and H_2CO [54], H_2 directly affects the oxidation state of the atmosphere. As such it strongly affects the possible O_2 content of the atmosphere, as well as speciation in the volatile nitrogen, carbon, and sulfur systems. In turn, this balance (e.g., the CH_4/CO_2 ratio, in the carbon system) affects the atmospheric radiation budget (through differing efficiencies of IR absorption), the potential for formation of photochemical hazes [55–57], and the formation and stability of simple organics. The latter process is particularly important in the context of prebiotic chemistry. For example, the synthesis of amino acids and other prebiotic organics by spark discharge has been shown to occur under either highly reducing conditions dominated by CH_4 and NH_3 [58,59], or approximately neutral mixtures of CO_2 and N_2 [14,60,61], but in both cases, the yield of organics requires the presence, and is dependent on the abundance, of H_2. Each of these factors—radiation budget, aerosol formation, and organic synthesis—helped to establish the context for the origins and early evolution of life, and each depended on atmospheric H_2 abundance.

Above and beyond its role in buffering the oxidation state of the atmosphere through chemical interactions, hydrogen can contribute to the irreversible oxidation of the planet by escape to space [52]. The flux to space can occur as H_2 or as dissociated atomic hydrogen [62], and need not originate with H_2. Any species capable of transporting atomic H into the upper atmosphere and undergoing photodissociation, can contribute to hydrogen escape. This process occurs at low rates in the modern, oxidized atmosphere [62], where the low abundance of reduced species like CH_4 and H_2 leaves H_2O as the principal H-carrying species. The upward flux of H_2O is stemmed by "trapping" in the cold regions of the atmosphere, leaving little capacity for H generation in the upper atmosphere [63]. A moderately reduced atmosphere, however, offers the possibility for transport of H by reduced compounds that are not subject to the atmospheric cold trap, so H escape due to, e.g., CH_4 or H_2 flux, could contribute significantly to the overall oxidation of the planet [64,65]. For further discussion, see Chapter 3, "Dioxygen Over Geological Time".

2.3.   Abiotic H_2 as an Energy Source for Photosynthesis-Independent Ecosystems

Microbial ecosystems based on the energy supplied by water–rock chemistry carry particular significance in the context of geo- and astrobiology. With no direct dependence on solar energy, lithotrophic microbes could conceivably penetrate a planetary crust to a depth limited only by temperature or pressure constraints. Given the average geothermal gradient, and the apparent ability of some microbes to grow up to $\sim 120°C$, it is conceivable that the upper 5–10 km of the Earth's crust represents habitable space [66]. The deep lithospheric habitat is thereby potentially much greater in volume than its surface counterpart. In addition, it could offer a stable refuge against inhospitable surface conditions related to climatic or atmospheric evolution, or even surface-sterilizing impacts [67]. H_2 is the chemical "fuel" most often envisioned to support a rock-hosted biosphere, because of its wide utilization in microbial metabolism and the existence of a variety of mechanisms for its production by the rock matrix (as above).

The possibility of a "deep biosphere" on Earth has received increasing attention during the last decade (e.g., [66,68–70]). Gold [66] allowed for the possibility that the deep biosphere could conceivably host a biomass equivalent to that at the Earth's surface. Microbes and microbial activity have been detected kilometers deep in oil reservoirs [71–73] and deep sea sediments [74,75]. Such habitats are fueled by organic matter that originates with photosynthesis at the Earth's surface. The magnitude and distribution of microbial communities that are supported solely by the chemistry of water–rock interactions (i.e., systems that are truly independent of the photosynthetic surface biosphere), has been much more difficult to characterize. Such communities have been inferred for basalt aquifers, where H_2 may be generated by the interaction of ferrous rocks with water [76,77], and granitic [78,79] and mixed mineralogy environments [45], where H_2 may be generated by radiolysis. The presence of possibly biogenic chemical signatures in hydrothermal vent effluents, and of microbes themselves in vent-associated fluids released by submarine eruptive events, also suggests the presence of rock-hosted communities within seafloor basalts [80–83]. However, difficulties in accessing such environments without contamination, and in measuring the extremely low rates of metabolic activity that might occur there, have hindered our ability to quantify the contribution of rock-hosted biomes to global biomass and productivity.

The potential that life can exist in a subsurface realm, independent of photosynthetic productivity, has raised the possibility that habitable niches may exist elsewhere in our solar system [84]. As defined by the long-term stability of liquid water, the Earth is at present the only body in the solar system capable of supporting a surface biosphere. Any biosphere elsewhere in the solar system would therefore be confined to the deep subsurface. Specifically, attention has been focused on Mars and Europa, both of which may support a vast volume of habitable space (as defined by the presence of liquid water at clement temperatures) within hundreds to thousands of meters of the surface [85–87]. Lacking access to sunlight and to the products of photosynthesis (i.e., abundant organic matter and high energy oxidants), any life in such environments would be limited to anaerobic chemotrophic metabolism, such as that supported by lithotrophic H_2 production. Indeed, the H_2-producing processes outlined earlier are expected to occur on any rocky body that experiences fluid circulation, including Mars and Europa. The Martian crust, even in the present day, is thought to consist largely of ultramafic rocks [88,89], so that serpentinization could lead to widespread H_2 production in the subsurface. On Europa, it is envisioned that processes similar to those occurring in terrestrial hydrothermal systems could contribute H_2 and other reduced components to the ocean [86,87]. In each case, the prospects for long-term survivability of any such biosphere would depend not only on the supply of H_2, but also on the continuous supply of suitable oxidants [90]. Mechanisms for creation of these oxidants, based on surface irradiation processes, have been proposed for both worlds.

3. H_2 CYCLING IN ANAEROBIC ECOSYSTEMS

The biogeochemistry of anoxic environments plays a key role in regulating the redox chemistry of our oceans and atmosphere [20], and frequently serves as the final filter on material passing into the rock record. The primary mediators of this chemistry (at temperatures within the biologically-tolerated range) are communities of anaerobic bacteria, which utilize molecular hydrogen as a widespread currency of electrons and energy. Nowhere is the significance of H_2 to microbial processes more clearly expressed than in anoxic ecosystems.

3.1. Interspecies H_2 Transfer

The reversible nature of H_2 biochemistry ($H_2 \rightleftharpoons 2H^+ + 2e^-$) permits its utilization as either an electron source or sink. By the late 1800s, it had been shown that H_2 is produced during fermentation of a wide variety of organic substrates [91–93]. Over the next 50 years, it became clear that H_2 is also consumed as a substrate in microbial processes such as methanogenesis [94] and acetogenesis [95]. It is now know that H_2 can serve as an electron donor in a broad spectrum of anaerobic reductions (Table 1), consistent with the widespread distribution of hydrogenase enzymes among anaerobic microorganisms [2,3].

3.1.1. Organic Decomposition in Anaerobic Ecosystems

In the absence of oxygen, the complete remineralization of complex organic molecules to CO_2 is a chemically difficult problem that cannot be accomplished by a single microbial metabolism. Instead, such substrates are degraded into successively smaller pieces via the activities of multiple groups of microorganisms ([19], Fig. 1). Most of these sequential steps are oxidation–reduction reactions

Table 1   Microbial Metabolism Involving H_2

Process	Role of H_2 [a]	Control by H_2 [b]	Sample reference
Reduction			
O_2	D	?	[233]
NO_3^-	D	U	[234]
Mn^{4+}	D	U	[235]
Fe^{3+}	D	U	[236]
As^{5+}	D	U	[237]
SO_4^{2-}, SO_3^{2-}, $S_2O_3^{2-}$, S^0	D	U	[238]
Olefins (C=C)	D	U	[239]
CO_2 (methanogenic)	D	U, R (?)	[240]
Fatty acid chain elongation	D	U	[241]
Fermentation			
Fatty acids	A	O	[103]
Saccharides	A	F	[107]
Alcohols	A	O	[112]
Amino acids	A, D	O, F	[108]
Acetate $\rightleftharpoons H_2/CO_2$	A, D	R	[125]
Formate $\rightleftharpoons H_2/CO_2$	A, D	R	[121]
Degradation of aromatics	A, D	?	[102]
Anaerobic methane oxidation	A	R (?)	[242]
Nitrogen fixation	A	?	[243]
$CH_3OH \rightarrow CH_4 + CO_2$	D, B	F	[136]
Acetate $\rightarrow CH_4 + CO_2$	B	F	[136]
Anoxygenic phototrophy	See Table 3		[24]
Oxygenic phototrophy	See Table 3		[23]

[a]D: electron donor, H_2 consumed; A: electron acceptor, H_2 produced from H^+; B: potential by-product.
[b]F: Facultative—H_2 shifts the products of metabolism; O: Obligate—H_2 production is obligate, H_2 partial pressures that are too high make the process thermodynamically unfavorable; U: Unfavorable—reaction becomes thermodynamically or kinetically unfavorable when H_2 partial pressures are too low, use of alternative electron acceptors may still be feasible; R: Reversal—H_2 can reverse the direction of metabolism from H_2 consumption to H_2 production or vice-versa.

that require transfer of electrons extracellularly from one group of microorganisms to another. This can be viewed as a series of redox half-reactions, each catalyzed by a different group of microbes: the overall reaction is not completed without transfer of electrons from one half-reaction to the other, or from one group of organisms to the other. Since free electrons cannot be transferred extracellularly through solution, a suitable molecular carrier is required. Several strategies employing various electron-transferring molecules have been documented, but none has the potential for more widespread utilization among a broad cross-section of the anaerobic microbial community than does H_2 transfer [12].

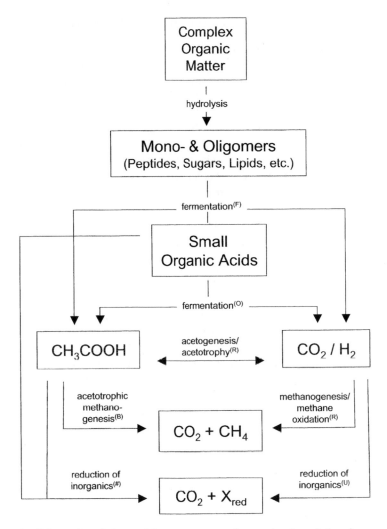

Figure 1 Schematic of the multi-step process of organic degradation in anaerobic systems, indicating the influence of H_2. Boxes indicate major chemical pools, arrows with text indicate microbially-mediated process. Superscripts on arrow text indicate the mode of control by H_2: (F)—facultative interspecies H_2 transfer; (O)—obligate interspecies H_2 transfer; (B)—H_2 can be a by-product, shifting the product yield; (U)—the reaction becomes thermodynamically unfavorable when H_2 concentrations are too low; (R)—the process is demonstrated or hypothesized to be reversible, based on H_2 concentrations. Note (#): important inorganic oxidants may include NO_3^-, Mn^{4+}, Fe^{3+}, and various oxidized sulfur species (e.g., SO_4^{2-}); X_{red} corresponds to the reduced form (e.g., H_2S). Adpated from Ref. [19].

3.1.2. Thermodynamics of Microbial H_2 Metabolism

To the extent that many of the microbial processes involved in anaerobic organic matter remineralization include H_2 as a product or reactant, they are energetically sensitive to variations in its concentration, in a fashion that can be quantified through thermodynamic calculations [96]. This is, of course, generally true for any chemical species involved in any chemical reaction:

Given a sample reaction, $aA + bB \rightarrow cC + dD$ (where uppercase letters are chemical species and lower case letters are stoichiometric coefficients), the Gibbs Free Energy of Reaction (which determines both whether the process can occur spontaneously and how much energy is available to bacteria mediating the reaction) under ambient conditions is given by,

$$\Delta G_{rxn} = \Delta G^\circ + R\,T \cdot \ln \frac{\{C\}^c \{D\}^d}{\{A\}^a \{B\}^b} \tag{5}$$

where ΔG° is the free energy change under standard conditions and unit activities; R is the universal gas constant; T is Kelvin temperature; and $\{\}$ indicates activity [97].

Hence, variations in the activity, or concentration, of any product or reactant influences the free energy yield of the overall reaction—with an exponential dependence on the reaction stoichiometry of the species in question. H_2 has a particularly strong influence in this regard, because it is generally produced or consumed in a high stoichiometric ratio relative to other products and reactants. For example, the anaerobic microbial processes of propionate fermentation and methanogenic CO_2 reduction involve H_2 in stoichiometric ratios of $3:1$ and $4:1$ (respectively) compared to the carbon substrates or products (by convention, water does not factor into energetic calculations for reactions occurring in aqueous solution):

$$CH_3CH_2COOH + 2H_2O \quad \longrightarrow \quad CH_3COOH + CO_2 + 3H_2 \tag{6}$$

$$CO_2 + 4H_2 \quad \longrightarrow \quad CH_4 + 2H_2O \tag{7}$$

Therefore, the energetics of these reactions are much more affected by variations in the concentration of H_2 than in concentrations of the other products and reactants. For example, to affect the same change in free energy yield that accompanies a doubling of the H_2 concentration, the CO_2 concentration would have to change by 8- or 16-fold (for propionate fermentation and methanogenesis, respectively). The sensitivity of such reactions to H_2 is enhanced in environmental settings, where H_2 residence times are frequently seconds or less (Section 3.2.1). H_2 concentrations in such systems can change very rapidly, which in turn can very rapidly (and significantly) change the energetics of H_2-producing or consuming processes. Table 2 illustrates this effect for a typical anoxic sediment system.

The realization that H_2 provides both a means of electron transfer between anaerobic microorganisms and a control on the energetics of those organisms led

Table 2 Relative Effectiveness of H_2 and Other Metabolites as Regulators of Microbial Energy Metabolism in a Natural Anoxic Ecosystem[a]

Species	Typical concentration (μM)[b]	Residence time (s)[c]	Δ[Concentration] for $\Delta(\Delta G) =$ 10 kJ mol^{-1} [d]	Time for $\Delta(\Delta G) =$ 10 kJ mol^{-1} (s)[e]
(Propionate fermentation: $CH_3CH_2COO^- + 3H_2O \rightarrow CH_3COO^- + H^+ + HCO_3^- + 3H_2)$[f]				
H_2	0.005	1.3	3.9 ×	5
$CH_3CH_2COO^-$	5	4×10^3	59 ×	2×10^5
CH_3COO^-	5	10^4	59 ×	6×10^5
HCO_3^-	60,000	4×10^7	59 ×	2×10^9
(Hydrogenotrophic methanogenesis: $4H_2 + HCO_3^- + H^+ \rightarrow CH_4 + 3H_2O)$[f]				
H_2	0.005	1.3	2.8 ×	3.6
CH_4	1,000	7×10^5	59 ×	4×10^7
HCO_3^-	60,000	4×10^7	59 ×	2×10^9

[a]Adapted from Ref. [12].
[b]Concentrations observed under methanogenic conditions in sediments from Cape Lookout Bight, NC, USA, at 22°C; data taken from Refs. [145,162,163].
[c]Time required for one complete replacement of the ambient pool, as calculated from measured turnover constants from each species, as reported in Refs. [145,162,163].
[d]Change in concentration (×-fold) required to alter the *in situ* free energy of reaction by ± 10 kJ mol^{-1} (depending on whether the species in question is a product or reactant).
[e]Time required for a concentration *increase* to raise the ambient concentration sufficiently to alter the *in situ* free energy of reaction by ± 10 kJ mol^{-1}, under the conditions described earlier. Note that a concentration *decrease* also yields an energetic change, but this is not treated here.
[f]Equations (6) and (7), modified to reflect HCO_3^- as the dominant species in the carbonate equilibrium series.

to the concept of "inter-species H_2 transfer". This concept holds that the biochemistry and bioenergetics of a microorganism involved in H_2 transfer can be affected by the activity of its partner, by virtue of the control the second organism may exert on H_2 concentrations [98]. The groundwork for this concept was laid by thermodynamic calculations showing that many H_2-producing fermentation reactions, which were known to occur in nature or in mixed culture, would be favorable only if H_2 concentrations were held at low levels (e.g., by the activities of other organisms [5,99]). Microbiologists generally credit Hungate [100] with being the first to demonstrate experimentally that variations in H_2 concentrations do, indeed, substantially influence the course of organic matter fermentation. In the decades since, numerous variations on the theme of interspecies H_2 transfer have been documented.

3.1.3. Obligate Interspecies H_2 Transfer

As the name implies, obligate interspecies transfer characterizes a microbial partnership in which efficient H_2 transfer is *required* to maintain conditions in which both partners can function. Typically, this applies to fermentations of small organic molecules that unavoidably generate H_2 as an end product, and are subject to thermodynamic inhibition by H_2 at low levels. Organisms catalyzing such reactions can only function in the presence of an H_2-consuming partner that can hold H_2 at the low levels required to make the fermentation thermodynamically favorable. The first and oft-cited example of obligate interspecies H_2 transfer is that of "*Methanobacillus omelianskii*", which was originally held to be an individual microorganism that catalyzed the methanogenic oxidation of ethanol:

$$2CH_3CH_2OH + CO_2 \longrightarrow 2CH_3COOH + CH_4 \tag{8}$$

The development of new cultivation techniques ultimately showed this culture to consist of two partner organisms that transferred H_2 between them [101]:

$$2CH_3CH_2OH + 2H_2O \longrightarrow 2CH_3COOH + 4H_2 \tag{9}$$

$$CO_2 + 4H_2 \longrightarrow CH_4 + 2H_2O \tag{7}$$

The ethanol-oxidizing organism could not be grown in pure culture (in the absence of its H_2-consuming partner), because the H_2 it generates would quickly accumulate to concentrations at which the reaction was rendered thermodynamically unfavorable. Subsequently, a variety of co-cultures were isolated in which the fermenting partner could only grow in the presence of a hydrogen-scavenging organism [102–105]. The list of compound types subject to obligate interspecies H_2 transfer now includes straight-chain saturated fatty acids [103,106,107], some amino acids [108–111], and small alcohols [101,107,112]. Interestingly, recent work has demonstrated that the H_2-consuming partner can, in some instances, be replaced by an efficient physical (non-biological) mechanism of H_2 removal [113]. This further demonstrates that H_2 concentration is among the key variables controlling the function of obligate H_2-transferring organisms, and raises the possibility that some of these organisms could be grown and studied in pure culture.

3.1.4.   Facultative Interspecies H_2 Transfer

During fermentation of larger and more complex substrates, a range of products—including more or less hydrogen—may be possible. The specific mixture of products formed in such reactions can sometimes be shifted towards the production of more H_2, if an H_2-consuming partner organism is present [98,114]. In the absence of a partner organism (e.g., in pure culture), the fermentation is still possible, but end products are shifted towards more reduced products and less (or no) H_2. In this sense, the interspecies H_2 transfer is facultative: the presence of an H_2 sink can affect the metabolism of fermentative organisms, but is not required for their continued function (as it is for obligate interspecies transfer). This process was first demonstrated by Iannotti et al. [115]. In pure culture, the fermenter *Ruminococcus albus* degrades glucose yielding both acetate and ethanol:

$$C_6H_{12}O_6 + 1.3H_2O \longrightarrow 1.3CH_3COOH + 0.7CH_3CH_2OH$$
$$+ 2CO_2 + 2.6H_2 \qquad (10)$$

When co-cultured with the H_2 consumer *Wolinella succinogenes*, however, the fermentation generates the more oxidized species acetate as the exclusive organic product, along with more H_2:

$$C_6H_{12}O_6 \longrightarrow 2CH_3COOH + 2CO_2 + 4H_2 \qquad (11)$$

The observed shift in products has a thermodynamic/energetic basis. The removal of H_2 by the partner organism creates a higher oxidation potential in the environment, which allows the fermenter to realize a higher energy yield through more complete oxidation of its substrate [115]. Similar product shifts resulting from facultative interspecies H_2 transfer were subsequently demonstrated for other fermentative organisms utilizing a variety of substrates [116–118]. The apparent energetic advantages of the interspecies transfer arrangement are borne out by studies showing significantly enhanced growth yields for fermenters cultured in the presence of H_2-consuming partners, vs. those grown in monoculture [119,120].

3.1.5. Enzymatic and Metabolic Reversal

Every chemical reaction is, in principle, reversible—regardless of whether that reaction occurs in a sterile beaker or a living organism, regardless of whether it results from a chance molecular encounter or is catalyzed by a complex enzyme. When ambient conditions change sufficiently that the sign of ΔG is reversed (becomes positive for the reaction in question), the reverse reaction becomes thermodynamically favored and will occur spontaneously provided kinetic barriers do not exist, and barring inputs of energy from other sources. The same holds true for the overall direction of a process consisting of multiple reaction steps. The examples of obligate and facultative interspecies transfer demonstrate that fluctuations in ambient H_2 concentrations can alter the energetic environment of H_2-generating reactions sufficiently to affect the products of microbial processes, or even to turn them on or off like a switch. For microbial processes that occur with free energy yields that are small under ambient conditions, it is therefore conceivable that a sufficiently large change in H_2 concentrations could reverse the sign of ΔG, and thereby make the reverse process thermodynamically favored. Exactly this behavior has been demonstrated in a few cases.

The H_2-based reversibility of biological reactions occurs at levels ranging from individual enzymes to multi-step processes comprising a complete, dissimilatory metabolism. The simplest example is the reversibility of hydrogenase enzymes themselves. A significant fraction of these enzymes can be made to either produce H_2 while oxidizing a biological electron carrier, or to reduce a biological electron carrier at the expense of H_2, depending on fluctuations in the ambient H_2 concentration (e.g., [2]). A similar behavior is observed for a coupled series of enzymatic reactions that, together, catalyze the complete reduction of CO_2 to formic acid:

$$CO_2 + H_2 \rightleftharpoons HCOOH \tag{12}$$

Wu et al. [121] demonstrated that the overall direction of reaction in this system of enzymes can be driven forward or reversed solely through the addition or removal of H_2. Because this formate-hydrogen lyase enzyme system is present in representatives of both fermentative [122] and terminal metabolic [123] organisms, it has been suggested that formate may serve as an agent of interspecies electron transfer in much the same way that H_2 does [122,124].

The highest order of demonstrated reversibility in a biological system occurs in an organism that can completely reverse its energy-harvesting, dissimilatory metabolism when ambient conditions change. Lee and Zinder [125] isolated an organism in pure culture that obtains energy via the acetogenic reduction of CO_2:

$$2CO_2 + 4H_2 \longrightarrow CH_3COOH + 2H_2O \text{ (acetogen in pure culture)} \tag{13}$$

However, when the organism was grown in co-culture with a methanogen that efficiently removed H_2 via interspecies transfer, it completely reversed this metabolism:

$$CH_3COOH + 2H_2O \longrightarrow 2CO_2 + \mathbf{4H_2} \text{ (acetogen in co-culture)} \tag{14}$$

$$CO_2 + \mathbf{4H_2} \longrightarrow CH_4 + 2H_2O \text{ (methanogen)} \tag{7}$$

Remarkably, the organism could grow while catalyzing either direction of metabolism.

A similar mechanism has been proposed to explain the anaerobic oxidation of methane in marine sediments. This process appears generally to be coupled to reduction of the terminal electron acceptor sulfate [126–128]:

$$CH_4 + SO_4^{2-} + 2H^+ \longrightarrow CO_2 + H_2S + 2H_2O \tag{15}$$

but no known sulfate reducers are capable of metabolizing methane. Accordingly, it was suggested that methanogenic organisms oxidize methane by operating in reverse, with suitable thermodynamic conditions maintained by interspecies transfer of H_2 to sulfate reducers [129]:

$$CH_4 + 2H_2O \longrightarrow CO_2 + \mathbf{4H_2} \text{ (methanogen)} \tag{16}$$

$$\underline{SO_4^{2-} + 2H^+ + \mathbf{4H_2} \longrightarrow H_2S + 4H_2O \text{ (sulfate reducer)}} \tag{17}$$

$$CH_4 + SO_4^{2-} + 2H^+ \longrightarrow CO_2 + H_2S + 2H_2O \text{ (overall)} \tag{15}$$

H_2 concentrations measured in a sulfate-containing and actively methane-oxidizing system were sufficiently low as to provide a thermodynamically favorable and bioenergetically feasible energy yield for both sulfate reducers and methanogens operating "in reverse" [129]. It has since been shown that anaerobic methane oxidation is often associated with consortia consisting of closely juxtaposed Archaea and sulfate reducers [130–132]. This strongly supports the idea of an electron-transferring consortium—and therefore, of a form of reversed methanogenesis—although the specific nature of the electron transfer agent (whether H_2, formate, acetate, or other) remains in question [129,133,134].

These examples of reversible metabolism represent the microbial equivalent of a human being that could survive either by oxidizing organic matter to CO_2 at the expense of oxygen (as we do) or, if conditions changed, by synthesizing organic matter and oxygen from CO_2 and water. Clearly this is a non-sensical example, mostly because the very large energy yields and complex biochemistry associated with human metabolism would make it totally unfeasible. But the low energy yields and relatively simple biochemistry characteristic of anaerobic microbial processes are strongly subject to energetic forcing, by H_2, even to the extent of complete metabolic reversal.

3.1.6. H_2 Leakage

The discussion thus far has focused on reactions that directly involve H_2. However, some reactions that do not directly involve H_2 are catalyzed by organisms that nonetheless possess hydrogenase enzymes. The presence of these enzymes makes it possible, and perhaps unavoidable, that the organism will produce H_2 if ambient levels are sufficiently low. An example is the methylotrophic methanogen *Methanosarcina barkeri* which, in pure culture, disproportionates acetate to methane and CO_2 (a reaction that does not involve H_2):

$$CH_3COOH \quad \longrightarrow \quad CH_4 + CO_2 \tag{18a}$$

In this process, electrons liberated by cleaving the carbon–carbon bond of acetate are transferred to a biological electron carrier, which subsequently reduces the methyl carbon to methane [135]. However, if the intermediate biological electron carrier can also serve as an electron donor for the organism's hydrogenase enzymes, maintenance of low ambient H_2 concentrations could create a thermodynamic driving force that would channel some of this reducing power into reduction of protons (production of H_2) rather than reduction of methyl carbon to methane. *M. barkeri* exhibits exactly such a behavior when co-cultured with the sulfate reducer *Desulfovibrio vulgaris*, which maintains a low ambient H_2 concentration within the culture:

$$CH_3COOH + 0.66H_2O \quad \longrightarrow \quad 0.66CH_4 + 1.33H_2 + 1.33CO_2 \ (M. \ barkeri) \tag{18b}$$

$$SO_4^{2-} + 4H_2 \quad \longrightarrow \quad S^{2-} + 4H_2O \ (D. \ vulgaris) \tag{19}$$

Thirty-three percent of the electrons that could be used by *M. barkeri* in the production of methane are instead channeled into production of H_2, which is lost from the cell [136]. When growing on methanol, rather than acetate, *M. barkeri* diverted nearly 42% of its electrons into H_2 production [136]. The same phenomenon has been inferred for natural ecosystems for acetate [137], methanol and methylamine [138], where the proportion of electrons lost to H_2 leakage may be higher still. Each of these is an example of a metabolism that does not outwardly seem to involve H_2, but may nonetheless be strongly affected by it. In theory, similar effects might be possible in any of the broad spectrum of organisms that contain hydrogenase enzymes.

3.2. Factors Controlling H_2 Concentrations

As suggested in the foregoing sections, many of the microbial populations and chemical processes in anaerobic ecosystems are tightly coupled through a shared metabolic and thermodynamic dependence on H_2 concentrations. Understanding how H_2 concentrations are themselves controlled in such systems can provide a quantitative framework in which to interpret and predict the responses of a wide variety of microbial processes to changing conditions.

3.2.1. The Production–Consumption Steady State

The ephemeral nature of H_2 as an intermediate during the decomposition of organic matter became clear when it repeatedly eluded detection in anaerobic environments despite clear expectations that it should be produced in large quantities [5,139]. Later work showed that H_2 is indeed present in such environments, but generally at nanomolar levels [140–143]. These concentrations are of the same order of magnitude as in co-cultures actively engaged in interspecies H_2 transfer [122,136]. At such levels, the residence time of H_2 in natural systems can be on the order of seconds or less [122,144,145]. The rapid turnover and low concentrations of H_2 indicate that rates of consumption closely match those of production. In this production–consumption steady state, the possible range of H_2 concentrations is ultimately bounded by the physiologic limitations (e.g., enzyme uptake limits, thermodynamic/bioenergetic requirements) of both partner organisms.

In practice, it is the influence of H_2 consumers that is most clearly expressed in the steady state H_2 levels of most anaerobic ecosystems. A demonstration of this effect was given by Lovley et al. [146], who showed that the steady-state H_2 level in otherwise identical sediment pore waters differed under methanogenic (sulfate absent) versus sulfate-reducing (sulfate present) conditions—strongly suggesting that methanogens and sulfate reducers (the "consumers" in the production–consumption balance) exerted the fine scale control on the H_2 steady state in that system. The same effect has been demonstrated for most anoxic sediments that have been so examined (e.g., [129,145,147–150]). These observations form the basis for understanding and predicting the factors that control environmental H_2 concentrations, in both a qualitative and quantitative sense.

3.2.2. Thermodynamic Controls

For microbial processes that offer relatively low energy yields, the chief physiologic limitation on the ability to consume H_2 appears to be a bioenergetic one. All known organisms require that free energy be available at a certain minimum level, corresponding to the minimum quantity necessary for generating ATP via molecular energy-conserving mechanisms [151,152]. As described earlier, the free energy yield of H_2-consuming processes is highly dependent on the H_2 activity. Lower H_2 concentrations correspond to smaller (less negative) free energy yields:

$$X_{ox} + nH_2 \longrightarrow X_{red} + yH_2O \tag{20}$$

$$\Delta G_{rxn} = \Delta G^\circ_{(T,P)} + RT \cdot \ln \frac{\{X_{red}\}}{(P_{H_2})^n \cdot \{X_{ox}\}} \tag{21a}$$

where ΔG_{rxn} is the overall free energy change for the reaction under ambient conditions (the energy yield available to the organism), $\Delta G^\circ_{(T,P)}$ is the standard free energy yield of the reaction adjusted for ambient temperature and pressure, P signifies partial pressure, $\{\}$ denotes activity, and X_{ox} and X_{red} represent the oxidized and reduced forms of the electron accepting partner for H_2 (e.g., $X_{ox} = CO_2$ or SO_4^{2-}, $X_{red} = CH_4$ or H_2S, for methanogenesis or sulfate reduction, respectively).

Hence, H_2 consumers may draw H_2 down to a level at which ΔG_{rxn} meets the minimum biological free energy requirement (ΔG_{min}), but no lower:

$$\Delta G_{rxn} = \Delta G^\circ_{(T,P)} + RT \cdot \ln \frac{\{X_{red}\}}{(P_{H_2})^n \cdot \{X_{ox}\}} \leq \Delta G_{min} \tag{21b}$$

(recalling that more negative values of ΔG correspond to more energetic processes).

The free energy available from a given reaction also depends on the temperature and pressure, and on activities of other products and reactants involved (and accordingly, on factors which may affect the speciation or activity of these species—e.g., pH and ionic composition/strength). If ΔG_{min} is known, the minimum partial pressure of H_2 that can be maintained by a given H_2-consuming microbe can therefore be related to a suite of environmental variables through rearrangement of Eq. (21b):

$$P_{H_2} \geq \frac{X_{ox}}{X_{red}} \cdot \exp\left(\frac{\Delta G_{min} - \Delta G^\circ_{(T,P)}}{RT}\right)^{1/n} \tag{22a}$$

The quantity $\Delta G^\circ_{(T,P)}$, the standard free energy change of the reaction, is a function of the particular electron acceptor that serves as the oxidant for H_2. Among electron acceptors that may be commonly available in natural systems, $\Delta G^\circ_{(T,P)}$ for hydrogen oxidation becomes more negative (the reaction becomes more energetically favorable) along the series: CO_2 (acetogenesis) > CO_2

(methanogenesis) $> SO_4^{2-} > Fe^{3+} \gg Mn^{4+} > NO_3^-$. Equation (22a) predicts that more negative values of $\Delta G_{(T,P)}^{\circ}$ should allow for lower levels of H_2. Indeed, a variety of experiments utilizing pure cultures and natural sediments have demonstrated that, when all other variables are held constant, H_2 partial pressures generally decrease along the spectrum of reactions from aceto-genesis to nitrate reduction [145,150,153–155]. Over the entire spectrum of processes, the variation in H_2 concentrations may be as much as four orders of magnitude [145,150].

The variation in H_2 with terminal electron acceptor is a key factor in the competition for substrates among terminal bacteria. By maintaining sufficiently low H_2 concentrations, terminal bacteria can thermodynamically inhibit competitors with less favorable reaction energetics [146,153]. In this way, one organism garners the entire H_2 supply until it has exhausted its electron acceptor, at which point the next most favorable process can occur. This effect provides a basis for biogeochemical zonation (e.g., the "stratification" of sediments into exclusive zones of sulfate reduction, methanogenesis, etc. [156]). Such zonation strongly affects the role a given ecosystem plays in biogeochemical cycling.

Further evidence of thermodynamic control on H_2 concentrations comes from an observed temperature effect that agrees with the predictions of Eq. (22a). In a series of studies, H_2 concentrations were consistently higher at higher temperatures in both methanogenic [129,145,157–159] and sulfate reducing [129,145] sediments. This corresponds to the decrease in available free energy as the temperature increases. The predicted effect of temperature on H_2 concentrations [from Eq. (22a)] can be significant for highly entropic reactions such as methanogenesis or sulfate reduction: a temperature variation from 0 to 30°C (such as might occur with seasonal variation in an environmental setting) corresponds to an approximately nine fold variation in H_2 concentrations; over a full biological range of temperatures (e.g., -10 to 130°C), H_2 is predicted to vary over four orders of magnitude (Fig. 2).

For reactions occurring in the gas phase, significant pressure changes can factor very prominently in reaction energetics, especially when the reaction in question entails a large molar volume change (e.g., as in the case of methanogenesis). For the present case, however, it should be borne in mind that biological reactions occur in aqueous solution. Partial molar volume changes associated with aqueous reactions are typically small, so large pressure changes would be required to have any significant effect on the overall reaction energetics of H_2-consuming processes, and therefore on the minimum H_2 levels that can be maintained. For example, Eq. (22a) predicts a minimum H_2 partial pressure for methanogenesis that is only $\sim$25% lower at the bottom of the Mariana Trench (at a pressure of $\sim$1100 bar) than at the ocean surface (given standard temperature of 298°K). Note that at present, this is purely a theoretical prediction, since it is not clear that any systematic studies concerning this effect have been undertaken.

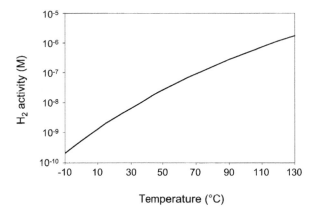

Figure 2 Theoretical temperature dependence of H_2 activity, when maintained by methanogenic organisms operating at a constant free energy yield: $HCO_3^- + 4\,H_2 + H^+ \rightarrow CH_4 + 3H_2O$ at $\Delta G_{mp} = -10\,kJ\,mol^{-1}$. Conditions: $\{HCO_3^-\} = 10^{-2}\,M$, $\{CH_4\} = 10^{-4}\,M$; $\{H^+\} = 10^{-7}\,M$, $P = 1.0\,bar$.

Variations in H_2 resulting from changes in the activities of other reactants and products (e.g., X_{ox} and X_{red}) in the H_2-consuming process have not been extensively examined. In at least one tested case, however, results closely followed the predictions of Eq. (22a). Under sulfate reducing conditions ($X_{ox} = SO_4^{2-}$, $X_{red} = H_2S$), increasing concentrations (activities) of sulfate translated to decreased levels of H_2 [145].

For both the sulfate experiments described above, and some of the temperature experiments, the variation in H_2 was *quantitatively* described by Eq. (22b):

$$P_{H_2} = \frac{X_{ox}}{X_{red}} \cdot \exp\left(\frac{\Delta G_{min} - \Delta G^\circ_{(T,P)}}{RT}\right)^{1/n} \tag{22b}$$

This requires that H_2-consuming organisms maintain H_2 at levels exactly sufficient to yield a constant free energy of reaction, regardless of whether the change in temperature or sulfate concentration would tend to increase or decrease the free energy of reaction. Hoehler et al. [160] have argued that this may result when H_2 consumers colonize discrete sources of H_2 (e.g., organic particles in sediments), and thereby form a microbial "net" capable of intercepting the outward flux of H_2. The impetus for further colonization of these sources should continue as long as the outgoing flux of H_2 persists at levels that are biologically useful, and cease at residual H_2 concentrations corresponding to ΔG_{min}. Accordingly, measurements of H_2 concentrations in systems exhibiting such behavior may be useful in assessing biological minimum free energy requirements, which in turn provide constraints on the distribution and activity of microorganisms in the environment [160].

3.2.3. Kinetic and Other Controls

Because all microorganisms are subject to minimum energy requirements, the bioenergetic/thermodynamic control on H_2 always represents the ultimate constraint on H_2 consumption in biological systems. Indeed, this appears to be the primary control expressed in methanogenic and sulfate reducing ecosystems. However, thermodynamic controls may be superceded by other considerations under a variety of circumstances. Thermodynamic control on H_2 concentrations in methanogenic and sulfate reducing systems likely results from two key characteristics; significant deviation from these characteristics may lead to control by kinetic or other factors:

1. The energy yields associated with these processes under naturally occurring conditions are very small, so that bioenergetic limitations are encountered when H_2 concentrations are still relatively high. However, for processes that offer much larger energy yields, the ability of enzymes to scavenge H_2 may become limiting at concentrations far higher than those corresponding to energetic limitations. An extreme example is the case of O_2. Aerobic H_2 oxidation offers a very large energy yield. Bioenergetic limitation would occur at H_2 concentrations of $\sim 10^{-44}$ M (less than one molecule of H_2 in all the world's oceans). Clearly, such a level could not be maintained by biological processes. This effect is demonstrated experimentally with the high-energy electron-acceptor nitrate, when added to methanogenic sediments at millimolar concentrations. The bioenergetic threshold for H_2 consumption under such circumstances is $\sim 10^{-33}$ M. And indeed, in the cases examined, H_2 levels decrease by several orders of magnitude upon addition of nitrate, consistent with the greater available free energy. However, the decrease is only to $\sim 10^{-11}$ M, still far above the bioenergetic threshold [145]. Hence, these levels more likely reflect enzyme uptake limitations, and thermodynamic control is qualitatively, but not quantitatively expressed.

2. The terminal electron acceptors for methanogenesis and sulfate reduction (CO_2 and SO_4^{2-}, respectively) occur predominantly in the dissolved form, rather than as solid precipitates. Because of this, methanogenic and sulfate reducing organisms can rely on diffusion to supply their electron acceptor, and can therefore localize in very close proximity to H_2 sources (e.g., organic particles). As described earlier, this spatial arrangement may be critical for the expression of bioenergetic control.

    Microorganisms are also capable of using predominantly insoluble electron acceptors, such as Fe^{3+} and Mn^{4+}, for the oxidation of H_2. But if the sources of these oxidants occur predominantly in particulate form, microorganisms must first localize themselves around these particles, and then rely on a diffusive flux of H_2 from the surrounding environment. Diffusion of H_2 in to the cell requires a gradient in concentration between the cell surface and the surroundings, so that even if intracellular concentrations of H_2 were held at the bioenergetic limit for these electron acceptors, the bulk medium would necessarily have significantly higher concentrations [137]. This has been illustrated in experiments in which particulate Fe and Mn were added to sulfate-reducing sediments. Although both Fe and Mn reduction offer higher free energy yields [which should translate to lower H_2 concentrations, according to Eq. (22a)], H_2 concentrations remained unchanged [145]. Thermodynamic control could be approximated in such systems if the sources of H_2 and metals were co-located (e.g., on the same particle), or if the concentration of metal particles was high, such that the diffusive regions around individual particles overlapped. In this case, H_2 concentrations would decrease towards the bioenergetic limit as the distance between particles decreased to zero. This effect may underlie instances in which the presence of Fe or Mn significantly affects H_2 concentrations [150] or excludes sulfate reduction (presumably, by lowering H_2 concentrations, [161]).

Lastly, thermodynamic/bioenergetic control over H_2 concentrations can be superceded when H_2 consumers are not able to closely match their consumption to the supply of H_2 from producers. Generally, this can occur if the population of consumers is not sufficiently developed in close proximity to the H_2 source. Hence, systems in which there is a flux and turnover of substrate and microbial biomass on time scales shorter than those required for development of a fine-tuned consumer population may exhibit H_2 concentrations significantly exceeding the bioenergetic minimum. Such systems may include, for example, animal digestive systems, sewage digesters, and continuous flow chemostats. Under these circumstances, it is possible that even the most thermodynamically favorable process can be superceded, if organisms catalyzing less favorable processes are capable of faster growth. Similar production–consumption imbalance may occur during the transient period in which a population of consumers grows to meet a newly available H_2 flux. In a number of systems, it has been observed that the transition from sulfate-reducing to methanogenic conditions (following the depletion of sulfate to the biologically useful limit) is accompanied by a transient pooling of fermentation products [162–165], including H_2. This likely corresponds to a time at which the methanogen population is growing to meet the supply of fermentation products, which were previously kept unavailable by the activities of sulfate reducers [163]. H_2 concentrations during such a period far exceed the bioenergetic minimum [165]. Finally, any chemical or physical limitation on the growth of H_2 consuming populations (e.g., chemical toxicity, inhospitable temperatures) will naturally also disrupt the production–consumption steady state, resulting in H_2 concentrations that deviate from thermodynamic predictions. Even in the most extreme cases, however, it should still be noted that thermodynamic/bioenergetic controls provide the ultimate boundary conditions on the possible H_2 concentrations within a system. Biologically controlled concentrations cannot drop below levels that offer useful energy yields to H_2 consumers, and cannot rise above levels that offer useful energy yields to H_2 producers.

3.3. Implications for Biogeochemistry

The biogeochemical significance of biological hydrogen cycling in anoxic systems is considerable. In the present day, anoxic conditions prevail in much of Earth's aquatic sediments and in the digestive systems of many animals. Collectively, these environments play an important role in regulating the redox chemistry of Earth's oceans and atmosphere [20]. They are significant sources of radiatively active trace gases, and represent the ultimate biological filter on material passing into the rock record, or into, e.g., oil producing horizons. The global significance of anoxic ecosystems was greater still during the early Archaean Era, when such systems were solely responsible for primary productivity, and for the biological transformation of organic matter. Our ability to decipher Earth's early history is linked to our ability to interpret the effects of anaerobic degradation processes on their environment. As we have seen, much of the community-level chemistry and metabolism in anoxic systems is dynamically and functionally dependent on H_2 cycling.

4. H$_2$ CYCLING IN PHOTOTROPHIC ECOSYSTEMS

4.1. H$_2$ in the Metabolism of Phototrophic Microorganisms

4.1.1. Photosynthetic Metabolism

Phototrophic microorganisms have evolved to free themselves from a dependence on chemical forms of energy, and instead depend on sunlight. Nonetheless, since these organisms arose from an anoxic world, it is perhaps not surprising that their metabolisms incorporate H_2 cycling at several levels (Table 3). The treatment of these processes here is necessarily brief. Thorough reviews are available on H_2 in the metabolism of cyanobacteria [22,23,25] and anoxygenic photosynthetic bacteria [24].

The photosynthetic bacteria can be functionally classed into oxygenic and anoxygenic representatives. Generically, both classes catalyze the same overall process, harnessing light energy to extract electrons from a donor molecule, and using those electrons to "fix" CO_2 into a more reduced organic form [166]:

$$X_{red} + h\upsilon \longrightarrow X_{ox} + ne^- \tag{23}$$

$$CO_2 + 4e^- + 4H^+ \longrightarrow CH_2O + H_2O \tag{24}$$

$$\overline{X_{red} + CO_2 + 4H^+ + h\upsilon \longrightarrow X_{ox} + CH_2O + H_2O \text{ (overall)}} \tag{25}$$

Table 3 H_2 Metabolism in Phototrophic Microorganisms[a]

Process	Catalyzed by	Conditions
H_2 Consumption		
Photoreduction of CO_2	AP, OP[b]	During anoxygenic photosynthesis
Reduction of O_2	AP, OP	Light or dark, oxic
Reduction of other inorganics	AP	APs: Dark, anoxic
(sulfate, thiosulfate, nitrate, nitrite)	OP	OPs: During anoxygenic photosynthesis
H_2 Production		
Photoproduction from organic electron donors	AP	Light, nitrogen limiting, high organics
Photoproduction from inorganic electron donors (sulfide,	AP	APs: Light, N limiting, high sulfide or thiosulfate
thiosulfate)	OP	OPs: Anoxygenic conditions, high H_2S
Water photolysis under CO_2 limitation	OP	High light, low CO_2
During nitrogen fixation	AP, OP	Light or dark, anoxic
Anaerobic fermentation	AP, OP	Dark, anoxic
$CO + H_2O \rightarrow H_2 + CO_2$	AP	Light or dark, anoxic, high CO

[a]See extensive reviews and Refs. [22–25].
[b]AP = anoxygenic phototrophs; OP = oxygenic phototrophs.

In the case of oxygenic photosynthesis, catalyzed in the bacterial world by cyanobacteria, the electron donor (X_{red}) is water, and the oxidized product (X_{ox}) is O_2. This is the same type of photosynthesis that is carried out by green plants, which indeed ultimately derived that ability from the cyanobacteria. Anoxygenic photosynthesis utilizes electron donors other than water (specifically, molecules for which less energy is required to extract electrons—e.g., hydrogen sulfide, ferrous iron), and accordingly, does not generate O_2 as a product.

Equation (25) describes a process of electron transfer from donor molecules to CO_2 (for both oxygenic and anoxygenic photosynthesis). Unlike the anaerobic processes described in Section 3, in which the electron transfer is frequently extracellular (between cells), the electron transfer in photosynthesis is intracellular. As such, it can occur completely via biological electron carriers, and without need for free electron carriers such as H_2. However, the presence of hydrogenase enzymes in representatives from both classes of phototrophs [2,24,25] makes H_2 metabolism possible in both accessory and primary roles.

4.1.2. H_2 Consumption

For both oxygenic and anoxygenic phototrophs, H_2 may serve as a supplemental or even principal electron donor for the reduction of CO_2 or other compounds. The ability to utilize H_2 as the electron donor for photoreduction of CO_2 is widespread among anoxygenic phototrophs, and some are capable of growing with H_2 as the sole electron source [24]. Some cyanobacteria are able to carry out anoxygenic photosynthesis, and may use likewise utilize H_2 for the photoreduction of CO_2 under those circumstances [167]. Reduction of other inorganic electron acceptors is also possible. Some anoxygenic phototrophs can utilize H_2 for the reduction of a variety of inorganic species, including sulfate, thiosulfate, sulfite, nitrite, nitrate, and fumarate, under dark and anoxic conditions [24]. Similarly, during anoxygenic phototrophic metabolism, some cyanobacteria can use H_2 to reduce nitrate or nitrite [168–170]. Both groups can use H_2 for the light-independent reduction of O_2 [171–174].

4.1.3. H_2 Production

4.1.3.1. Photoproduction: Phototrophic microorganisms exhibit a wide variety of H_2 producing mechanisms. Perhaps the best studied, due to its potential to serve as an alternative energy source, is the photoproduction of H_2. High light levels and an abundant supply of electron donor [X_{red} in Eq. (23)] may build up excess reducing power within the electron transport system of the phototrophs. Given enzymes capable of converting biological reducing equivalents to molecular hydrogen, this excess may "bleed off" in the form of H_2 [175,176]. Anoxygenic phototrophs exhibit this process with a wide variety of organic electron donors and, in a few cases, with sulfide and thiosulfate as inorganic electron donors [177–179]. In this case, nitrogenase enzymes (the enzymes that catalyze nitrogen fixation) are chiefly responsible for the generation of H_2 [180,181]. Because of this, photoproduction of H_2 occurs at higher rates when the system is limited by N_2 and combined nitrogen (when nitrogenase enzymes are present, but are not otherwise engaged in N_2 fixation). Similarly, an abundant supply of organic or inorganic electron donor, combined with high light levels (to rapidly generate electrons by photolysis), tend to drive the process to maximal rates. Cyanobacteria exhibit a similar process, generating H_2 from photolysis of H_2S [167,182,183] or water [22,23]. In these organisms, photoproduction of H_2 may derive from both nitrogenase and hydrogenase enzymes [173,184].

4.1.3.2. Nitrogen fixation: Most representatives of the oxygenic and anoxygenic phototrophs are capable of nitrogen fixation [185], a process that generates H_2 as an apparently unavoidable by-product [186,187]:

$$N_2 + 8H^+ + 8e^- \longrightarrow 2NH_3 + H_2 \tag{26}$$

For every N_2 fixed, one H_2 is liberated. The typical (Redfield) C:N ratio for biomass is about 6.6:1. Hence, in a system completely free of combined nitrogen

(N in forms other than N_2), one molecule of N_2 must be fixed—and therefore, one molecule of H_2 produced—for every 13 CO_2 molecules fixed into organic matter. Maximally, then, nitrogen fixation could generate H_2 at $\sim 8\%$ of the rate of CO_2 fixation in a given system. As sources of combined nitrogen become increasingly abundant, the need for nitrogen fixation decreases accordingly [188,189], and the rate of H_2 production is expected to follow.

Due to O_2-sensitivity of the responsible enzymes, the conduct of nitrogen fixation requires low ambient O_2 levels [190,191]. For anoxygenic phototrophs, this is not a problem, as they can simply localize to anoxic niches. However, the very metabolic activity of oxygenic phototrophs (O_2 production) creates an environment unsuitable for nitrogen fixation. The cyanobacteria have developed two main approaches for dealing with this problem [191–195]. The first is development of specialized "heterocysts" within a filament of "normal" cells. These heterocysts have thick membranes that limit the diffusion of oxygen in from the surrounding environment, and do not themselves photosynthesize (do not produce O_2) [191,192]. They are therefore able to maintain internal conditions suitable for the conduct of nitrogen fixation during daylight hours, when the rest of the cells are actively photosynthesizing.

A second strategy for avoiding oxygen poisoning of nitrogen fixation is to conduct the process in the dark (when photosynthesis is not active), in structures that limit the influx of oxygen from the surrounding environment [193]. Several taxa of cyanobacteria accomplish this by constructing biofilms or mats, dense networks of organisms and polysaccharide matrix material in which transport of solutes such as O_2 is limited to molecular diffusion [195–197]. Because the mat matrix limits the transport of O_2 from the surrounding environment, and respiration actively consumes O_2 within the mat, these systems frequently develop anoxic conditions (suitable for nitrogen fixation) within a few hours of the onset of darkness. Thus, while heterocystous cyanobacteria can fix nitrogen during daylight hours, most non-heterocystous mat-forming organisms typically exhibit very low rates of N_2 fixation during the day, followed by elevated rates at night (e.g., [198]). The potential for H_2 generation associated with nitrogen fixation varies accordingly.

Nitrogen fixation is overall a reducing process with respect to N_2, so that liberation of reducing power as H_2 represents a biochemical inefficiency. Perhaps to stem this loss, nitrogen-fixing phototrophs possess "uptake hydrogenases", which recapture H_2 and cycle reducing power back into the system— thereby limiting release to the environment [22,23,25]. For this reason, strategies for optimizing photobiological H_2 production frequently seek to eliminate the activity of uptake hydrogenases [11].

4.1.3.3. Fermentation: Some phototrophic microorganisms (both oxygenic and anoxygenic) are capable of fermenting organic molecules (e.g., saccharides) to simple organic acids and hydrogen, as a way of releasing energy [199–204]. Phototrophs fuel cellular energy demands principally by harnessing solar radiation during daylight hours. A portion of this energy is stored in a chemical form, via the accumulation of polysaccharides or osmoregulants. Under dark conditions, this stored energy can be released by breaking down the polysaccharides. Oxygenic and some anoxygenic phototrophs do this preferentially by oxidizing the storage compounds using O_2, because aerobic respiration affords a large energy yield per molecule of saccharide oxidized. In the absence of O_2, organisms can still release stored energy via the anaerobic fermentation (decomposition) of storage polysaccharides or osmoregulants. H_2 can be generated in these fermentations, along with organic acids (lactate, acetate, formate, and propionate) and ethanol [203,205–207]. As with the anaerobic fermentations discussed in Section 3, it is expected that fermentations conducted by phototrophic bacteria should be thermodynamically sensitive to ambient H_2 concentrations.

The capacity for fermentative metabolism appears to be constitutive in cyanobacteria, and these organisms initiate fermentation rapidly in response to anoxic conditions. A significant fraction of stored carbohydrates may be used in fermentation on a daily basis: Nold and Ward [208] demonstrated $\sim 75\%$ consumption of polysaccharides by *Synechococcus* sp. in 12 h under dark, anoxic conditions. On the other hand, some organisms (e.g., *Oscillatoria terebriformis* and *Microcoleus* sp.) can apparently support fermentation for up to several days based on available stores of glycogen [204,209]. In either case, the potential for H_2 generation is significant. A typical glucose fermentation in the presence of a hydrogen sink (e.g., reaction [11]) yields about four molecules of H_2 per glucose fermented [107,115]. Photosynthetic formation of glucose yields six molecules of O_2:

$$6CO_2 + 6H_2O \quad \longrightarrow \quad C_6H_{12}O_6 + 6O_2 \tag{27}$$

Thus, fermentation in an oxygenic photosynthetic system is conceivably capable of producing H_2 at two-thirds of the rate at which it produces O_2—or even more, if the fermentation yields a higher proportion of H_2 (to an absolute maximum of two times the O_2 production rate, if glucose was fermented completely to CO_2/H_2). Similarly, anoxygenic phototrophs may be capable of quantitatively significant H_2 production via fermentation [24].

4.2. H_2 Cycling in Photosynthetic Microbial Mats

Photosynthetic microbial mats are densely packed, highly complex, and highly structured assemblages of microorganisms representing a diverse array of metabolic strategies [197,210]. Broadly, such communities contain photosynthetic primary producers in close association with chemotrophic organisms, with the latter supported by the flux of photosynthetically-fixed carbon. Especially in well-developed systems, the physical matrix of the mat limits the exchange of materials with the environment to molecular diffusion, which may serve to promote more efficient internal recycling of nutrients and fixed carbon

[195–197]. However, this also means that chemical conditions within the mats are strongly affected by microbial metabolic activity. As light-driven systems, microbial mats experience dramatic daily fluctuations in metabolic processes, and therefore in the resulting chemical microenvironment [211,212]. It is against this backdrop of diverse metabolic capabilities and dramatically oscillating physical and chemical conditions that the cycling of H_2 within microbial mats must be considered.

The extremely close physical proximity of metabolically diverse organisms within this variable matrix [197,210], coupled with a frequent inability to grow constituent organisms in isolation, suggests that consortial and feedback interactions among community members may be of first order importance in many aspects of microbial mat biogeochemistry [213]. Given the H_2-metabolizing capacities of such a broad variety of mat microorganisms, it seems likely that H_2 cycling could serve as the basis for many of these interactions. By virtue of their dominant role in chemical cycling within the mats, and their capacity to variably generate or consume large quantities of H_2 depending on ambient conditions, photosynthetic microorganisms seem poised to exert the overriding influence on the H_2 economy within the mat. Much of a mat's chemical contribution to the environment may depend on how the function of accessory microorganisms, such as the anaerobes associated with organic matter degradation, is pushed and pulled by the H_2-cycling activities of the phototrophs.

The relatively few studies that have examined H_2 in natural phototrophic microbial mat communities include both heterocystous [214] and non-heterocystous [65,215,216] cyanobacterial assemblages. In all cases, H_2 cycling is distinctly light dependent. Light periods are characterized by low H_2 levels at the mat surface ($P_{H_2} < 10 \, \text{Pa}$) [65], by little or no flux of H_2 to the environment [65,214], and by H_2 consumption when concentrations are initially high [65,214,215]. Skyring et al. [215] found evidence for both light- and O_2-dependent mechanisms of H_2 consumption in their non-heterocystous mats, while Oremland [214] reported a distinctly O_2-dependent consumption mechanism in heterocystous assemblages. Light-dependent H_2 consumption may stem from photoreduction of CO_2 (although this presumably requires anoxygenic conditions). O_2-dependent consumption of H_2 could result either from the activities of cyanobacterial uptake hydrogenases, or independent aerobic H_2-oxidizing organisms.

Dark periods are characterized by H_2 production, which appears to result principally from fermentation activities [65,214–216]. The H_2 production is, in all determined cases, distinctly sensitive to O_2, which supports either a fermentative or N-fixing mechanism. However, in the heterocystous mats examined, nitrogen fixation occurred almost exclusively during the light period, while H_2 flux occurred principally in the dark [214]. Similarly, H_2 production occurred in non-heterocystous mats whether or not nitrogen fixation was active [65,217]. Additionally, organic fermentation products were liberated simultaneously with H_2 in at least two mat systems [218]. Thus, fermentation appears to be quantitatively the most important process of H_2 production in these systems during the dark period. While the ultimate source of fixed carbon for fermentation is most likely productivity by phototrophs within the system, it is not clear whether cyanobacteria or other microorganisms actually carry out the process.

It is perhaps not surprising that nitrogen fixation is a quantitatively less important mechanism of H_2 production than fermentation, given the stoichiometry of the two processes. Recall that H_2 production coupled to nitrogen fixation could occur at up to 8% of the rate at which carbon is fixed *into biomass*, in a completely nitrogen-limited system (Section 4.1.3). However, if nitrogen fixation occurs during the day (as in heterocystous cyanaobacteria), much of the H_2 is expected to be subsequently reclaimed by cyanobacterial "uptake hydrogenases" [23]. If nitrogen fixation occurs at night, then much of the carbon fixed during the day would have to be respired in order to meet the high energy demand of the process (16–24 ATP per N_2 fixed [219]). Under dark, anoxic conditions, fermentation is the only source of energy available to fuel this demand. Fermentation of one molecule of glucose [as in reaction (11)] yields about 2–3 ATP [115], so that roughly eight molecules of glucose would have to be fermented to provide the energy required for fixation of one molecule of N_2. Each glucose molecule fermented may yield 4 H_2 [107,115], giving a ratio of about 32 H_2 produced by fermentation for every N_2 fixed. Recall that the direct production of H_2 by nitrogen fixation occurs in a stoichiometry of $1:1$ (H_2 produced per N_2 fixed). Thus, the rate of H_2 production that is *directly* attributable to nitrogen fixation is perhaps only a few percent of that deriving from fermentation. Nonetheless, nitrogen fixation may still factor prominently in H_2 cycling if its energy demand drives higher levels of fermentation at night. Such a mechanism was believed partially responsible for an orders-of-magnitude difference in night time H_2 production in nitrogen-fixing (higher H_2 production) vs. non-nitrogen-fixing mats [65].

Sulfate reduction serves as an important moderator of H_2 accumulation and emission from the mat matrix under dark, anoxic conditions, when sulfate is present at millimolar levels [215]. Since sulfate reduction may also occur in the upper layers of microbial mats under light, aerobic conditions [220], the process could also provide a daytime H_2 sink. Methanogenesis appears to be a considerably less important sink for H_2, even when sulfate is limited to sub-millimolar concentrations [221]. Despite the presence of microbial H_2 consumers, mats can still generate a substantial flux of H_2 at night, especially under anoxic conditions. Under dark, anoxic conditions, up to 16% of the reducing equivalents fixed by *Lygnbya* spp. mats during the day were subsequently liberated as H_2 at night [65]. The associated pooling of H_2 to partial pressures of almost 10^4 Pa can dramatically change the redox potential within the mat, and therefore potentially affect a wide range of H_2-sensitive microbial processes (shifting towards more reduced products) [12]. Similarly, the release of significant quantities of H_2 to the environment could have important consequences for oceanic and atmospheric redox chemistry, in a system dominated by microbial mat productivity (e.g., on the early Earth [65], see below).

4.3. Implications for Biogeochemistry

In terms of integrated historical influence, photosynthetic microbial mat communities have been among the most important arbiters of global biogeochemical change. Organisms capable of oxygenic photosynthesis appeared on Earth at least 2.7 billion years ago [222], and perhaps much earlier [223]. Anoxygenic photosynthesis is believed to be older still [224,225]. The advent of these processes (particularly oxygenic photosynthesis) as a metabolic strategy made available an abundant new energy source that may have enhanced the productivity of the biosphere by 2–3 orders of magnitude relative to its non-photosynthetic forerunner [21]. Abundant fossil evidence suggests that much of the new photosynthetic productivity during this era was localized in cyanobacterial mats and stromatolites [26], which remained globally widespread until the evolution of effective competitors or predators less than a billion years ago. Hence, photosynthetic microbial mat communities may have served as the principal biological arbiters of global geochemical change for 2 billion years or more of Earth's history. Similarly, these mat communities may have been an important crucible for microbial evolution, combining a dynamic environment of physical and chemical challenges with a dense packing of functionally diverse constituents. Many aspects of microbial mat chemistry, particularly the regulation of redox state and the coupling of elemental cycles, are tied to H_2 cycling among the constituent organisms. Hence, this process may have contributed significantly to the coupled biological and geochemical evolution of our planet for much of its early history.

5. SUMMARY

Hydrogen has had an important and evolving role in Earth's geo- and biogeochemistry, from prebiotic to modern times. On the earliest Earth, abiotic sources of H_2 were likely stronger than in the present. Volcanic out-gassing and hydrothermal circulation probably occurred at several times the modern rate, due to presumably higher heat flux [17]. The H_2 component of volcanic emissions was likely buffered close to the modern value by an approximately constant mantle oxidation state since 3.9 billion years ago [51], and may have been higher before that, if the early mantle was more reducing [226]. The predominantly ultramafic character of the early, undifferentiated crust [32] could have led to increased serpentinization and release of H_2 by hydrothermal circulation, as in modern ultramafic-hosted vents [227]. At the same time, the reactive atmospheric sink for H_2 was likely weaker [49]. Collectively, these factors suggest that steady state levels of H_2 in the prebiotic atmosphere were 3–4 orders of magnitude higher than at present [14–16], and possibly higher still during transient periods following the delivery of Fe and Ni by large impact events [36]. These elevated levels had direct or indirect impacts on the redox state of the atmosphere, the radiation budget, the production of aerosol hazes, and the genesis of biochemical precursor compounds.

The early abiotic cycling of H_2 helped to establish the environmental and chemical context for the origins of life on Earth. The potential for H_2 to serve as a source of energy and reducing power, and to afford a means of energy storage by the establishment of proton gradients, could have afforded it a highly utilitarian role in the earliest metabolic chemistry. Some origin of life theories suggest the involvement of H_2 in the first energy-generating metabolism [10,38], and the widespread and deeply-branching nature of H_2-utilization in the modern tree of life [2,12] suggests that it was at least a very early biochemical innovation. The abiotic production of H_2 via several mechanisms of water–rock interaction could have supported an early chemosynthetic biosphere. Such processes offer the continued potential for a deep, rock-hosted biosphere on Earth (e.g., [66,68–70]) or other bodies in the solar system [85].

The continued evolution of metabolic and community-level versatility among microbes led to an expanded ability to completely exploit the energy available in complex organic matter. Under the anoxic conditions that prevailed on the early Earth, this was accomplished through the linked and sequential action of several metabolic classes of organisms [19]. By transporting electrons between cells, H_2 provides a means of linking the activities of these organisms into a highly functional and interactive network. At the same time, H_2 concentrations exert a powerful thermodynamic control on many aspects of metabolism and biogeochemical function in these systems [145]. Anaerobic communities based on the consumption of organic matter continue to play an important role in global biogeochemistry even into the present day. As the principal arbiters of chemistry in most aquatic sediments and animal digestive systems, these microbes affect the redox and trace-gas chemistry of our oceans and atmosphere [20], and constitute the ultimate biological filter on material passing into the rock record. It is in such communities that the significance of H_2 in mediating biogeochemical function is most strongly expressed.

The advent of phototrophic metabolism added another layer of complexity to microbial communities, and to the role of H_2 therein. Anoxygenic and oxygenic phototrophs retained and expanded on the utilization of H_2 in metabolic processes [22–25]. Both groups produce and consume H_2 through a variety of mechanisms. In the natural world, phototrophic organisms are often closely juxtaposed with a variety of other metabolic types, through the formation of biofilms and microbial mats [197]. In the few examples studied, phototrophs contribute an often swamping term to the H_2 economy of these communities [65,214,215], with important implications for their overall function—including regulation of the redox state of gaseous products, and direct release of large quantities of H_2 to the environment [12,65]. As one of the dominant sources of biological productivity for as much as 2 billion years of Earth's history, these communities have been among the most important agents of long-term global biogeochemical change [21,228].

On the modern Earth, H_2 is present at only trace levels in the atmosphere and oceans. Nonetheless, its function as an arbiter of microbial interactions and chemistry ensures an important role in biogeochemical cycling. The significance of H_2 in a global sense may soon increase, as the search for alternative fuels casts attention on the clean-energy potential of hydrogen fuel cells (e.g., [229,239]). Already, H_2 utilization plays an important role in all three phylogenetic domains of life. Humans may soon add an important new term to this economy. Considerable research is focused on the H_2-producing capacities of phototrophic and other microorganisms as potential contributors in this regard [11]. Regardless of source, the large scale utilization of H_2 as an energy source could carry important consequences for biogeochemistry [230,231].

ACKNOWLEDGMENTS

The author acknowledges support from the NASA Exobiology Program and the NASA Astrobiology Institute. I am grateful to Mitch Schulte, Kevin Zahnle, and Norm Sleep for helpful discussions on aspects of this manuscript.

REFERENCES

1. Fenchel T, Finlay BJ. In: May RM, Harvey PH, eds. Ecology and Evolution in Anoxic Worlds. Oxford Series in Ecology and Evolution. Oxford: Oxford University Press, 1995.
2. Vignais PM, Billoud B, Meyer J. FEMS Microbiol Rev 2001; 25:455–501.
3. Schwarz E, Friedrich B. In: Dworkin M, ed. The Prokaryotes: An Evolving Electronic Resource for the Microbiological Community. New York: Springer Verlag, 2003.
4. Mitchell P. Nature 1961; 191:144–148.
5. Zobell CE. Bull. Am Assoc Pet Geochemists 1947; 31:1709–1751.
6. Neal C, Stanger G. Earth Planet Sci Lett 1983; 66:315–320.
7. Greenland LP. In: Decker RW, Wright TL, Stauffer PH, eds. Volcanism in Hawaii. Washington, D.C.: U.S. Geologic Survey, 1987.
8. Elderfield H, Schulz A. Annu Rev Earth Planet Sci 1996; 24:191–224.
9. Wächtershauser G. Proc Natl Acad Sci 1990; 87:200–204.
10. Wächtershauser G. Pure Appl Chem 1993; 65:1343.
11. Nandi R, Sengupta S. Crit Rev Microbiol 1998; 24:61–84.
12. Hoehler TM, Albert DB, Alperin MJ, Bebout BM, Martens CS, Des Marais DJ. Antonie van Leeuwenhoek 2002; 81:575–585.
13. de Duve C. Blueprint for a Cell. Burlington, NC: Neil Patterson, 1991.
14. Pinto JP, Gladstone GR, Yung YL. Science 1980; 210:183–185.
15. Kasting JF, Brown LL. In: Brack A, ed. The Molecular Origins of Life: Assembling the Pieces of the Puzzle. New York: Cambridge University Press, 1998:35–56.
16. Kasting JF, Pavlov AA, Siefert JL. Origins Life Evol Biosphere 2001; 31:271–285.
17. Turcotte DL. Earth Planet Sci Lett 1980; 48:53–58.

18. Farquhar J, Bao HM, Thiemens M. Science 2000; 289:756–758.
19. Schink B. In: Zehnder AJB, ed. Biology of Anaerobic Microorganisms. New York: Wiley-Interscience, 1988:771–846.
20. Henrichs S, Reeburgh WS. Geomicrobiol J 1987; 5:191–237.
21. Des Marais DJ. In: Banfield J, Nealson K, eds. Geomicrobiology. Washington, DC: Mineralogical Society of America, 1997:429–445.
22. Lambert GR, Smith GD. Biol Rev Cambr Philos Soc 1980; 56:589–660.
23. Houchins JP. Biochim Biophys Acta 1984; 768:227–255.
24. Vignais P, Colbeau A, Willison JC, Jouanneau Y. In: Rose AH, Tempest DW, eds. Advances in Microbial Physiology. Vol. 26. London: Academic Press, 1985:155–234.
25. Tamagnini P, Axelsson R, Lindberg P, Oxefelt F, Wunschiers R, Lindblad P. Microbiol Molec Biol Rev 2002; 66:1–20.
26. Walter MR, ed. Stromatolites. Amsterdam: Elsevier, 1976.
27. Walter MR, Grotzinger JP, Schopf JW. In: Schopf JW, Klein C, eds. The Proterozoic Biosphere: A Multidisciplinary Study. New York: Cambridge University Press, 1992:253–260.
28. Embley TM, van der Giezen M, Horner DS, Dyal PL, Foster P. Philos Trans Biol Sci 2003; 358:191–203.
29. Homann PH. Photosynth Res 2003; 76:93–103.
30. Reeburgh WS. In: Lidstrom ME, Tabita FR, eds. Microbial Growth on C_1 Compounds. Dordrecht: Kluwer Academic Publishers, 1996:334–342.
31. Moody JB. Lithos 1976; 9:125–138.
32. Nisbet EG. The Young Earth: An Introduction to Archaean Geology. Cambridge: Cambridge University Press, 1987.
33. Abrajano TA, Sturchio NC, Kennedy BM, Lyon GL, Muehlenbachs K, Bohlke J. Appl Geochem 1990; 5:625–630.
34. Sturchio NC, Abrajano TA, Murdwchick J, Muehlenbachs K. Tectonophysics 1989; 168:101–107.
35. Früh-Green GL, Connolly JAD, Plas A, Kelley DS, Grobety B. J Geophys Res. In press.
36. Kasting JF. Origins Life Evol Biosphere 1990; 20:199–231.
37. Drobner E, Huber H, Wächtershauser G, Rose D, Stetter KO. Nature 1990; 346:742–744.
38. Wächtershauser G. System Appl Microbiol 1988; 10:207–210.
39. Freund F. J Non-Cryst Solids 1985; 71:195–202.
40. Freund F, Dickinson JT, Cash M. Astrobiology 2002; 2:83–92.
41. Kita I, Masuo S, Wakita A. J Geophys Res 1982; 87:10789–10795.
42. Spinks JWT, Woods RJ. An Introduction to Radiation Chemistry. New York: Wiley, 1964.
43. Vovk IF. In: Fritz P, Frape SK, eds. Saline Water and Gases in Crystalline Rocks. Vol. 33. Toronto: Geological Association of Canada, 1987:197–210.
44. Sherwood-Lollar B, Frape SK, Fritz P, Macko SA, Whelan JA, Blomquist R, Lahermo PW. Geochim Cosmochim Acta 1993; 57:5073–5085.
45. Lin LH, Onstott TC, Lippmann J, Ward JA, Hall J, Sherwood-Lollar B. paper presented at the 12th Annual V. M. Goldschmidt Conference, Davos, Switzerland, 2002.
46. Carmichael ISE, Turner FJ, Verhoogen J. In: Krauskopf K, ed. Igneous Petrology. McGraw-Hill International Series in the Earth and Planetary Sciences. New York: McGraw-Hill, 1974.

47. Conrad R. Adv Microbiol Ecol 1988; 10:231–283.
48. Levine JS. In: Levine JS, ed. The Photochemistry of Atmospheres. Orlando: Academic Press, 1985:518.
49. Atkinson R, Darnall KR, Lloyd AC, Winer AM, Pitts JN Jr. Adv Photochem 1979; 11:375–488.
50. Holland HD. In: Engle AEJ, James HL, Leonard BF, eds. Petrologic Studies: A Volume in Honor of A. F. Buddington. New York: Geological Society of America, 1962.
51. Delano JW. Origins Life Evol Biosphere 2001; 31:311–341.
52. Walker JCG. Evolution of the Atmosphere. New York: Macmillan, 1977.
53. Hart MH. Origins Life 1979; 9:261–266.
54. Canuto VM, Levine JS, Augustsson TR, Imhoff CL, Giampapa MS. Nature 1983; 305:281–286.
55. Zahnle KJ. J Geophys Res—Atmos 1986; 91:2819–2834.
56. Kasting JF, Zahnle KJ, Pinto JP, Young AT. Origins Life Evol Biosphere 1989; 19:95–108.
57. Sagan C, Chyba C. Science 1997; 276:1217–1221.
58. Miller SL. Science 1953; 117:528–529.
59. Miller SL, Urey H. Science 1959; 130:245–251.
60. Schlesinger G, Miller SL. J Mol Evol 1983; 19:376–382.
61. Schlesinger G, Miller SL. J Mol Evol 1983; 19:383–390.
62. Hunten DM. J Atmosph Sci 1973; 30:1481–1494.
63. Holland IID. Chemical Evolution of the Atmosphere and Oceans. Princeton: Princeton University Press, 1984.
64. Catling DC, Zahnle KJ, McKay CP. Science 2001; 293:839–843.
65. Hoehler TM, Bebout BM, DesMarais DJ. Nature 2001; 412:324–327.
66. Gold T. Proc Natl Acad Sci 1992; 89:6045–6049.
67. Sleep N, Zahnle K. J Geophys Res 1998; 103:28529–28544.
68. Pedersen K. Earth Sci Rev 1993; 34:243–260.
69. Frederickson JK, Onstott TC. Sci Am 1996; 275:68–73.
70. Fyfe WS. Science 1996; 273:448.
71. Davydovacharakhchyan IA, Kuznetsova VG, Mityushina LL, Belyaev SS. Microbiology 1992; 61:202–207.
72. Fardeau ML, Cayol JL, Magot M, Ollivier B. FEMS Microbiol Lett 1993; 113:327–332.
73. Stetter KO, Huber R, Blochl E, Kurr M, Eden RD, Fielder M, Cash H, Vance I. Nature 1993; 365:743–745.
74. Parkes RJ, Cragg BA, Bale SJ, Getliff JM, Goodman K, Rochelle PA, Fry JC, Weightman AJ, Harvey SM. Nature 1994; 371:410–413.
75. Kormas KA, Smith DC, Edgcomb V, Teske A. FEMS Microbiol Ecol 2003; 45:115–125.
76. Stevens TO, McKinley JP. Science 1995; 270:450–454.
77. Chapelle FH, O'Neill K, Bradley PM, Methe BA, Ciufo SA, Knobel LL, Lovley DR. Nature 2002; 415:312–315.
78. Kotelnikova S, Pedersen K. FEMS Microbiol Rev 1997; 20:339–349.
79. Pedersen K. FEMS Microbiol Rev 1997; 20:399–414.
80. Deming JW, Baross JA. Geochim Cosmochim Acta 1993; 57:3219–3230.
81. Holden JF, Summit M, Baross JA. FEMS Microbiol Ecol 1998; 25:33–41.

82. Summit M, Baross JA. Deep-Sea Res Part II: Topical Stud Oceanogr 1998; 45:2751–2766.
83. Huber JA, Butterfield DA, Baross JA. FEMS Microbiol Ecol 2003; 43:393–409.
84. Brack A. Adv Space Res 1999; 24:417–433.
85. Boston PJ, Ivanov MV, McKay CP. Icarus 1992; 95:300–308.
86. Kargel JS, Kaye JZ, Head III JW, Marion GM, Sassen R, Crowley JK, Ballesteros OP, Grant SA, Hogenboom DL. Icarus 2000; 148:226–265.
87. Chyba CF, Phillips CB. Origins Life Evol Biosphere 2002; 32:47–67.
88. Longhi J, Knittle E, Holloway JR, Waenke H. In: Kieffer HH, Jakosky BM, Snyder CW, Matthews MS, eds. Mars. Tucson, AZ: University of Arizona Press, 1992:184–208.
89. Singer RB, McSween HY Jr. In: Lewis JS, Matthews MS, Guerrieri ML, eds. Resources of Near-Earth Space. Tucson, AZ: University of Arizona Press, 1993:709–736.
90. Gaidos EJ, Nealson KH, Kirschvink JL. Science 1999; 284:1631–1633.
91. Popoff L. Arch Ges Physiol 1875; 10:113–146.
92. Fitz A. Ber Dtsch Chem Ges 1876; 9:1348–1352.
93. Hoppe-Seyler F. Plügers Arch Ges Physiol 1876; 12:1–17.
94. Söhngen NL. Rec Trav Chim 1910; 29:238–244.
95. Fischer R, Lieske R, Winzer K. Biochem Z 1932; 245:2–12.
96. Thauer RK, Jungermann K, Decker K. Bacteriol Rev 1977; 41:100–180.
97. Atkins PW. Physical Chemistry. 4th ed. New York: Oxford University Press, 1990.
98. Wolin MJ, Miller TL. ASM News 1982; 48:561–565.
99. Johns AT, Barker HA. J Bacteriol 1960; 80:837–841.
100. Hungate RE. Archiv für Mikrobiologie 1967; 59:158–164.
101. Bryant MP, Wolin EA, Wolin MJ, Wolfe RS. Arch Microbiol 1967; 59:20–31.
102. Ferry JG, Wolfe RS. Arch Microbiol 1976; 107:33–40.
103. McInerney MJ, Bryant MP, Pfennig N. Arch Microbiol 1979; 122:129–135.
104. Boone DR, Bryant MP. Appl Environ Microbiol 1980; 40:626–632.
105. Mountfort DO, Bryant MP. Arch Microbiol 1982; 133:249–256.
106. Stieb M, Schink B. Arch Microbiol 1985; 140:387–390.
107. Dolfing J. In: Zehnder AJB, ed. Biology of Anaerobic Microorganisms. New York: Wiley-Interscience, 1988:417–468.
108. Barker HA. Ann Rev Biochem 1981; 50:23–40.
109. Stams AJM, Hansen TA. Arch Microbiol 1984; 137:329–337.
110. Nanninga HJ, Gottschal JC. FEMS Microbiol Lett 1985; 31:261–269.
111. McInerney MJ. In: Zehnder AJB, ed. Biology of Anaerobic Microorganisms. New York: Wiley-Interscience, 1988:373–415.
112. Laanbroek HJ, Abee T, Voogd JL. Arch Microbiol 1982; 133:178–184.
113. Valentine DL, Reeburgh WS, Blanton DC. J Microbiol Meth 2000; 39:243–251.
114. Wolin MJ. In: Schlegel HG, Gottschalk G, Pfennig N, eds. Microbial Formation and Utilization of Gases. Göttingen: E. Goltze KG, 1976.
115. Iannotti EL, Kafkewitz D, Wolin MJ, Bryant MP. J Bacteriol 1973; 114:1231–1240.
116. Scheifinger CC, Linehan B, Wolin MJ. Appl Microbiol 1975; 29:480–483.
117. Chen M, Wolin MJ. Appl Environ Microbiol 1977; 34:756–759.
118. Tewes FJ, Thauer RK. In: Gottschalk G, Pfennig N, Werner H, eds. Anaerobes and Anaerobic Infections. Stuttgart: Gustav Fischer, 1980:97–104.
119. Chung KT. Appl Environ Microbiol 1976; 31:342–348.

120. Traore AS, Fardeau ML, Hatchikan EC, LeGall J, Belaich JP. Appl Environ Microbiol 1983; 46:1152–1156.
121. Wu W.-M, Hickey RF, Jain MK, Zeikus JG. Arch Microbiol 1993; 159:57–65.
122. Boone DR, Johnson RL, Liu Y. Appl Environ Microbiol 1989; 55:1735–1741.
123. Schauer NL, Ferry JG. J Bacteriol 1982; 150:1–7.
124. Thiele JH, Zeikus JG. Appl Environ Microbiol 1988; 54:20–29.
125. Lee MJ, Zinder SH. Appl Environ Microbiol 1988; 54:124–129.
126. Devol AH, Ahmed SI. Nature 1981; 291:407–408.
127. Alperin MJ, Reeburgh WS. Appl Environ Microbiol 1985; 50:940–945.
128. Iversen N, Jørgensen BB. Limnol Oceanogr 1985; 30:944–955.
129. Hoehler TM, Alperin MJ, Albert DB, Martens CS. Global Biogeochem Cycles 1994; 8:451–463.
130. Hinrichs K.-U, Hayes JM, Sylva SP, Brewer PG, DeLong EF. Nature 1999; 398:802–805.
131. Boetius A, Ravenschlag K, Schubert CJ, Rickert D, Widdel F, Gieske A, Amann R, Jørgensen BB, Witte U, Pfannkuche O. Nature 2000; 407:623–626.
132. Orphan VJ, House CH, Hinrichs KU, McKeegan KD, DeLong EF. Science 2001; 293:484–487.
133. Valentine DL, Reeburgh WS. Environ Microbiol 2000; 2:477–484.
134. Sørensen KB, Finster K, Ramsing NB. Microbial Eco 2001; 42:1–10.
135. Thauer RK. Microbiology 1998; 144:2377–2406.
136. Phelps TJ, Conrad R, Zeikus JG. Appl Environ Microbiol 1985; 50:589–594.
137. Hoehler TM, Ph.D. University of North Carolina, 1998.
138. Finke N, Hoehler TM, Jørgensen BB. Appl Environ Microbiol submitted.
139. Carroll EJ, Hungate RE. Arch Biochem 1955; 56:525.
140. Sørensen J, Christensen D, Jørgensen BB. Appl Environ Microbiol 1981; 42:5–11.
141. Lilley MD, Baross JA, Gordon LI. Deep-Sea Res 1982; 29:1471–1484.
142. Conrad R, Argano M, Seiler W. Appl Environ Microbiol 1983; 45:502–510.
143. Scranton MI, Novelli PC, Loud PA. Limnol Oceanogr 1984; 29:993–1003.
144. Zinder SH. In: Ferry JG, ed. Methanogenesis. New York: Chapman & Hall, 1993:128–206.
145. Hoehler TM, Alperin MJ, Albert DB, Martens CS. Geochim Cosmochim Acta 1998; 62:1745–1756.
146. Lovley DR, Dwyer DF, Klug MJ. Appl Environ Microbiol 1982; 43:1373–1379.
147. Lovley DR, Klug MJ. Appl Environ Microbiol 1983; 45:187–192.
148. Conrad R, Lupton FS, Zeikus JG. FEMS Microbiol Ecol 1987; 45:107–115.
149. Novelli PC, Scranton MI, Michener RH. Limnol Oceanogr 1987; 32:565–576.
150. Lovley DR, Goodwin S. Geochim Cosmochim Acta 1988; 52:2993–3003.
151. Schink B. In: Finn RK, Präve P, eds. Biotechnology, focus 2. Munich: Hanser Publishers, 1990:63–89.
152. Schink B. Microbiol Mol Biol Rev 1997; 61:262–280.
153. Lovley DR, Phillips EJP. Appl Environ Microbiol 1987; 53:2636–2641.
154. Cord-Ruwisch R, Seitz H.-J, Conrad R. Arch Microbiol 1988; 149:350–357.
155. Chapelle FH, Lovley DR. Ground Water 1992; 30:29–36.
156. Zehnder AJB, Stumm W. In: Zehnder AJB, ed. Biology of Anaerobic Microorganisms. New York: Wiley-Interscience, 1988:1–38.
157. Conrad R, Schutz H, Babbel M. FEMS Microbiol Ecol 1987; 45:281–289.
158. Conrad R, Wetter B. Arch Microbiol 1990; 155:94–98.

159. Westermann P. FEMS Microbiol Ecol 1994; 13:295–302.
160. Hoehler TM, Alperin MJ, Albert DB, Martens CS. FEMS Microbiol Ecol 2001; 38:33–41.
161. Thamdrup B. In: Schink B, ed. Advances in Microbial Ecology. Vol. 16. New York: Kluwer Academic/Plenum, 2000:41–84.
162. Sansone FJ, Martens CS. Geochim Cosmochim Acta 1982; 46:1575–1589.
163. Alperin MJ, Blair NE, Albert DB, Hoehler TM, Martens CS. Global Biogeochem Cycles 1992; 6:271–291.
164. Shannon RD, White JP. Limnol Oceanogr 1996; 41:435–443.
165. Hoehler TM, Albert DB, Alperin MJ, Martens CS. Limnol Oceanogr 1999; 44:662–667.
166. Overmann J, Garcia-Pichel F. In: Dworkin M, ed. The Prokaryotes: An Evolving Electronic Resource for the Microbiological Community. New York: Springer-Verlag, 2000.
167. Belkin S, Shahak Y, Padan E. Meth Enzymol 1988; 167:380–386.
168. Peschek GA. Arch Microbiol 1978; 119:313–322.
169. Peschek GA. Biochim Biophys Acta 1979; 548:187–202.
170. Bothe H, Neuer G, Kalbe I, Eisbrenner G. In: Stewart WDP, Gallon JR, eds. Nitrogen Fixation: Proceedings of the Phytochemical Society of Europe Symposium, Sussex, September 1979. London: Academic Press, 1980:83–112.
171. Madigan MT, Gest H. J Bacteriol 1979; 137:524–530.
172. Siefert E, Pfennig N. Arch Microbiol 1979; 122:177–182.
173. Howarth DC, Codd GA. J General Microbiol 1985; 131:1561–1569.
174. Kämpf C, Pfennig N. J Basic Microbiol 1986; 26:517–531.
175. Omerod JG, Omerod KS, Gest H. Arch Biochem Biophys 1961; 94:449.
176. Gest H, Omerod JG, Omerod KS. Arch Biochem Biophys 1962; 97:21.
177. Kondratieva EN, Gogotov IN. Adv Biochem Eng Biotechnol 1983; 28:139–191.
178. Sasikala K, Ramana CV, Rao PR, Kovacs KL. Adv Appl Microbiol 1993; 38:211–295.
179. Vignais PM, Toussaint B, Colbeau A. In: Blankenship RE, Madigan MT, Bauer CE, eds. Anoxygenic Photosynthetic Bacteria. Dordrecht: Kluwer Academic Publishers, 1995:1175–1190.
180. Gest H, Kamen MD. Science 1949; 109:558.
181. Bulen WA, Burns RC, Le Combe JR. Proc Natl Acad Sci 1965; 53:532.
182. Belkin S, Padan E. FEBS Lett 1978; 94:291–294.
183. Fry I, Robinson AE, Spath S, Packer L. Biochem Biophys Res Commun 1984; 123:1138–1143.
184. Howarth DC, Codd GA, Stewart WD. Soc General Microbiol Quarterly 1981; 8:153.
185. Martinez-Romero E. In: Dworkin M, ed. The Prokaryotes: An Evolving Electronic Resource for the Microbiological Community. New York: Springer-Verlag, 2000.
186. Schubert KR, Evans HJ. Proc Natl Acad Sci 1976; 73:1207–1211.
187. Evans HJ, Harker AR, Papen H, Russell SA, Hanus FJ, Zuber M. Annu Rev Microbiol 1987; 41:335–361.
188. Newton W, Postgate JR, Rodriguez-Barrueco C. Recent Developments in Nitrogen Fixation. London: Academic Press, 1977.
189. Stal LJ. In: Packer L, Glazer AH, eds. Methods in Enzymology. Vol. 167. San Diego: Academic Press, 1988:475–485.
190. Fay P, Cox RP. Biochim Biophys Acta 1967; 143:562–569.

191. Fay P. Microbiol Rev 1992; 56:340–373.
192. Wolk CP, Ernest A, Elhai J. In: Bryant DA, ed. The Molecular Biology of Cyanobacteria. Dordrecht: Kluwer Academic Publishers, 1994:769–823.
193. Bergman B, Gallon JR, Rai AN, Stal LJ. FEMS Microbiol Rev 1997; 19:139–185.
194. Mulholland MR, Capone DG. Trends in Plant Science 2000; 5:148–153.
195. Stal LJ. In: Whitton BA, Potts M, eds. The Ecology of Cyanobacteria. Dordrecht: Kluwer Academic Publishers, 2000:61–120.
196. Krumbein WE, Cohen Y, Shilo M. Limnol Oceanogr 1977; 22:635–656.
197. Cohen Y, Castenholz R, Halvorson HO. Microbial Mats. New York: Alan R. Liss, 1984.
198. Bebout BM, Paerl HW, Bauer JM, Canfield DE, DesMarais DJ. In: Stal LJ, Caumette P, eds. Microbial Mats: Structure, Development, and Environmental Significance. Vol. G-35. Berlin: Springer-Verlag, 1994:265–271.
199. Uffen RL, Wolfe RS. J Bacteriol 1970; 104:462–472.
200. Schön G, Biedermann. Biochim Biophys Acta 1973; 304:65–75.
201. Yen HC, Marrs B. Arch Biochem Biophys 1977; 181:411–418.
202. Van Der Oost J, Bulhuis BA, Feitz S, Krab K, Kraayenhof R. Arch Microbiol 1989; 152:415–419.
203. Moezelaar R, Bijvank SM, Stal LJ. Appl Environ Microbiol 1996; 62:1742–1758.
204. Stal LJ, Moezelaar R. FEMS Microbiol Rev 1997; 21:179–211.
205. Heyer H, Krumbein WE. Arch Microbiol 1991; 155:284–287.
206. De Philippis R, Margheri MC, Vincenzini M. Arch Hydrobiol Algological Studies 1996; 83:459–468.
207. Aoyama KL, Uemura I, Miyake J, Asada Y. J Ferment Bioeng 1997; 83:17–20.
208. Nold SC, Ward DM. Appl Environ Microbiol 1996; 62:4598–4607.
209. Richardson LL, Castenholz RW. Appl Environ Microbiol 1987; 53:2142–2150.
210. Stal LJ, Caumette P, eds. Microbial Mats: Structure, Development and Environmental Significance. Berlin: Springer-Verlag, 1994.
211. Revsbech NP, Jørgensen BB, Blackburn TH. Limnol Oceanogr 1983; 28:1062–1074.
212. Revsbech NP, Jørgensen BB. Adv Microbial Ecol 1986; 9:293–352.
213. Paerl HW, Pinckney JL, Steppe TF. Environ Microbiol 2000; 2:11–26.
214. Oremland RS. Appl Environ Microbiol 1983; 45:1519–1525.
215. Skyring GW, Lynch RM, Smith GD. In: Cohen Y, Rosenberg E, eds. Microbial Mats: Physiological Ecology of Benthic Microbial Communities. Washington, D.C.: American Society for Microbiology, 1989:170–179.
216. van der Meer MTJ, Schouten S, Bateson MM, Nübel U, Wieland A, Kühl M, De Leeuw JW, Sinninghe Damsté JS, Ward DM. Appl Environ Microbiol submitted.
217. Skyring GW, Lynch RM, Smith GD. Geomicrobiol J 1988; 6:25–31.
218. Albert DB, Hoehler TM, Des Marais DJ, Bebout BM. Abstracts of the First Astrobiology Science Conference, Moffett Field, CA 2000.
219. Madigan MT, Martinko JM, Parker J. Brock Biology of Microorganisms. 8th ed. Upper Saddle River, NJ: Prentice Hall, 1996.
220. Canfield DE, DesMarais DJ. Science 1991; 251:1471–1473.
221. Bebout BM, Hoehler TM, Thamdrup B, Albert DB, Carpenter S, Hogan ME, Turk KA, Des Marais DJ. Geobiology 2004; 2:87–96.
222. Summons RE, Jahnke LL, Hope JM, Logan GA. Nature 1999; 400:554–557.
223. Rosing MT, Frei R. Earth Planet Sci Lett 2004; 217:237–244.

224. Xiong J, Fischer WM, Inoue K, Nakahara M, Bauer CE. Science 2000; 289:1724–1730.
225. Des Marais DJ. Science 2000; 289:1703–1705.
226. Kasting JF, Eggler DH, Raeburn SP. J Geol 1993; 101:245–257.
227. Kelley D, Karson JA, Blackman D, Früh-Green G, Butterfield DA, Lilley MD, Schrenk M, Olson EO, Roe K, Lebon J. Ridge Events 2001; 11:3–9.
228. Des Marais DJ. In: Jones JG, ed. Advances in Microbial Ecology. Vol. 14. New York: Plenum Press, 1995.
229. Barreto L, Makihira A, Riahi K. Int J Hydrogen Energy 2002; 28:267–284.
230. Ritter JA, Ebner AD, Wang J, Zidan R. Mater Today 2003; 6:18–23.
231. Tromp TK, Shia R-L, Allen M, Eiler JM, Yung YL. Science 2003; 300:1740–1742.
232. Schulz MG, Diehl T, Brasseur GP, Zittel W. Science 2003; 302:624–627.
233. Argano M. In: Balows A, Trüper HG, Dworkin M, Harder W, Schleifer KH, eds. The Prokaryotes. New York: Springer Verlag, 1992:3917–3933.
234. Tiedje JM. In: Zehnder AJB, ed. Biology of Anaerobic Microorganisms. New York: Wiley-Interscience, 1988:179–244.
235. Lovley DR, Phillips EPJ, Lonergan DJ. Appl Environ Microbiol 1989; 55:700–706.
236. Balashova VV, Zavarzin GA. Microbiol (Engl Trans Mikrobiologiya) 1979; 48:635–639.
237. Stolz JF, Oremland RS. FEMS Microbiol Rev 1999; 23:615–627.
238. Widdel F. In: Zehnder AJB, ed. Biology of Anaerobic Microorganisms. New York: Wiley-Interscience, 1988:469–586.
239. Wolin MJ, Wolin EA, Jacobs NJ. J Bacteriol 1961; 81:911–917.
240. Ferry JG. Crit Rev Biochem Mol Biol 1992; 27:473–503.
241. Sharak-Genthner BR, Davis CL, Bryant MP. Appl Environ Microbiol 1981; 42:12–19.
242. Zehnder AJB, Brock TD. Appl Environ Microbiol 1980; 39:194–204.
243. Mortenson LE. Biochimie 1978; 60:219–223.

3

Dioxygen over Geological Time

Norman H. Sleep

Department of Geophysics, Stanford University,
Stanford, California 94305, USA

1.	Introduction	50
2.	Long-Term Crustal Reservoirs and Cycles	51
3.	Modern and Ancient Biological Redox Cycles	53
	3.1. Thermodynamics of Life	54
	3.2. The Geological Record of the Modern Cycle	55
	3.3. Ancient Ecosystems	56
4.	Mantle Cycle	59
	4.1. Coupled Sulfur and Carbon Cycles and the Net Mantle Flux	59
	4.2. Oxidation of Basaltic Crust at Midocean Ridges	61
	4.3. Mantle Oxidation and Volcanic Gases	62
	4.4. Subduction of Sediments	64
	4.5. Serpentinite	65
	4.6. Hydrogen Escape to Space	65
5.	Carbon Burial	66
	5.1. Coupled Organic Carbon and Carbonate Surface Cycles	66
	5.2. Sites for Organic Carbon Burial	67
	5.3. Phosphorus and Organic Carbon Burial	67
6.	Conclusion: Oxygen Build-Up Over Geological Time	68
	6.1. Hydrogen Escape and the Crustal Oxygen Reservoir	68
	6.2. Transition to Oxygen-Rich Atmosphere	69

    6.3.   Controls on Reservoir Sizes   70
    6.4.   Exportable Biological Implications   70
Notes added in Proof   71
Acknowledgments   71
Abbreviations   71
References   71

1. INTRODUCTION

Oxygen and air are almost synonymous in common speech. There is oxygen in air and the tendency is to give the matter little thought until the oxygen gets dangerously depleted within the small volume that one has to breathe. Only a chemist would call O_2 "dioxygen" in normal conversation. I do not further use it in this paper.

In contrast, if we look further, back in time or to the other planets, the presence of oxygen in large quantities in our air becomes anomalous. The Earth is the only solar planet to have significant free oxygen in its atmosphere. Earth lacked this component throughout much of its history, before 2 billion years ago. Even now, the oxygen in our air is grossly out of equilibrium with the ferrous iron and sulfides in common crustal rocks as well as buried organic matter.

Given the widespread biological and geological interest, numerous scientists have recently reviewed the literature on the origin of oxygen in the air and the oxygen cycle. Canfield and Raiswell [1] and Lenton [2] provide general treatments. Berner et al. [3] discuss the Phanerozoic Era [the last 600 million years (m.y.)]. Hansen and Wallmann [4] discuss the last 135 m.y. Anbar and Knoll [5] discuss the Proterozoic Era ($\sim$2500–600 m.y. ago). Des Marais [6], Kasting and Siefert [7], Holland [8], and Nisbet and Sleep [9,10] discuss the early Earth. I concentrate on long-term geological aspects of the problem in this review. Ideally, one would like constraints that one can extrapolate to other planets, like Mars and any extrasolar planets to be imaged by the Terrestrial Planet Finder [11] (see Ref. [12] for a general discussion of planetary habitability). Other chapters in the volume cover the short-term aspects of cycles strongly coupled to oxygen: hydrogen (T. M. Hoehler, Chapter 2), sulfur (O. Klimmek, Chapter 5), phosphorus (B. Schink, Chapter 6), and iron (J. H. Street and A. Paytan, Chapter 7).

The short-term behavior and maintenance of oxygen in the air is available for observation. Green plants and marine plankton use light to photosynthesize O_2 and organic matter. The reaction is written schematically in Eq. (1):

$$CO_2 + H_2O + h\nu \implies O_2 + CH_2O \qquad (1)$$

About 8.4×10^{15} mol of this reaction occur every year. The photosynthetic organisms themselves, heterotrophic organisms in the food chain, including decay, consume all but 0.1% of this bounty. The remaining 10^{13} mol of oxygen goes into the air and sediments bury the remaining 10^{13} mol of organic matter. As there are 38×10^{18} mol of oxygen in the air plus the ocean, the residence time of free oxygen with respect to burial and weathering is ~ 4 m.y. The residence time with respect to passing through photosynthesis is 4000 years.

Both residence times are long on the scale of human lifetimes. We do not have to thank the green plant of today, as the vast size of the reservoir alone provides short- and medium-term stability. In contrast, stability does not prevail in regions temporarily isolated from the main reservoir. The winterkill of fish where dissolved O_2 levels drop in shallow ice-covered temperate lakes is one example.

However, the mentioned 4 m.y. is a tiny fraction of the ~ 4.5 billion year age of the Earth and even of the 600 m.y. period where oxygen-consuming large (metazoan) animals have roamed. In this paper, I concentrate on these long-term aspects. That is, how and when did O_2 become significant in the Earth's air? What are the long-term controls on the O_2 abundance? I begin by reviewing the crustal cycle and reservoirs of "available" oxygen in Section 2. I then examine the biological controls on oxygen and alternative biological cycles in Section 3. Thereafter, I consider the descent of available oxygen and organic carbon into the mantle and the escape of H_2 to space in Section 4. I examine controls on the long-term burial of carbon in sediments on the Earth in Section 5 and I conclude that several processes led to the build up of O_2 in our air in Section 6.

2. LONG-TERM CRUSTAL RESERVOIRS AND CYCLES

Rocks play a major role in the long-term cycle of oxygen. I define reservoirs and cycles in line with model geochemical and tectonic concepts. Oxygen is the most common element in the Earth's crust and mantle, but almost all of it is fixed to cations that have only one oxidation state under the prevailing conditions: silicon, magnesium, aluminum, calcium, sodium, and potassium [13]. I am concerned with the much smaller reservoir of "available" oxygen. This includes free oxygen and oxygen fixed to iron, carbon, sulfur, and hydrogen. These reservoirs formed ultimately at the expense of mantle-derived igneous rocks. At depth, the iron was about $1/10$ ferric and $9/10$ ferrous. The carbon was CO_2; the hydrogen was water, and the sulfur was sulfide. In a global compilation, we ignore minor fluxes with other oxidation states, like reduced carbon in diamonds.

For bookkeeping convenience, I include the Earth's ocean, atmosphere, and sediments as parts of its continental crust. The oceanic crust (excluding its sediments) almost always gets subducted into the mantle and is conveniently considered part of the mantle. I refer to the high-grade metamorphic rocks and igneous rocks of the continental crust as "hard rocks" to distinguish them from sediments, "soft rocks". Subduction includes any process that has the net effect of putting sediments and basaltic crust into the mantle. With these definitions, all the reactants in Eq. (1) have long-term continental crustal cycles and significant crustal reservoirs. In addition, O_2 reacts with sulfide in rocks to form sulfates and with ferrous iron to form ferric iron. Sulfate in sediments and excess ferric sediments and ferric iron in hard rocks are additional available oxygen reservoirs.

I begin with CO_2 and need not consider water, as it is unlikely to ever have been a globally limiting reactant on the Earth. The pre-industrial concentration of CO_2 in the air is $\sim$300 ppm or 0.055×10^{18} mol globally [14]. This reservoir is in dynamic equilibrium with 3.3×10^{18} mol in the global ocean [14]. The current rate of carbon burial would consume the combined ocean and air CO_2 reservoir in $\sim$300,000 years. Carbonates (limestones and dolomites) on continental platforms are the main continental crustal CO_2 reservoir of 6000×10^{18} mol [14], which would be consumed by carbon burial in 600 m.y., still a small fraction of the age of the Earth. Carbonates tend to break down releasing CO_2 at high-grade metamorphic conditions and do not form in common igneous rocks. It is thus unlikely that globally significant amounts exist within the hard rocks of the continental crust.

The air and the ocean (38×10^{18} mol as O_2) are the most obvious but not the most massive reservoir of available oxygen. There are 40×10^{18} mol of sulfate in the ocean and 140×10^{18} mol of sulfate in sediments [14]. It required 80×10^{18} and 280×10^{18} mol of O_2, respectively, to form these reservoirs from sulfide derived from igneous rocks. The inventory of excess ferric iron in sediments, 50×10^{18} mol of Fe_2O_3, required 25×10^{18} mol of O_2 to form from FeO derived from igneous rocks [15]. The inventory of excess ferric iron in hard rocks, 4000×10^{18} mol of Fe_2O_3, required 2000×10^{18} mol of O_2 to form from FeO over geological time [15]. Much of this hard rock reservoir is over 2 billion years old.

The accuracy of estimates of the sizes of these reservoirs varies greatly. The ocean and the air are available for precise sampling. Sedimentary rocks have been extensively sampled by drilling, particularly for petroleum exploration. The estimates of sediment reservoirs are hence better than a factor of 1.5.

The estimate for excess ferric iron in hard rocks is problematic, as we do not have good global sampling of the middle and deep crust. The estimates compiled by Lecuyer and Ricard [15] extrapolate near surface ferric/ferrous ratios [equivalent to an excess of 3% Fe_2O_3 (by mass)] downward to 35 km depth. This may lead to an overestimate as the fraction of rocks that have been exposed to oxidation at the surface should decrease with depth.

More subtlety, the reference percentage of Fe_2O_3 from which to define "excess" is not obvious. The existence of some excess is obvious, for example, the ferric iron in red granites. However, earlier treatments (e.g., including that of Wolery and Sleep [16]) effectively defined the global excess as zero. This vestige of preplate tectonic thinking slipped through. That is, the continents were thought to be a primary feature of the Earth that came with its observed ferric/ferrous ratio. A better reference for defining excess is the low $\sim 1/10$ ferric to total iron ratio in the mantle-derived igneous rocks (mid-oceanic ridge basalt, MORB, is a good standard) that formed the crust. Arc basalts are a less optimal choice. They are somewhat oxidized relative to MORB. They contain subducted water along with other material that has recently been at the surface (e.g., [17,18]). These foreign materials plausibly oxidized the magma. In any case, the composition of the continental crust is poorly enough known that using any mantle basalt is effectively assuming that the bulk of the estimated 3% Fe_2O_3 in crustal rocks is excess.

The important reduced carbon reservoirs have been extensively sampled. The living biomass (0.05×10^{18} mol of C equivalent [14]) is a trivial reservoir. Sediments (1200×10^{18} mol of C [14]) are the only significant reservoir that is well constrained by drilling. The residence time of this reservoir with respect to carbon burial is only 120 m.y. This is about the expected lifetime for sediments between deposition and being uplifted and eroded [19]. There is significant carbon of sedimentary origin in graphitic gneisses. It is very difficult to get an accurate estimate of their global abundance because of sparse sampling. I ignore this reservoir.

I add up the available oxygen reservoirs and compare them using reaction (1) with the buried organic carbon to show that an imbalance exists. I attribute this imbalance to the crustal cycle not being closed. That is, some H_2 escapes to space, some O_2 is subducted from sulfate and O_2 reacting with ferrous iron in the oceanic crust, and some organic carbon is subducted mainly with sediments. Taking the numbers at face value, there is an excess of $\sim 1200 \times 10^{18}$ mol of O_2 equivalent on the oxygen side, consistent with H_2 loss to space and/or net subduction of organic matter. The rate of loss to generate this over the geological time of 4 billion years is 0.3×10^{12} mol per year, $1/30$ of the O_2 release to the air by carbon burial.

Alternatively one could force a balance by adjusting the amount of Fe_2O_3 in hard rocks to $\sim 2000 \times 10^{18}$ mol. The equivalent thickness of oxidized hard rocks then becomes 17.5 km. Given that there are few samples of the lower crust, I do not exclude this situation. However, I do use my preferred imbalance of $\sim 1200 \times 10^{18}$ mol in subsequent examples.

3. MODERN AND ANCIENT BIOLOGICAL REDOX CYCLES

The typical ecology presentation starts with photosynthetic primary producers and continues with the food chain dependent on it. It is hard to find biota on the Earth that do not (at least indirectly) depend on photosynthesis. For example, hydrothermal vent organisms that oxidize sulfide use O_2 from photosynthesis. A counter example is that deep methanogens in principle can be independent of photosynthesis, using CO_2 from the interior of the Earth and H_2 from the local reaction of ferrous iron in the rock with water [20,21].

It is unlikely that life on the Earth began with photosynthesis. The root (the last universal common ancestor) may have been a chemoautothrophic thermophile organism (see Refs. [22–25] for discussion). Still photosynthesis is ancient. It occurs independently in Archaea and bacteria at a midlevel in the branches [1,26]. Oxygenic photosynthesis is of later (but still ancient) origin from anoxygenic photosynthesis. It is more complex and occurs only on one branch (cyanobacteria including the chloroplasts in green plants).

It is still hard to imagine global ecology without photosynthesis. We have to try to do it if we want to see how oxygen arose in our air.

3.1. Thermodynamics of Life

Before going into the geological record, I review the thermodynamics of hetero-trophic (and chemoautotrophic) life with the intent of obtaining generalized features. The generic form of these biological process involves a (free) energy producing reaction of say sulfate with H_2 to make sulfide and a coupled energy consuming reaction that the organism needs to go about its business, like making ATP. The free energy change of the combined reaction is relevant to the minimum concentrations of the reactants for biota to use them. The free energy of the reaction depends on the logarithm of the ion product of the reactants and products.

In a typical local environment, one of the reactants is present in excess of the other. Given time, biota consume the reactants until the limiting one is almost used up to the amount expected from the ion product and the free energy of the reaction. Hoehler et al. [27,28] observed this behavior in natural environments and laboratory analogs. For example, sulfate in the presence of significant H_2 gets drawn down to ~ 10 μM depending on temperature. H_2 in the presence of sulfate gets drawn down to ~ 1 nM, again depending on temperature. Life competes with the abiotic reaction, which has a larger free energy because it does not produce ATP. The abiotic reaction can in principle draw down the reactants below the level needed to sustain life.

The free energy limit does not apply to reactions involving O_2. Rather the free energy of the biological reaction is large enough that its predicted oxygen fugacity would imply less than one molecule in any reasonable volume (T. Hoehler, personal comunication, 2003). (The same is true with a mineral assemblage including minerals with ferrous iron [A. Hessler, personal communication, 2003].) Rather, a very small concentration of oxygen would be maintained by the finite time for a molecule to get from its source to its sink.

A general lesson is that an ecological system, like the modern cycle with photosynthesis and abundant O_2, may also work with O_2 being a trace component and say abundant H_2, organic matter, and methane and somewhat different organisms. That is, oxygenic photosynthesis could have been important without producing significant atmospheric oxygen and hence evidence for oxygen in the geological record (e.g., [29]).

Ancient sediments are not likely to yield evidence of the presence of trace amounts of oxygen in the water at the time of their deposit. Microbes consumed most of the oxygen produced by photosynthesis. Only minute amounts of oxygen reacted with sulfides and ferrous iron in the rock. The mass balance is like that of a bathtub: the water level can be quite low with the tap full on if the drain works.

3.2. The Geological Record of the Modern Cycle

The geological record indicates that on average the Earth's surface environments have become more oxidized with time. It is much harder to document whether the changes are effectively steps between different buffered plateaus, a gradual increase, or a trend with significant fluctuations. To show some of the issues, I start with the recent events and work back to less oxidized conditions on the Earth in its ancient past. At each stage, there was a workable global ecology and carbon cycle. The tendency of evolution to make organisms fit for their environment limits what we can learn about any long-term biological buffers on the oxidation state of the Earth's surface.

Starting with the present, humans are obviously quite fit to breathe the 21% O_2 that we find in our air and quite unfit if the oxygen concentration gets far below that. It is tempting to invoke the anthropic principle that there is 21% O_2 in our air because that is what is necessary to support intelligent life forms. However, this is a science stopper. As noted already, the oxygen concentration in the air changes slowly so there is time for evolution. Some kinds of animals do inhabit suboxic environments so it is plausible that intelligence could have originated there, especially if these environments were more common than aerobic ones. The anthropic argument can be easily turned around to that we breathe oxygen because aerobic environments were quite common when (and since) the biological innovation of control genes for diverse cell types occurred and our multicellular clade colonized them.

It is hard though more productive to use the macrostructure of fossil animals (and plants) to get constraints on the O_2 concentration of the air because evolution makes animals (to use Darwin's word) serviceable in their environments. For example, Berner et al. [3] regard the large size of Devonian dragonflies as evidence for high values of O_2 (35%) at the time. Alternatively, these large insects filled a niche that is now occupied by birds and that there is selective pressure from birds that keeps the size of modern dragonflies in check. Careful physiology and flight dynamics are needed to resolve these possibilities.

A still better, related argument is that the charcoal from ancient forest fires and the existence of ancient forests provide limits on the O_2 concentration in the air. At the low end, the geological argument has much weight. Even dry tinder will not ignite if there is too little O_2 (below $\sim 15\%$ [2,3]). (This is physics not biology. Cutting off the air is a time-honored method of putting out fires.) On the upper end, trees have evolved to tolerate and benefit from fire (like jack pines) or to be fire resistant to some extent. They have done so most recently in the current atmosphere. Strongly fire resistant trees, like Ohia on the Big Island of Hawaii, the site of frequent lava flows, do exist. Ohia does not ignite when near lava flows or after being hit by hot volcanic ash. It is killed by direct contact with lava flows but even then the trunk may not burn. Berner et al. [3] discuss how one might recognize fire-resistant high-O_2 plants in the geological record.

3.3. Ancient Ecosystems

The geological record indicates that there were at least modest amounts ($\sim 3\%$) of O_2 in the air after 2 billion years ago (e.g., [1,2,6,30]). That is, there was enough dissolved oxygen in rain and river water to have macroscopic effects on exposed rock. Studies of paleosols indicated that FeO was oxidized rapidly enough to be immobile and widespread redbeds indicate that much FeO did get oxidized.

A much more precise constraint comes from multiple sulfur isotope systematics. Sulfur has four stable isotopes, 32, 33, 34, and 36. In most geological situations, the isotopes fractionate proportionally to their atomic masses. However, ultraviolet radiation in the atmosphere produces mass-independent fractionation by photolysis. This fractionation can be preserved in the geological record only if there is too little oxygen in the air to oxidize the sulfur species to sulfate. Otherwise, the soluble sulfate mixes into the massive reservoir of the ocean. The critical concentration of oxygen below which mass-independent sulfur isotope anomalies can be preserved is 2 ppm [31]. Studies of ancient sediments indicate that the transition to an atmosphere with <2 ppm oxygen to a more oxygen-rich one lasted from ~ 2.45 to ~ 2.0 billion years ago [32] (see Ref. [33] for a field study). The oxygen cycle after the transition was basically like the modern one. As at present, the tendency of organic matter to back react with oxygen provided a stabilizing buffer. There is, however, no obvious biological process to buffer O_2 at a given value.

Going back before ~ 2.8 billion years ago and back to the end of a good sedimentary rock record ~ 3.5 billion years ago, there were only trace amounts of O_2 in the water and there may have been significant amounts of methane in the air. (Trace O_2 does not persist in air with methane because of photolysis [34].) Macroscopic evidence includes mobile iron in paleosols and survival of easily oxidizable detrital minerals like pyrite in sediments.

There is evidence that oxygenic photosynthesis occurred at this time, but typically produced only trace amounts of O_2 in the water. Molecular fossils indicate oxygen-producing microbes as does the tendency of carbon isotopic fractionation to behave like that with modern photosynthesis [29,30].

Lead isotope anomalies indicate locally oxic water back to 3.75 Ga [35]. Physically, the isotopes ^{235}U, ^{238}U, and ^{232}Th decay to ^{207}Pb, ^{206}Pb, and ^{208}Pb, respectively. Measurement of the Pb isotopes can indicate U–Th fractionation and constrain when it occurred. Geochemically U and Th do not fractionate in reduced environments on the scale of the source region for a sedimentary rock.

However, in oxic environments oxidized U(VI) is soluble and fractionates from Th. Later, it reacts often with organic matter to produce insoluble U(IV) compounds. The net effect is that the U/Th ratio in the organic-rich sediments is much greater than that of the source rocks.

During that interval and still further back, a highly productive anoxic sulfuretum system based on sulfur is conceivable (e.g., [29]). The net photosynthetic and heterotrophic back reaction is given in simple form below:

$$2H_2O + H_2S + 2CO_2 + h\nu \implies 2H^+ + SO_4^{2-} + 2CH_2O \tag{2}$$

Extant organisms of ancient lineage do the reaction each way [1,26,30]. As with the modern cycle, the obvious limit on primary productivity is the rate by which the heterotropic back reaction restores the reactants on the left side.

One might suppose that the low solubility of iron sulfides in near neutral pH water would inhibit the cycle in Reaction (2). This, however, seems unlikely. First, complexing keeps total ferrous iron at a finite amount in sulfide-rich water [36]. Second, iron sulfide on a shallow bottom could act as a storage bin of both S and Fe reactant. Finally, I note in analogy that having a small concentration of a reactant need not be fatal. Modern photosynthesis works fine even though CO_2 is a trace gas (300 ppm in the air) and even though most of the crustal carbonate and organic carbon reservoirs are buried out of reach of the surface ecology at any one time.

In any case, a global ecology based on iron, and not sulfur, is possible with extant organisms. Diverse kinds of bacteria do photosynthesis that in idealized form is given in Eq. (3) [37,38]:

$$H_2O + 4FeO + CO_2 + h\nu \implies 2Fe_2O_3 + CH_2O \tag{3}$$

Species of bacteria and Archaea get energy from the back reaction [39,40]. These extant species use a variety of ferric iron sources including crystalline hematite. Their genes to do this may predate those for sulfate respiration. A vigorous ancient biospheric cycle based on Reaction (3) thus seems likely. Such photosynthesis could produce banded iron formations on the Earth [38] and hematite deposits on Mars [41] (see Ref. [42] for discussion of these rocks). The biosphere was upside-down from the modern one [43]. The ferric iron was immobile and built up if it is in a different place than the buried organic matter. One would not expect to find organic matter with ferrous iron formation as biota would have quickly consumed it by reacting it with an excess of hematite.

An iron-based ecosystem is particularly likely in regions where the surface waters are dominated by basalt. There is no shortage of FeO as it is $\sim 10\%$ by mass of the rock. In contrast S is only 0.1%. This situation seems likely on Mars and within basaltic regions on land on the early Earth.

Photosynthesis using H_2 occurs with extant organisms (see the works of Ehrenreich and Widdel [37] and Zaar et al. [44] for studies of particular microbes). The idealized reaction is indicated below:

$$2H_2 + CO_2 + h\nu \iff CH_2O + H_2O \tag{4}$$

This reaction is exothermic and the light acts like a catalyst. A heterotrophic ecosystem based on the back reaction is not possible. On the modern Earth, this reaction as well as sulfur and iron photosynthesis often depends on reduced organic matter associated with oxygenic photosynthesis.

Before the advent of other types of photosynthesis, productivity depended on the H_2 flux from volcanoes and hydrothermal vents and was orders of magnitude lower than present. I note that little H_2 reached the photic zone as methanogens in the dark crevices of hydrothermal systems and in the deep ocean above submarine volcanoes got first crack at the reactants. Modern methanogens can draw H_2 down to $\sim$13 nM [20,21].

The niche of hydrogen-based photosynthesis may have served to preadapt organisms for the more bountiful iron, sulfur, and eventually oxygenic systems. It could have even started as a facultative mode in an organism that could do methanogenesis in the dark. As it is exothermic, it could have been quite inept.

To reiterate, the Earth's surface environments have become more oxidized with time. The first cellular microbes were chemoautotrophs and lithoautotrophs dependent on hydrothermal vents and the solid and gaseous products of igneous activity. Eventually, hydrogen-based photosynthesis evolved as a minor niche. These photosynthetic organisms evolved into the more bountiful niche of sulfide photosynthesis. The global ecology slowly moved from a low-sulfate sulfuretum to a high-sulfate sulfuretum. Cyanobacteria show genomic evidence of having evolved during this time in sulfide-rich seawater [36]. As long as sulfide was abundant, oxygenic photosynthesis was not globally significant. The selective advantage of not being dependent on sulfide and the feedback that sulfide decreased as O_2 built up from a trace gas to modern levels led to the dominance of cyanobacteria. An iron-based system may have prevailed between hydrogen and sulfur.

The deep water may well have remained anoxic after the shallow water became oxic [5,36,45]. In their formulations, the bottom water is ferrous iron rich before 2.5 billion years ago and becomes sulfide rich after that. Upwelling of these waters would have triggered high productivity based on iron and sulfide photosynthesis, respectively. That is, the organic sediments produced beneath the anoxic upwellings would show molecular evidence of anoxygenic photosynthesis even though most of the shallow ocean was oxic.

Organisms existed for geological lengths of time without becoming dominant on the Earth or obvious in the geological record. For example, O_2 built up in small bodies of stagnant productive water as indicated by Pb isotopes [35], much the way methane saturates water in modern swamps. Organisms persist after their times of global dominance. At present, anoxygenic photosynthesis persists as a minor niche. Anaerobic organisms are relegated to local anoxic environments.

It is clear that this global change required biological innovations. It is obvious that the tendency of oxidation of organic matter to go faster when oxygen (or sulfate) is abundant provides some negative feedback. No hard biological buffers are obvious. Positive biological feedback, like with O_2 reacting with sulfide to quench the sulfuretum, may have even hastened some changes. It is not evident that the living organisms set the conditions of their own rises and falls from grace, rather than responding to processes in the Earth's mantle, crust, and upper atmosphere. In the next two sections, I examine these controls on the redox state of the Earth's surface.

4. MANTLE CYCLE

To the first order, the oxygen cycle can be viewed as steady state because the residence time for oxygen in the air and even buried organic carbon is short compared to the age of the Earth. We need to recognize subtle effects that allow the cycle to be slightly out of steady state. In particular, oxygenic photosynthesis produces oxygen and organic matter in equal amounts. Simply speeding up the productivity of it or any of the anoxygenic modes of photosynthesis does not necessarily lead to a build-up of oxygen if the back reaction is fast. Rather, geological processes must separate the oxygen and organic carbon for geological periods of time. Put in another way, the net growth of the excess oxygen and buried carbon reservoirs is a tiny fraction, a few percent, of the instantaneous burial and exhumation rates.

I consider carbon burial in the next section. In this section, I consider the mantle cycle. Mid-oceanic ridges act as oxygen sinks. Reaction with basalt consumes dissolved oxygen and sulfate. Organic carbon in sediments subducts into the mantle. Reduced volcanic gasses vent on land and beneath the ocean. I consider the loss of H_2 to space in this section as it too moves a crustal reactant out of sight.

One can approach the problem in several ways. I first constrain the sulfur and oxygen fluxes by examining the long-term coupled sulfur and carbon cycles. I next review more direct current estimates of sulfur and oxygen fluxes at ridges. I then examine fluxing of gasses and the change of redox state of the mantle over time, the subduction of sedimentary carbon, and the oxygen flux associated with serpentinization. Finally I review H_2 escape to space.

4.1. Coupled Sulfur and Carbon Cycles and the Net Mantle Flux

The carbon, oxygen, and sulfur cycles are strongly coupled. Study of this coupling provides independent information on how closed (or open) the crustal cycle is over periods of several hundred million years (see Refs. [14,16]). The O_2 from photosynthesis reacted with sulfide to form sulfate that is now a major reservoir for free oxygen. The net effect of photosynthesis is the idealized Reaction (2) with a minor correction for FeS_2 in the buried product. That is, an increase in the sulfate reservoir implies an increase in the organic carbon reservoir.

Both carbon and sulfur isotopes fractionate during this process. Organic matter is depleted in ^{13}C relative to marine carbonates and bacteriogenic sulfides are depleted in ^{34}S relative to marine sulfates. If the crustal system is in fact closed, the total mass of carbon and its average $\delta^{13}C$ stays constant as summarized in Eq. (5),

$$\delta^{13}C_{carb}M_{carb} + \delta^{13}C_{org}M_{org} = \delta^{13}C_{total}M_{total} \tag{5}$$

where M is the mass of a reservoir, the subscript "carb" indicates carbonates, the subscript "org" indicates organic matter, and the subscript "total" is that of the crustal system. Similarly, for a closed sulfur system Eq. (6) holds,

$$\delta^{34}S_{ide}M_{ide} + \delta^{34}S_{ate}M_{ate} = \delta^{34}S_{total}M_{total} \tag{6}$$

where the subscript "ide" indicates sulfides and the subscript "ate" indicates sulfate. Study of sediments indicates that the global average $\delta^{13}C_{total}$ is about -5 per mil (relative to a standard) similar to that of mantle derived igneous rocks [14]. This implies that there has been no preferential subduction of carbonates or organic matter.

Similarly, the total ^{34}S anomaly in sediments is about 0 per mil that of the ultimate igneous source [14]. This implies no preferential loss of sulfate or sulfide to the mantle.

Analysis is simplified because the carbon fractionation $\delta^{13}C_{carb} - \delta^{13}C_{org}$ and sulfur fractionation $\delta^{34}S_{ide} - \delta^{34}S_{ate}$ did not change much in the Phanerozoic and may be considered constants. The carbon fractionation and mass balance [Eq. (5)] provide two equations with two unknowns, $\delta^{13}C_{org}$ and M_{org}/M_{total} ($M_{carb}/M_{total} + M_{org}/M_{total} = 1$), if $\delta^{13}C_{carb}$ is measured. Similarly, one can get $\delta^{34}S_{ide}$ and M_{ide}/M_{total}, if $\delta^{34}S_{ate}$ is measured. This allows one to focus on easily measurable marine carbonate and marine sulfate deposits that sampled a well-mixed reservoir that evolved gradually.

A long-term correlation of sulfur and carbon ratios is implied if the system remains closed. From Reaction (2), the fraction of carbonate in the carbon reservoirs is low when the fraction of sulfur in the sulfate reservoir is high. The mass balances in Eqs. (5) and (6) imply high $\delta^{13}C_{carb}$ at times of low $\delta^{34}S_{ate}$. The slope of the correlation depends on the relative size of the reservoirs with the small reservoir sulfur changing the most.

Extensive studies of marine carbonates and marine sulfates do yield the predicted correlation for a closed system with some scatter [14]. This indicates that the sedimentary sulfur and carbon systems are more or less closed on the time scale of the isotopic fluctuations, namely a few 10^8 years.

This reasoning constrains the closure of cycle with the smaller reservoir, sulfur. The total sedimentary plus ocean sulfur reservoir is 350×10^{18} mol of S [14]. For example, Wolery and Sleep [16] assumed that the crustal residence time (for exchange with the mantle) of sulfur is 800 m.y., more than the length of the Phanerozoic record over which the cycle appears to be closed. They got an upper limit of the sulfur flux into the mantle of 0.45×10^{12} mol per year of sulfur (or 0.9×10^{12} mol as O_2 equivalent). Considering direct estimates of the fluxes in the next section indicates that this is a tolerable estimate of the actual flux.

4.2. Oxidation of Basaltic Crust at Midocean Ridges

Hydrothermal fluids circulate through the oceanic crust. One can constrain the oxygen flux associated with these processes by sampling the fluids or the before and after composition of the oceanic crust.

The oceanic crust is potentially a huge oxygen sink. New seafloor forms at a rate of 3 km^2 year^{-1}, which would cover the Earth's entire surface in 170 m.y. The basaltic rocks of the crust have a density of 3000 kg m^{-3} and contain $\sim 10\%$ FeO by mass. Oxidizing this material to 1 km depth to magnetite, Fe$_3$O$_4$ would consume 2×10^{12} mol of O_2 per year. Oxidation of the whole 6 km of crust would consume 12×10^{12} mol of O_2 equivalent per year, more than the flux from carbon burial. Compilations of the composition of the oceanic crust indicate a flux of $(0.5-1.9) \times 10^{12}$ mol of O_2 equivalent per year indicating that less than a kilometer of equivalent section gets oxidized [15]. Studies of hydrothermal fluids are in agreement with the lower end of this range, but poorly constrained [4]. I use the rounded estimate of 10^{12} mol of O_2 equivalent per year in my examples.

I discuss the mass balance and physics of oxidation of the oceanic crust so that I can extrapolate the present situation into the past. Today, sulfate is the main oxidant in seawater, followed by minor fluxes from dissolved oxygen and the water itself that produces H_2 from reaction with ferrous iron. Carbon dioxide precipitates as carbonates and is not a globally significant oxidant under these conditions. Most of the circulating sulfate that enters hydrothermal systems, however, does not become reduced. Rather, it precipitates as anhydrite (CaSO$_4$) at temperatures below 150°C. Later when the crust has spread away from the ridge axis and cooled, the anhydrite dissolves and the sulfate returns to the ocean. Sulfate reduction is quite slow at the temperatures where anhydrite is sequestered (below 150°C).

A fraction of $\sim 10\%$ of the sulfate that originally entered the hydrothermal system from the ocean continues downward and reacts with ferrous iron to form sulfide and magnetite. Some of the sulfide precipitates in the rock and the rest vents at the seafloor. The vent fluid leaches additional sulfide from the rock and carries it into the ocean. The vented sulfides then react with dissolved oxygen in the water to form sulfate. The net effect is to replace one batch of sulfate ions in the water with another that came from the rock and to transfer oxygen into oceanic crust. The oxygen sinks from oceanic crust being reduced at depth and vented sulfide being oxidized in the ocean are comparable. I set them to 0.5×10^{12} mol of O_2 equivalent of per year in my examples.

The rate of oxygen loss depends on the rate that seafloor forms, currently 3 km^2 per year, since most of the alteration occurs within crust younger than a million years. In the past, the seafloor production rate depended on the square of the heat flow from the mantle. I follow Sleep and Zahnle [46] and use the present rate in some examples and a rate 10 times present for the early Earth in another example. The oxygen flux from marine sulfate reduction was then 5×10^{12} mol per year if sulfate was at least $1/10$ the present concentration. More likely, this potent sink buffered sulfate at low levels until the rate of plate tectonics waned.

I am now ready to discuss the ancient Earth. At a time when the ocean was anoxic and constrained little sulfate, the oceanic crust was not a significant oxygen sink. There was little sulfate in the circulating seawater to react with the basalt and no oxygen to oxidize vented sulfides.

Later the concentration of sulfate gradually increased in the seawater. Initially its concentration was too low to form anhydrite and all the sulfate entering the hydrothermal system got reduced. This formed a stabilizing buffer on the sulfate concentration in seawater because the sulfate flux into the oceanic crust was proportional to its concentration.

Still later the sulfate concentration in seawater reached $\sim 10\%$ of the current value. Anhydrite precipitated in the intake regions of the hydrothermal systems. At that time, the sulfate concentration reaching the deep hot parts of the hydrothermal systems approached the present value. The flux of sulfate reduced per area of seafloor also approached its present value. The vent systems then ceased to be stabilizing buffers on seawater sulfate concentration. The subsequent increases in sulfate concentration resulted in more anhydrite sequestration but did not affect the amount of sulfate that got reduced.

Eventually the deep ocean became oxic. The vented sulfide then got reduced. This process at first formed a stabilizing buffer that kept oxygen concentration in seawater down. Once the oxygen concentration reached the level that all of the sulfide was oxidized, the buffer was overwhelmed and the oxygen sink was independent of its concentration. This situation prevails today.

4.3.   Mantle Oxidation and Volcanic Gases

The mantle composition changes with time as oxidized oceanic crust and organic carbon in sediments subduct into the mantle. Conversely, volcanic rocks erupt from the mantle and gases vent from these rocks.

One speculation is that the build up of oxygen in the air occurred because volcanic gases have become more oxidized with time [8,47]. Biological processes then partitioned the gases into carbonates, sulfates, sulfides, and organic matter. If one assumes that the partition of carbon is fixed at $1:4$ organic to carbonate, then a subtle change in the gas composition can flip the system into having sulfate and oxygen [8]. There are problems with this approach.

The first is that it ignores the igneous rock that brought the gases up in the first place. Much of the rock erodes quickly if it comes out of a volcano on land or later if it is intrusive. The ferrous iron in the igneous rock then becomes available to be oxidized to ferric iron. Eventually, much of the ferric iron formed in this way ends up in sedimentary rocks that get metamophosed or melted into hard rocks. As noted in Section 2, the excess ferric iron in hard rocks is the largest crustal available oxygen reservoir. The sink involved in its formation from mantle-derived igneous rocks cannot be ignored in favor of considering only volcanic gases.

The second problem is cause and effect. The continental crust and mantle are more plausibly oxidized by the surface environments than the other way around. For example, arc volcanics contain surface water and ferric iron from altered oceanic crust.

Crustal melts form various parts of the crust. As the crust is highly hetero-geneous, the mean composition of volcanic gases does not depend simply on the mean composition of the crust. Rather, the fraction of a gas component depends exponentially on its chemical potential in the local region that actually melts. The bulk of the more reduced gases then comes from the most reduced graphitic domains.

Finally, one can see if the mantle has actually become more oxidized over time by looking at the trace element (particularly Cr and V) variations of rocks of different ages. Overall the data indicate that the terrestrial mantle composition has not changed within resolution from the time of a good igneous rock record (3.6 billion years ago) and possibly back to 3.9 billion years ago [48–50]. I note that Warren and Kallemeyn [51] show that the mantle of Mars became more oxidized between 4.5 billion years ago, the age of Mars rock ALH-84001, and the recent (1.3 billion years and younger) SNC rocks. The younger Martian rocks may have a source region enriched in subducted materials including platinum-group elements from meteorites.

This behavior is what one would expect from the relative sizes of the crust and mantle reservoirs. The available oxygen reservoirs in the mantle are water, ferric iron, and CO_2. The mantle water reservoir is a significant fraction of the mass of the oceans (40% or 20000×10^{18} O_2 equivalent [52]; 30%, 15000×10^{18} O_2 equivalent [53]). Ferric iron is 15000×10^{18} O_2 equivalent [52]. Carbon dioxide is 6500×10^{18} O_2 equivalent [53] (see Ref. [46] for an earlier review of carbon in the mantle). As these reservoirs are an order of mag-nitude greater than the crustal organic carbon and available oxygen reservoirs, one would not expect crustal processes to greatly affect the chemical potential of mantle oxygen. This is compatible with the observation that arc-derived mantle rocks are only occasionally more oxidized than typical mantle rocks [50]. Note that FeO and Fe_2O_3 are in solid solution in various mantle minerals, so that oxygen chemical potential varies continuously with composition. There are no sharp buffers where a minor change in composition causes the chemical potential to jump, like metallic iron–FeO and FeO–magnetite.

Still, the Earth's mantle is more oxidized than the other solar system bodies for which we have information from samples, the asteroid Vesta, the Moon, and Mars. The Earth's mantle most likely became oxidized during its accretion (for review see Ref. [52]). The climax was the collision of a Mars-sized body with the Protoearth that left the present Earth–Moon system in its wake. During this event (and after subsequent asteroid impacts) metallic iron mixed into the mantle and consumed hydrogen equivalent to that in the present oceans. A comparable amount of hydrogen was consumed making ferric iron in the mantle.

Immediately after the Moon-forming impact, a rock–vapor atmosphere surrounded the Earth for a few thousand years and a stream atmosphere then surrounded the Earth for a few million years [54]. These violent surface conditions subsequently returned for thousands of year periods to the surface after asteroid impacts. One interesting effect is that an iron-rich asteroid may have transiently created a reducing atmosphere leading to the origin of life [55,56].

4.4. Subduction of Sediments

There is a net excess of oxygen in my crustal mantle balance. Subduction of organic carbon is a conceivable way to create this imbalance.

At present, most sedimentary deposits on oceanic crust accumulated slowly in aerated seawater [57]. They thus contain globally insignificant amounts of organic carbon. The organic carbon that does get subducted comes from near-coast sediments and the arc rocks removed by tectonic erosion. Neither process is particularly exposed to observation.

Diamond pipes exhume samples of ancient rocks from $\sim$200 km depth beneath continents. Carbon and nitrogen isotopic anomalies in the diamonds and sulfur isotope anomalies within diamond inclusions provide evidence for ancient subduction and the existence of an ancient active surface biochemical cycle. Farquhar et al. [58] present data on 2.9 billion-year-old diamonds and a brief review of the literature on the topic. Given the rarity of diamond pipes, these data relate to the existence of processes and give little information on fluxes into the mantle.

It is useful to look at the volcanic gases in island arcs. Typically helium and carbon isotopes are variable indicating a mixture of mantle, sedimentary carbonate, and sedimentary organic matter source [59]. This established the reality of carbon subduction, but it is hard to constrain the absolute flux. Sleep and Zahnle [46] review direct estimates from volcanic gases and sediment entering subduction zones. The data are consistent with carbonate subduction being in crude balance with the source at the ridge axis. They estimate that the ridge source is $(1-2.5) \times 10^{12}$ mol per year (1×10^{12} mol per year is consistent with the mantle CO_2 reservoir discussed earlier). Their total subducted carbon flux is $(1-3) \times 10^{12}$ mol per year, of which $3/4$ makes it into the deep mantle and $1/4$ degases in arc volcanoes.

I use carbon isotopes to constrain the long-term partition of subducted carbon between carbonate and organic carbon. Following Des Marais [6], the mantle response to the surface cycle is sluggish given the large size of the mantle reservoir. Currently, it takes 7200 m.y. to cycle the mantle through the ridge axis. Still the crust and mantle reservoirs are comparable and both have on average the same average $\delta^{13}C$. This indicates that carbonate and organic carbon have on average been subducted proportional to their crustal abundance, i.e., $4:1$.

The subducted organic carbon that returns to the surface as CO_2 is an oxygen source as is the deeply subducted organic carbon that was ultimately derived from igneous CO_2 as oxygen source. The current organic carbon subduction flux is $(0.2-0.6) \times 10^{12}$ mol per year, $1/5$ of the total carbon subduction flux of $(1-3) \times 10^{12}$ mol per year.

The subduction flux depends on the rate by which organic matter enters trenches. That is, it is a complicated function of the details of biology and the details of erosion and sediment transport. Note that ferric iron could have been preferentially subducted in sediments relative to organic carbon at times when iron photosynthesis dominated.

4.5. Serpentinite

The mantle of the Earth outcrops on the seafloor at ultraslow spreading ridges [60]. Hydrothermal circulation alters the rock to a mixture of hydrous magnesium silicates and magnesium hydroxide called serpentinite. Some ferrous iron substitutes for magnesium, but much of it is oxidized to magnetite. The oxygen to do this comes directly from the water. (Serpentinite is much more reducing than basalt and liberates lots of hydrogen.) This provides a potential hydrogen source (oxygen sink) that does not require seawater to contain dissolved oxygen or sulfate.

At present, significant mantle rock is exposed along very slow and ultraslow ridges which produce ~ 0.3 km^2 of seafloor per year [60]. (It will be much more common on a planet where plate tectonics has died.) The deep crust has not been extensively sampled, but surface sampling and seismic data [61] indicate that 50% serpentinization to a depth of 2 km is a reasonable average (Henry Dick, personal communication, 2003).

The oxygen flux is modestly significant. For example, if half of the 10% of FeO in the rock (with density of 3000 kg m^{-3}) gets oxidized to magnetite to a depth of 2 km, the global flux of O_2 is only 0.2×10^{12} mol per year.

4.6. Hydrogen Escape to Space

Currently, hydrogen escape to space is only a minor global flux. The physics are simple. Water is the main hydrogen species in the air. Only trace quantities make it above the cold trap at the top of the troposphere.

Catling et al. [62] point out that the situation was quite different on the early Earth. Like now, organic processes made methane and H_2 (Chapter 2) in massive quantities. Unlike now, there was little oxygen and further back little sulfate in the ocean with which the methane and hydrogen could quickly react. It is likely that methane built up to levels of hundreds of parts per million in the air and was a significant greenhouse gas.

Unlike water, there is no cold trap for methane or H_2 on an earthlike planet. Rather, methane and H_2 diffuse to the top of the atmosphere and disassociate from photolysis. The H then escapes to space. The remaining carbon makes it back to the surface where it eventually reacts to form more methane. The hydrogen to do this ultimately comes from water. The net reaction is equivalent to disassociating water and having the hydrogen escape to space.

The fluxes for this process can be large $(1-10) \times 10^{12}$ mol per year, comparable to modern carbon burial [62]. This process is thus very attractive for getting an O_2 atmosphere started. The high-end flux would build up the net excess of oxygen in the crustal reservoirs in only 120 m.y., the low end in 1200.

5. CARBON BURIAL

Carbon burial separates oxygen from carbon. It takes time for the reactants to get back together. Processes that increase the burial flux or increase the duration of burial thus increase the amount of oxygen in the air. I discuss the coupling of the carbonate cycle with the organic carbon cycle before discussing more traditional topics of the sites for carbon burial and the effect of phosphorus.

5.1. Coupled Organic Carbon and Carbonate Surface Cycles

I begin with the weathering of organic carbon and carbonates from outcropping rocks. These weathering products immediately mix in the surface cycle. For simplicity, I let the burial time be the same for sedimentary organic carbon and sedimentary carbonate. The burial and the exhumation rates are thus proportional to the size of the reservoirs: four carbonate to one organic carbon. The major unburied organic carbon reservoir is by definition the biomass, 0.05×10^{18} mol of C [14]. Its residence time with respect to photosynthesis is this quantity divided by global productivity of 8.4×10^{15} mol per year or 6 years. The main reservoir of CO_2 is in the ocean, 3.3×10^{18} mol [14]. Its residence time with respect to photosynthesis is 400 years. (This is shorter than the mixing time of the ocean. At any one time, part of the CO_2 is in the photic zone.)

The global rate of carbon burial is 50×10^{12} mol per year, four parts carbonate and one part organic carbon. The residence time of the combined surface carbon and carbonate reservoirs is 67,000 years. During this time, a carbon atom is in CO_2 for 66,000 years and in the biomass for 1000 years. It gets oxidized and reduced 165 times. This example indicates that organic carbon burial cannot be considered independently of carbonate burial as both processes compete for the same carbon atoms. The carbon is oxidized most of the time and ends up in carbonate deposits 4/5 of the time.

In fact, a negative feedback exists. Reduction of carbonate to organic carbon leaves the divalent cations Mg^{2+} or Ca^{2+} in its wake. An excess of these cations speeds carbonate burial. [The Urey cycle of stabilizing climate is based on this effect (see Ref. [46] for review). Weathering releases more divalent cations when the climate is warm leading to an enhanced CO_2 sink.] A dearth of divalent cations retards the burial of carbonate.

5.2. Sites for Organic Carbon Burial

Organic carbon burial requires that it be separated from free oxygen and marine sulfate. This process depends on the relative mobilities of the reactants and the tendency productivity to be highly variable in space and time. Dissolved sulfate is the most massive oxidant in seawater. Oxygen is present in the air in massive quantities and in modest quantities in seawater. However, neither are particularly mobile in stabilized stratified basins (where chemical diffusion is the mixing mechanism) or in fine-grained sediments. Organic matter is relatively immobile once it has sunk to the seafloor or other surface. Nutrients, water, and sunlight locally make productivity highly variable in time and space. Organic carbon piles up in regions with high productivity and ineffective aeration.

The presence of oceans and continents on the Earth plays a fundamental part in burial processes. Erosion of the continent supplies sediments that collect in the ocean basins, rift valleys and other intermontane basins, and platform basins. Weathering releases nutrients necessary for life, like phosphorus. We have the right amount of water so that the freeboard of the continents is small enough that shallow seas often cover platform regions like the interior of North America. Long-lived sediments accumulate where the platforms slowly subside. The sea going in over the platforms and then out is a major part of historical geology. The life in the shallow seas is a major part of paleontology.

However, one should not conclude that a planet with shallow platforms is necessary for carbon to get buried and oxygen to build-up. Rather, a diverse variety of sites for carbon burial exist on the Earth. The carbon atoms find their way into these traps unless they become part of carbonates first.

5.3. Phosphorus and Organic Carbon Burial

Modelers (e.g., [4]) have correctly focused on phosphorus as a limiting medium-term nutrient for carbon burial. Phosphorus comes from continental weathering. The ridge axis and ocean crust in general are significant sinks [63]. Phosphorus and hence continental weathering control carbon burial. That is, the burial ratio is approximately the Redfield Ratio $C : P \sim 116$ (atomic) of the dying organisms. This situation is approximately true in many environments [64].

The trouble may be more in visualizing the biological innovations than the intrinsic difficulty of these innovations. There appears to be considerable flexibility under water and inefficiency on land that argues against extrapolating a rigid phosphorus control to other planets or even to the early Earth.

Phosphorus reenters the water column in anoxic environments where organic carbon is >2% (by weight) in the sediment [64,65]. The C:P ratio of organically derived material making it into the sediments gets as high as 400. This lets three times as much organic carbon get buried as would occur if P did not recycle.

On land, life does not now efficiently mobilize phosphorus for its use. Only one P atom out of ten gets mobilized. [The C:P ratio in an average sediment is about 13 with 0.5% (by weight) organic carbon in the sediment and andesite (see Table 21-2 of Ref. [13]) as a source rock.] The rest ends up in sediments where it is inaccessible to marine biota.

6. CONCLUSION: OXYGEN BUILD-UP OVER GEOLOGICAL TIME

Photosynthesis supplied and maintains the oxygen we breathe. We must look to the Earth's interior and to space to see why this bounty exists. One must look at the other reservoirs, as the oxygen in the air is only 1.6% of the total available oxygen. Like with the American Revolution, it is futile to search for a single simple cause. Rather, there is interplay of multiple processes, all of which are expected on an earthlike planet. I begin with the long-term cycles.

At present, significant fluxes of available oxygen into the mantle are marine sulfate reduction at ridge axes, 0.5×10^{12} mol per year, oxidation of vented sulfide, 0.5×10^{12} mol per year, and serpentinization, 0.2×10^{12} mol per year. They are partly offset by the oxygen source from subduction of organic carbon of $(0.20-0.6) \times 10^{12}$ mol per year. The net loss of oxygen is $(0.5-1.0) \times 10^{12}$ mol per year.

The size of the crustal reservoirs allows an imbalance of this magnitude to persist on a 100 m.y. time scale with limited observable effect. For example, it would take 1260–2670 m.y. for this oxygen loss to produce the buried organic carbon reservoir. Given that my estimates have some slop and that subduction of organic carbon involves the vagaries of tectonics and biology, I am not particularly concerned about an instantaneous modern imbalance.

6.1. Hydrogen Escape and the Crustal Oxygen Reservoir

The massive reservoir of available oxygen as ferric iron in hard rocks played a major role in the build-up to an oxygen atmosphere. Hydrogen escape to space before 2.45 m.y. produced this reservoir. Biological methane carried hydrogen to the top of the atmosphere and biological hydrogen escaped. This process has not detectably affected the oxidation state of the massive mantle reservoir.

The flux of hydrogen to space need not have been particularly large. For example, a flux of 1×10^{12} mol per year of O_2 equivalent (at the lower end of the range quoted by Catling et al. [62]) would have produced the reservoir of 2000×10^{18} mol over the geological time before 2.45 billion years ago. The actual flux would have been greater as some O_2 entered the mantle in serpentinites. This flux would have been significant only if slow spreading ridges existed. There was also a significant but unconstrained flux from the subduction of ferric iron in sediments.

Storage of oxygen of the hard rocks in the crust required vigorous tectonics. Igneous rocks became oxidized when they erupted to the surface or were exposed to erosion. Later tectonics took these supracrustal rocks back into the deep crust where they metamorphosed and melted into hard rocks. This process did not require free oxygen in the air. Rather the ferrous iron was directly oxidized by iron-based photosynthesis and indirectly oxidized by sulfate derived from photosynthesis.

The oxygen from methane-controlled hydrogen escape filled the continental reservoir to the point that its oxidation was no longer a massive sink. Tectonic processes contributed to this decline. Continental tectonics have waned over geological time. Much of the hard rock in the continental crust stayed put over the last 2 billion years. Over time, the concentration of ferrous iron in exposed hard rocks waned as granites became more common relative to basalts. Physically, granites form by partial melting in hydrous conditions where ferrous iron-rich minerals stay in the deep source region. In addition, iron-rich rocks are more dense than granites and mechanically end up deeper in the crust.

Hydrogen escape to space got the Earth only part way to its present state. Methane ascent requires that there is little O_2 or sulfate around to consume it, biologically in the water or by photolysis in the air. The major sink for the oxygen released by methane ascent waned once the exposed hard rocks were oxidized. Further oxygen produced sulfate and eventually free oxygen. This occurred $\sim$2.45 billion years ago, killing hydrogen escape.

6.2. Transition to Oxygen-Rich Atmosphere

Methane-induced hydrogen escape left the Earth on the threshold of an oxygen atmosphere. As the hard rock sink for oxygen waned, carbon burial charged the modest sulfate and free-oxygen reservoirs. This process eventually overwhelmed two oxygen buffers. The first was sulfate reduction by basalt at ridge axes which reached its current flux per area of new seafloor when sulfate concentration in seawater reached $\sim$10% of its current amount. The second was oxidation of vented sulfide, which reached its current flux per area when the deep sea became aerobic.

There were positive and negative feedbacks and complicated lag times in the transition from an oxygen-poor atmosphere to an oxygen-rich one between 2.0 and 2.45 billion years ago. For example, it takes time for evolution to adapt heterotrophs to a new array of photosynthetic producers. (It has not escaped me that the evolution of land plants is an analogous case.) There were strong variations in the fraction of carbon that got buried as organic carbon vs. as carbonate during the transition [6]. I have already noted instability where that reaction of O_2 and sulfide suppresses sulfide-based photosynthesis. Oxidation reduces the concentration of the greenhouse gas methane in the air. Fluctuation in and out of ice ages is likely.

6.3.   Controls on Reservoir Sizes

Then why do we have 21% oxygen in our air? There is no obvious long-term process that leads to this specific value. Rather, it is a typical value in a dynamic fluctuating system that is buffered to some extent. Clearly, abundant free oxygen makes it more difficult to bury organic carbon.

Such negative feedback has kept O_2 fluctuations in check over the last 600 m.y. There have been significant swings in the last 600 m.y. [3–4]. The size of the buried carbon reservoir fluctuated during these excursions.

Why is 1/5 of the sedimentary carbon organic? Competition by sedimentary carbonates for surface carbon is likely to be important. The amount of available divalent cations limits the carbonate reservoir. Instantaneously, reduction of carbonates frees divalent cations speeding up carbonate burial. Oxidation of reduced carbon leaves a surfeit of CO_2 over the available cations.

Globally, the amount of divalent cations in sedimentary rocks has build up over geological time. There is an excess of calcium over any reasonable igneous source rock [16]. The availability of divalent cations tends to stabilize the great size of the carbonate reservoir and hence, the size of the smaller carbonate reservoir. Feedback occurs since the rate at which divalent cations get trapped in the subsurface during metamorphism while CO_2 vents to the surface depends on the size of the carbonate reservoir. The vented CO_2 is prone to being trapped in reduced sediments until weathering of silicates liberates more divalent cations. Conversely, reaction of buried organic carbon with buried ferric iron and sulfate to liberate CO_2 during metamorphism depends on the size of the organic carbon reservoir.

A simple but incomplete hypothesis is that carbon in sediments was approximately 1 : 4 organic to carbonate at 2 billion years ago when the atmosphere became oxic. Tectonic processes have not changed a lot since then. These include uplift and weathering that release divalent cations, metamorphism that traps them, and the formation and exhumation of sedimentary basins.

6.4.   Exportable Biological Implications

The physical processes that have allowed oxygen in our air are those of an earthlike silicate planet. I extrapolate in generalities from biology. Life can evolve biochemical pathways for multiple modes of photosynthesis.

Evolution has allowed several global ecologies on the Earth. These include rock-based life with no photosynthesis and hydrogen-based, iron-based, and sulfide-based photosynthesis. The currently dominant oxygenic photosynthesis existed in limited environments before it became dominant and did not immediately produce oxygen-rich air when it did.

I do not extrapolate the details of the timing of innovations and their environments. For example, vacancy signs hung on available niches for vast periods of time. It took ~1.4 billion years for metazoa to fill the niche for animals after there was significant oxygen around!

NOTES ADDED IN PROOF

Press releases on the *Opportunity* Mars Rover make it obvious that the hematite deposits are not direct products of photosynthesis. See [66] for more on the rise of atmosphere oxygen at 2.32 Ga and [67] for the presence of sulfate reducers at 3.47 Ga.

ACKNOWLEDGMENTS

I thank Kelvin Zahnle, Henry Dick, Mak Saito, Tori Hoehler, and Angela Hessler for helpful discussions. This research was partly supported by the National Science Foundation. This material is based upon work supported by the National Aeronautics and Space Administration through the NASA Astrobiology Institute under Cooperative Agreement No. CAN-00-OSS-01 and issued through the Office of Space Science.

ABBREVIATIONS

ATP	adenosine 5′-triphosphate
Ga	billion years before present
m.y.	million years
MORB	mid-oceanic ridge basalt
SNC	Shergottite–Nakhlite–Chassignite, three classes of stony meteorites regarded to have come from Mars

REFERENCES

1. Canfield DE, Raiswell R. Am J Sci, 1999; 299:697–723.
2. Lenton TM. The coupled evolution of life and atmospheric oxygen. In: Rothschild LJ, Lister AM, eds. Evolution on Planet Earth The Impact of the Physical Environment, Amsterdam: Academic Press, 2003:35–53.
3. Berner RA, Beerling DJ, Dudley R, Robinson JM, Wildman RA Jr. Annu Rev Earth Planet Sci, 2003; 31:108–134.
4. Hansen KW, Wallmann K. Am J Sci, 2003; 303:94–148.

5. Anbar AD, Knoll AH. Science, 2002; 297:1137–1142.
6. Des Marais, DJ. Rev Mineral Geochem, 2001; 43:555–578.
7. Kasting JF, Siefert JL. Science, 2002; 296:1066–1068.
8. Holland HD. Geochim Cosmochim Acta, 2002; 66:3811–3826.
9. Nisbet EG, Sleep NH. Nature, 2001; 409:1083–1091.
10. Nisbet EG, Sleep NH. The physical setting for early life. In: Rothschild LJ, Lister AM, eds. Evolution on Planet Earth The Impact of the Physical Environment, Amsterdam: Academic Press, 2003:4–24.
11. Wolstencroft RD, Raven JA. Icarus, 2002; 157:535–548.
12. Kasting JF, Catling D. Annu Rev Astron Astrophys, 2003; 41:429–463.
13. Krauskopf KB, Bird DK. Introduction to Geochemistry, 3rd ed. New York: McGraw-Hill, 1995:647.
14. Holser WT, Schidlowski M, Mackenzie FT, Maynard JB. In: Gregor CB, Garrels RM, Mackenzie FT, Maynard JB, eds. Chemical Cycles in the Evolution of the Earth, New York: John Wiley, 1988:105–173.
15. Lecuyer C, Ricard Y. Earth Planet Sci Lett, 1999; 165:197–211.
16. Wolery TJ, Sleep NH. Interactions of geochemical cycles with the mantle. In: Gregor CB, Garrels RM, Mackenzie FT, Maynard JB, eds. Chemical Cycles in the Evolution of the Earth. New York: John Wiley, 1988:77–103.
17. Stopler E, Newman S. Earth Planet Sci Lett, 1994; 121:293–325.
18. Morris JD, Gosse J, Brachfeld S, Tera T. Rev Mineral Geochem, 2002; 50:207–270.
19. Veizer J. The evolving exogenic cycle. In: Gregor, CB, Garrel RM, Mackenzie FT, Maynard JB, eds. Chemical Cycles in the Evolution of the Earth, New York: John Wiley, 1988, 175–222.
20. Chapelle FH, O'Neill K, Bradley PM, Methé BA, Ciufo SA, Knobel LL, Lovley DR. Nature, 1997; 1202:312–315.
21. Kral TA, Brink KM, Miller SL, McCay CP. Origins Life Evol. Biosphere, 1998; 28:311–319.
22. Brinkman H, Philippe H. Mol Bio Evol, 1999; 16:817–825.
23. Di Giulio M. J Theor Biol, 2003; 221:425–436.
24. Forterre P, de la Tour CB, Philippe H, Duguet M. Trends Genet, 2000; 16:152–154.
25. Forterre P, Brochier C, Philippe H. Theor Popul Biol, 2002; 61:409–422.
26. Xiong J, Fischer WM, Inoue K, Nakahara M, Bauer CE. Science, 2000; 289: 5485, 1724–1730.
27. Hoehler TM, Alperin MJ, Albert DB, Martens CS. Geochim Cosmochim Acta, 1998; 62:1745–1756.
28. Hoehler TM, Alperin MJ, Albert DB, Martens CS. FEMS Microbiol Ecol, 2001; 38:33–41.
29. Grassineau NV, Nisbet EG, Bickle MJ, Fowler CMR, Lowry D, Mattey DP, Abell P, Martin A. Proc R Soc Lond Ser B—Biol Sci, 2001; 268:113–119.
30. Nisbet EG, Fowler CMR. Proc R Soc Lond Ser B—Biol Sci, 1999; 266:2375–2382.
31. Pavlov AA, Kasting JF. Astrobiology, 2002; 2:27–41.
32. Farquhar J, Wing BA. Earth Planet Sci Lett, 2003; 213:1–13.
33. Ono S, Eigenbrode JL, Pavlov AA, Kharecha P, Rumble III D, Kasting JF, Freeman KH. Earth Planet Sci Lett, 2003; 213:15–30.
34. Pavlov AA, Brown LL, Kasting JF. J Geophys Res, 2001; 106:23267–23287.
35. Rosing MT, Frei R. Earth Planet Sci Lett, 2004; 217:237–244.
36. Saito MA, Sigman DM, Morel FMM. Inorg Chim Acta, 2003; 356:308–318.

37. Ehrenreich A, Widdel F. Appl Environ Microbiol, 1994; 60:4717–4526.
38. Kappler A. Newman DK. Geochim Cosmochim Acta, in press.
39. Schröder I, Johnson E, de Vries S. FEMS Microbiol Rev, 2003; 27:427–447.
40. Luu Y-S, Ramsay JR. World J Microbiol Biotechnol, 2003; 19:215–225.
41. Catling DC, Moore JA, Icarus, 2003; 293:839–843.
42. Christensen PR, Morris RV, Lane MD, Bandfield JL, Malin MC. J Geophys Res, 2001; 106:23873–23885.
43. Walker JCG. Nature, 1987; 329:710–712.
44. Zaar A, Fuchs G, Golecki JR, Overmann J. Arch Microbiol, 2003; 197:174–183.
45. Canfield DE. Nature, 1998; 396:450–453.
46. Sleep NH, Zahnle K. J Geophys Res, 2001; 106:1373–1399.
47. Kump LR, Kasting JF, Barley ME. Geochem Geophys Geosys, (2001).
48. Delano JW. Origins Life Evol Biosphere, 2001; 31:311–341.
49. Canil D. Earth Planet Sci Lett, 2002; 195:75–90.
50. Lee C-TA, Brandon AD, Norman M. Geochim Cosmochim Acta, 2003; 67:3045–3064.
51. Warren PH, Kallemeyn GW, Meteoretics Planet Sci, 1996; 31:97–105.
52. Kuramoto K, Matsui T. J Geophys Res, 1996; 101:14909–14932.
53. Saal AE, Hauri EH, Langmuir CH, Perfitt MR. Nature, 2002; 419:451–455.
54. Sleep NH, Zahnle K, Neuhoff PS. Proc Nat Acad Sci USA, 2001; 98:3666–3672.
55. Kasting JF. Origins Life Evol Biosphere, 1990; 20:199–231.
56. Miyakawa S, Yamanashi H, Kobayashi K, Cleaves HJ, Miller SL. Proc Nat Acad Sci, USA, 2002; 99:14628–14631.
57. Plank T, Langmuir CH. Chem Geol, 1998; 145:325–394.
58. Farquhar J, Wing BA, McKeegan KD, Harris JW, Cartigny P, Thiemens MH. Science, 2002; 298:2369–2372.
59. Sano Y, Williams SN. Geophys Res Lett, 1996; 23:2749–2752.
60. Dick HJB, Lin J, Schouten H. Nature, 2003; 426:405–412.
61. Jokat W, Ritzmann O, Schmidt-Aursch MC, Drachev S, Gauger S, Snow J. Nature, 2003; 423:962–965.
62. Catling DC, Zahnle KJ, McKay CP. Science, 2001; 293:839–843.
63. Wheat CG, Feely R, Mottl MJ. Geochim Cosmochim Acta, 1996; 60:3593–3608.
64. Anderson LD, Delaney ML, Faul KL. Global Biogeochem Cycles, 2001; 15:65–79.
65. Ingall E, Jahnke R. Marine Geol, 1997; 139:219–229.
66. Hannah JL, Bekker A, Stein HJ, Markey RJ, Holland HD. Earth Planet. Sci. Lett., 2004; 225:43–52.
67. Shen YN, Buick R. Earth Sci. Rev., 2004; 64:243–272.

4

The Nitrogen Cycle: Its Biology

Marc Rudolf and Peter M. H. Kroneck

Universität Konstanz, Mathematisch-Naturwissenschaftliche Sektion,
Fachbereich Biologie, Postfach M665, D-78457 Konstanz, Germany

1.	Introduction	76
2.	Nitrogen Fixation	78
	2.1. Chemical Aspects	78
	2.2. Biological Aspects	79
	2.3. Nitrogenase: Three-Dimensional Structure and Reaction Mechanism	81
	2.3.1. Structural Aspects	81
	2.3.2. Mechanistic Aspects	81
	2.3.3. The P-Cluster and the FeMo-Cofactor	84
3.	Respiratory Processes: Energy Conservation with Inorganic Nitrogen Compounds	85
	3.1. Denitrification: Reductive Transformation of Nitrate to Dinitrogen	85
	3.1.1. Nitrate Reductase	85
	3.1.2. Nitrite Reductase	86
	3.1.3. NO Reductase	89
	3.1.4. N_2O Reductase	90
	3.2. Nitrite Ammonification: Reductive Transformation of Nitrite to Ammonia	92
	3.2.1. From Nitrate to Nitrite	92
	3.2.2. From Nitrite to Ammonia	93

3.3.   Nitrification and Nitrite Oxidation: Oxidative
       Transformation of Ammonia   94
4. Assimilatory Processes: Building the Essential Molecules
    of Life   95
5. Outlook   96
Acknowledgments   97
Abbreviations   97
References   98

1. INTRODUCTION

The biological nitrogen cycle has received considerable attention over the past decades because of its global importance (Fig. 1) [1–4]. As a consequence of the increase of the world population agricultural technologies had to be improved significantly to provide food for a rapidly growing number of people. Following the original discovery of the nature and value of mineral fertilization by Liebig,

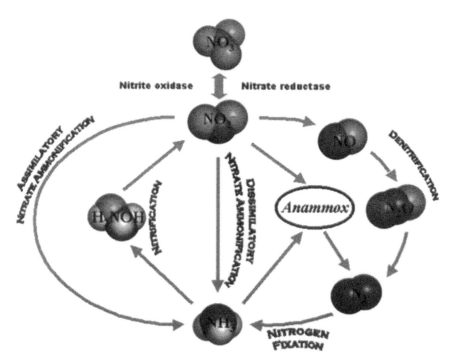

Figure 1   The biological nitrogen cycle. For the Anammox process see text and Eq. (1).

nitrogen compounds were used in increasing quantities as an ingredient of mineral fertilizers [5].

Why is nitrogen of such central importance? The amount of nitrogen, in comparison to hydrogen, carbon, and oxygen, is only minor in living organisms (roughly 3% in humans). The three basic elements hydrogen, carbon, and oxygen are readily bioavailable in form of water and carbon dioxide for plants which are the key players in our food chain [6]. Dinitrogen, though the major constituent of the atmosphere, has to be converted first to ammonia by a complex process called nitrogen fixation. Nitrate reduction plays a key role and has important agricultural, environmental, and public health implications. Assimilatory nitrate reduction, performed by bacteria, fungi, algae, and higher plants, is one of the most fundamental biological processes. The tremendous use of fertilizers in agricultural activities led to a significant nitrate accumulation in groundwater, and consumption of drinking water with high levels of nitrate had been associated with severe diseases due to formation of highly toxic N-nitroso compounds. Nitrogen oxides produced by denitrification are also linked with the greenhouse effect and the depletion of stratospheric ozone. Dinitrogen monoxide (N_2O, *laughing gas*) is recognized as a greenhouse gas with a tropospheric concentration of ≈ 300 ppb by volume [7].

Nitrogen is a basic element for life because it is a component of essential biomolecules, such as amino acids, proteins, and nucleic acids. Nitrogen exists in the biosphere in several oxidation states, ranging from $N(5+)$ to $N(3-)$. Interconversions of these nitrogen species constitute the global biogeochemical nitrogen cycle which is sustained by biological processes, with bacteria playing a predominant role (Fig. 1) [8–18]. Although multicellular organisms such as fungi, algae, higher plants, and humans have made a significant mark on the geochemistry of the earth, averaged over geological time, it is clear that the most important geochemical agents by far have been unicellular microorganisms, e.g., *Bacteria*, *Archaea*, and single-celled *Eucarya* [19]. Microbes have altered the chemistry of the atmosphere via oxygenic photosynthesis, nitrogen fixation, and carbon sequestration. Most remarkably, they perform these many chemical reactions in every nook and cranny from the near surface to the depths, including even the most extreme environments [20].

Inorganic nitrogen is converted to a biologically useful form by dinitrogen fixation ($N_2 \rightarrow NH_3$) or nitrate assimilation ($NO_3^- \rightarrow NH_3$) and the incorporation of ammonia into organic molecules. Inorganic nitrogen compounds are recycled from the environment by nitrification ($NH_3 \rightarrow NO_3^-$), the oxidative conversion of ammonia to nitrate, denitrification ($NO_3^- \rightarrow N_2$) whereby nitrate is successively transformed to nitrite, nitrogen monoxide (NO), dinitrogen monoxide, and dinitrogen, and nitrate ammonification ($NO_3^- \rightarrow NH_3$), with nitrite as intermediate (Fig. 1). Nitrate plays a key role in the biogeochemical nitrogen cycle which is contributed to by both prokaryotic and eukaryotic organisms. It is a source of nitrogen for assimilation, whereby the key enzymatic reaction is the reduction of nitrate to nitrite. Nitrite is further reduced to ammonium, which can be further assimilated into organic nitrogen. The same series of reactions, though often catalyzed by distinct enzymes, can also occur as part of a nitrate respiration process in some enteric and sulfate-reducing bacteria, whereby nitrate serves as a terminal electron acceptor during anaerobic metabolism [15].

It was thought for a long time that in the ocean denitrification was the only mechanism of N_2 production. However, most recently it was shown by two independent groups that large scale conversion of fixed inorganic nitrogen (nitrate, nitrite, ammonium) to N_2 can occur through another important route, the so-called Anammox process [Eq. (1)] [21].

$$NH_4^+ + NO_2^- \rightleftharpoons N_2 + 2H_2O \qquad (1)$$

In this context, a new autotrophic member of the order *Planctomycetales* was isolated earlier and identified which possessed the Anammox pathway [22]. The process of Anammox has to be considered to be one of the most innovative technological advances in removal of ammonia nitrogen from wastewater.

Very likely CO_2, CO, and N_2 were among the main components by the time of the origin of life within the Earth's early atmosphere. In addition, NO, H_2, H_2O, and sulfuric gases could also have been present, and probably trace amounts of O_2 [23]. Thus, it was not too surprising that several molecular links do exist between the evolution of denitrification enzymes and cytochrome *c* oxidase, the terminal oxidase of aerobic respiration [24]. Perhaps the most important is the homology between nitric oxide reductase and heme/copper cytochrome oxidases, and the presence of the mixed-valence $[Cu^{1.5+}(S_{Cys})_2Cu^{1.5+}]$ CuA electron-transfer center in nitrous oxide reductases, the quinol-dependent NO reductase from the gram-positive *Bacillus azotoformans*, and heme/copper cytochrome oxidases [24–27].

In this chapter, we will discuss recent advances in the field of the biological nitrogen cycle, with emphasis on the metalloenzymes involved in the various oxidative and reductive interconversions of nitrogen compounds. Physiological, structural, and mechanistic aspects of these enzymes and their active sites will be described. Note that all nitrogen transformations involve metalloenzymes, many of them with complex multimetal catalytic sites with unusual chemical and electronic properties which have attracted the interest of chemists and spectroscopists. In view of the data accumulating in the field, we recommend as primary references the *Handbook on Metalloproteins* [28], the *Handbook of Metalloproteins* [29], and the special issue of *Chemical Reviews* on Biomimetic Chemistry [30]. The following comprehensive reviews are also recommended for further reading [9–12,18,24,31–33].

2. NITROGEN FIXATION

2.1. Chemical Aspects

The industrial production of ammonia from N_2 and H_2 by the Haber-Bosch procedure is a high energy consumption process. Due to the volume reduction and exothermic reaction, the conversion has to take place at high pressures and low temperature. Unfortunately, the iron oxide/alkaline metal oxide catalysts need high temperatures for effective conversion rates, the conditions of the process are pressures around 200 bar and temperatures around 500°C [34,35].

Ammonia synthesis proceeds according to the following reaction [Eq. (2)]:

$$0.5N_2 + 1.5H_2 \rightleftharpoons NH_3 \quad \Delta H_{298} = -46.22 \, kJ/mol \tag{2}$$

The yield is nearly 11%, but the unreacted compounds reenter the cycle leading to an overall yield of 100%. Nearly 100% of the ammonia synthesized worldwide is produced by the Haber-Bosch process. Even though this technology is expensive, no low cost alternatives have been developed so far.

To lower the energy consumption in terms of economic and environmental aspects, much effort has been invested to design new catalysts that work under ambient conditions. A decisive hint concerning effective catalysts might come from investigating microorganisms that will convert N_2 under physiological conditions of temperature and pressure [35]. The first N_2 fixation catalyst working under ambient conditions employed a ruthenium center [36]. Another strategy aims at modeling the active site of the enzyme nitrogenase. Recently, a biomimetic complex was presented based on the substitution of iron sulfur clusters with molybdenum and vanadium as heterometal atoms, respectively. These authors concluded that the synthetic and native cluster cores approach congruency (Fig. 2) [37].

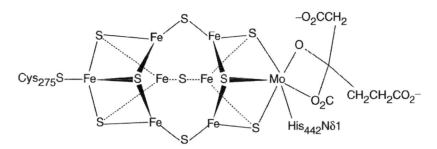

Figure 2 Schematic drawing of the structure of the nitrogenase FeMo-cofactor model [43].

2.2. Biological Aspects

Biological nitrogen fixation, a process found only in some prokaryotes, is catalyzed by the nitrogenase complex. Bacteria containing nitrogenase occupy an indispensable ecological niche, supplying fixed nitrogen to the global nitrogen cycle. Due to this inceptive role in the nitrogen cycle, so-called diazotrophs are present in virtually all ecosystems, with representatives in aerobic soils (e.g., *Azotobacter* species), in the ocean surface layer (*Trichodesmium*), and in specialized nodules in legume roots (*Rhizobium*) [38]. In any ecosystem, diazotrophs must respond to varied environmental conditions to regulate the tremendously taxing nitrogen fixation process. All characterized diazotrophs regulate nitrogenase at the transcriptional level. A smaller set also possesses a fast-acting post-translational regulation system. Although there is little apparent variation in the sequences and structures of nitrogenases, there appear to be almost as many nitrogenase-regulating schemes as there are nitrogen fixing species.

Biological nitrogen fixation provides the dominant route for the transformation of atmospheric N_2 into a bioavailable form, ammonia [39–42]. The overall stoichiometry of biological nitrogen fixation reaction [Eq. (3)]

$$N_2 + 8H^+ + 8e^- + 16Mg\text{-}ATP \rightleftharpoons 2NH_3 + H_2 + 16Mg\text{-}ADP + 16P_i$$

(3)

reflects the requirements for reducing equivalents, Mg-ATP, and protons, in addition to the nitrogenase complex constituted by the Fe-protein and the MoFe-protein (Fig. 3) [43]. So far, high resolution X-ray structures of nitrogenase have been obtained for the enzymes from *Azotobacter vinelandii* [43–48], *Clostridium pasteurianum* [49,50], and *Klebsiella pneumoniae* [51]. The fundamental significance of the process and its intrinsic chemical formidability have prompted extensive, extended studies to attain a molecular description of this complicated enzyme. These efforts have met with minimal success, and even simple molecular details such as the minimal requisite reaction stoichiometry, the location of substrate binding, and the oxidation states of the metals involved remain either unknown or in debate [32].

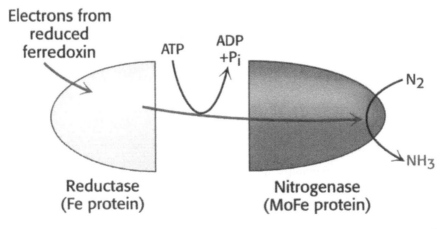

Figure 3 Schematic drawing of the nitrogenase complex consisting of the Fe-protein and the MoFe-protein.

2.3. Nitrogenase: Three-Dimensional Structure and Reaction Mechanism

2.3.1. Structural Aspects

The nitrogenase complex consists of two metalloproteins, highly conserved in sequence and structure throughout nitrogen-fixing bacteria [52]. The protein containing the site of substrate reduction is the molybdenum–iron protein (MoFe-protein), also known as dinitrogenase or component I. The obligate electron donor to the MoFe-protein is an iron protein (Fe-protein), also known as dinitrogenase reductase or component II (Fig. 3). The genes encoding both the MoFe-protein and the Fe-protein, as well as accessory genes for electron transfer proteins, metal cluster synthesis and regulation comprise the *nif* regulon [52].

The MoFe-protein is organized as a $\alpha_2\beta_2$ tetramer of the *nifD* and *nifK* gene products [38]. It contains two copies each of two unique polynuclear metal clusters designated the P-cluster and the FeMo-cofactor with a total molecular mass of ≈ 240 kDa [43]. The tertiary structures of the α and β subunits are quite similar [50]. The Fe protein is a dimer of two identical subunits of the *nifH* gene product [38]. It coordinates a single [Fe_4S_4] cluster [53] with a total molecular mass of 60 kDa. Each subunit folds as a single α/β type domain, which together symmetrically ligate the surface exposed [Fe_4S_4] cluster through two cysteines (Cys97 and Cys132) from each subunit. A single bound ADP molecule is located in the interface region between the two subunits. Because the phosphate groups of this nucleotide are ≈ 20 Å from the [Fe_4S_4] cluster away, it is unlikely that ATP hydrolysis and electron transfer are directly coupled. Instead, it appears that interactions between the nucleotide and cluster sites must be indirectly coupled by allosteric changes occurring at the subunit interface. Each Fe-protein dimer can bind two nucleotide molecules, distal from the [Fe_4S_4] active site [39]. Binding of Mg-ATP at these sites causes a conformational change in the Fe-protein. The two subunits rotate toward each other, extruding the [Fe_4S_4] cluster towards the protein surface (and surmised interaction with the P-cluster of the MoFe-protein) by 4 Å [54]. This conformational change is thought to be a key step in the catalytic cycle of nitrogenase.

2.3.2. Mechanistic Aspects

The mechanism of the reduction of N_2 to ammonia catalyzed by the nitrogenase complex is one of the continuing mysteries of chemistry and biology. Since the first N_2 complex was discovered [55] and the basic structure of the active site of conventional molybdenum nitrogenases was unraveled [43,44,48,53], efforts have been made to combine chemistry and biology to explain the mechanism of biological nitrogen fixation at the atomic level [56,57].

The homodimeric Fe-protein couples ATP hydrolysis to intermolecular electron transfer and is the only known mechanistically competent source of electrons for the catalytically active component. The reduction of N_2 by nitrogenase involves three basic types of electron transfer steps: (i) reduction of the Fe-protein by electron carriers such as ferredoxin or flavodoxin; (ii) transfer of single electrons from the Fe-protein to the MoFe-protein in a Mg-ATP dependent process, with a minimal stoichiometry of two molecules of Mg-ATP hydrolyzed per electron transferred; and (iii) electron and proton transfer to the substrate, which is almost certainly bound to the FeMo-cofactor [53]. Because all known nitrogenase substrates including protons and acetylene [38] are reduced by at least two electrons, the first and the second step must be repeated until sufficient electrons have been provided for complete substrate reduction. Each electron transfer between the Fe-protein and the MoFe-protein appears to involve a cycle of obligatory association or dissociation of the protein complex, with the dissociation step having been identified as rate determining for the overall reaction [58]. The three-dimensional structure of the associated protein complex was solved recently [47].

But how does the FeMo-cofactor reduce N_2 to ammonia, and where does the substrate N_2 bind? Several reaction pathways are discussed which employ a wide range of oxidation states of the Mo center [57,59,60]. Very little of the reactivity observed to date is likely to occur at a nitrogenase active site. New insights in the speculative synthetic chemistry and the *nitrogenase problem* were reviewed recently. Note that without N_2 as substrate, the enzyme spontaneously reduces protons to H_2 [32].

In one model the reduction of N_2 starts with Mo in the zero-valent state [56]. This cannot be the case in nitrogenase because biological reducing agents are probably not strong enough to bring this about. Several proposals have invoked N_2 binding to the iron atoms of the FeMo-cofactor, but outside the cluster [61]. No synthetic cluster has yet been shown to interact with N_2 in a way to model the enzyme binding. Therefore, empirical support for such proposals is lacking.

Most recently, a molybdenum complex was found [57] that can convert N_2 to ammonia in a series of steps that involve no oxidation state lower than Mo(III) [57]. In their model complex, the Mo ion is enclosed in a ligand that restricts access to the metal so that only small reactants such as N_2 and protons can reach it. The cycle starts with N_2 bound to Mo(III). After stepwise addition of two protons and oxidizing the metal center to Mo(VI), the next protonation step leads to free ammonia and a Mo(VI)–(N) species. Nowadays, such a species is discussed to be the resting state of the MoFe-protein of nitrogenase (Fig. 4) [48,61,62]. The next three protons convert this bound nitrogen atom to the second molecule of ammonia with the coupled reduction of Mo(VI) to Mo(III).

The structure of the active site may change during the eight-electron N_2 reduction cycle. The molybdenum is connected to the Fe–S-cluster by three sulfide ions, and is linked to the protein by nitrogen coordination via His442. The remaining two coordination sites are occupied by an unusual ion, homocitrate, which is bidentate. The Mo is thus six-coordinate and seems not, at least

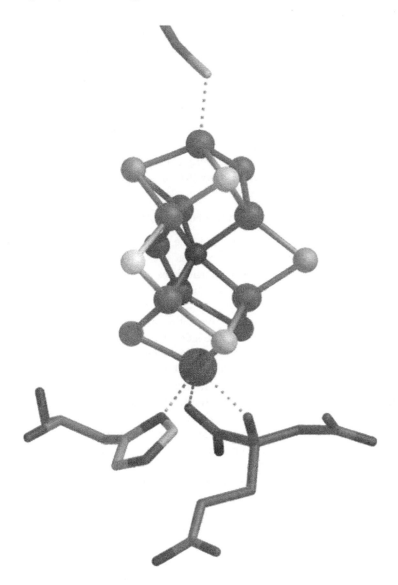

Figure 4 Stick-and-ball model of the nitrogenase FeMo-cofactor showing the μ_6-interstitial atom in the center of the Fe–S moiety [48,63] (see Fig. 2 for a schematic drawing).

in the crystal, to have an obvious available coordination site for N_2 [43,44,48,53]. It seems that the bidentate homocitrate has exactly the correct size and shape to disconnect from one coordination site of the Mo and form a hydrogen bond to His442. A mutant *nifV* nitrogenase that carries citrate rather than homocitrate can fix N_2 with <10% of the activity of the wild-type, but the size and shape of citrate do not allow it to form a hydrogen bond to His442 [64,65]. Consequently, it is less able to help open up a coordination site on the molybdenum. Reactivity studies on isolated cofactors of wild-type and of the mutant nitrogenase support this interpretation.

Coupling the results from the Mo model complex [57] with the explanation for the presence and possible action of homocitrate, these results provide the best model to date for N_2 reduction by conventional molybdenum nitrogenases. Vanadium-dependent nitrogenases probably undergo similar reactions, but iron-only nitrogenases [66] and the unique molybdenum-dependent nitrogenase from *Streptomyces thermoautrophicus* [67] present further mysteries.

2.3.3. The P-Cluster and the FeMo-Cofactor

The electron-transfer P-cluster and the catalytic FeMo-cofactor of the MoFe-protein of nitrogenase have been assembled only by biosynthesis. Whereas the P-cluster most likely participates in interprotein electron transfer, the FeMo-cofactor is the active site of substrate binding and reduction [43–46,48–51].

The FeMo-cofactor, with a composition of [Mo : 7Fe : 9S] : homocitrate, is coordinated to the protein through the side chains of only two residues bound to Fe and Mo sites located at opposite ends of the cluster. In the case of the terminal Fe the residue is a cysteine, for Mo it is a histidine (Figs. 2 and 4). Perhaps the most unusual feature of the cofactor in these structures is the trigonal prismatic arrangement of the six central iron atoms. They are located on the surface of a sphere with a radius of 2.0 Å from the cofactor center and are each coordinated to three sulfide anions. Furthermore, all nine sulfur atoms of the FeMo-cofactor are themselves equidistant from the center on a second sphere with a radius of 3.3 Å. The high-resolution crystallographic analysis (1.16 Å) of the MoFe-protein revealed a μ_6-interstitial atom at the center of the FeMo-cofactor [48]. Although the new structure refinement did not allow identification of this ligand with certainty, the electron density and bond distances are consistent with a light (2p) element, perhaps carbon, nitrogen, or oxygen. The resolution dependence of the electron density best matches the curve generated by a nitrogen atom. These considerations plus the universal acceptance of the FeMo-cofactor as the site of nitrogen reduction have led to the tentative assignment of nitride as the newfound core ligand. The most obvious consequence of this finding stems directly from its impact on cluster environment. The revelation of the internal ligand makes all iron centers in the FeMo-cofactor four-coordinate and (distorted) tetrahedral [32]. The question whether Fe or Mo is the key player in substrate binding during nitrogenase catalysis still has to be answered. Arguments for both possibilities have been developed from consideration of the catalytic features of different forms of the FeMo-cofactor, from chemical considerations based on Fe- or Mo-based model compounds, and from theoretical calculations [61].

The structure of the P-cluster, with composition [8Fe:7S], can be considered composed of two [Fe$_4$S$_3$] subclusters that are bridged by a hexacoordinate S atom, with the overall assembly coordinated to the protein through six cysteine ligands. The arrangement in the native state (P^N) consists of two [Fe$_4$S$_3$] cuboidal halves, vertex fused through a μ_6-sulfide and further interconnection by two μ_2-cysteinate bridges. After oxidation (P^{OX}), the cluster was found rearranged to a more open, asymmetric geometry, with additional protein ligation provided by a serine side chain oxygen and a backbone amide nitrogen [46].

3. RESPIRATORY PROCESSES: ENERGY CONSERVATION WITH INORGANIC NITROGEN COMPOUNDS

3.1. Denitrification: Reductive Transformation of Nitrate to Dinitrogen

In denitrification, NO$_3^-$ is reduced via NO$_2^-$ to the gaseous products NO, N$_2$O, and N$_2$. The biological process comprises several steps including the formation of the N−N bond in N$_2$O and the cleavage of the N−O bond in N$_2$O. The biology and microbial ecology of the denitrification pathway has been reviewed [9,14,68–70]. Denitrifying bacteria utilize the positive redox potentials of nitrate [E$_0'$ (NO$_3^-$/NO$_2^-$) = +0.43 V], nitrite [E$_0'$ (NO$_2^-$/NO) = +0.35 V], nitric oxide [E$_0'$ (NO/N$_2$O) = +1.175 V], and nitrous oxide [E$_0'$ (N$_2$O/N$_2$) = +1.355 V] [71] for energy conservation by chemiosmotic coupling of ATP synthesis and electron transport. The terminal oxidoreductases of denitrifying bacteria obtain the reducing equivalents via quinones [9]. Recently, a novel species of the *Ectothiorhodospira* clade was isolated that grew anaerobically with As(III) as the electron donor [72].

3.1.1. Nitrate Reductase

Nitrate is a major source of inorganic nitrogen for plants, fungi, and many species of bacteria. Animals cannot assimilate nitrate. Assimilation involves three pathway-specific steps: uptake followed by reduction to nitrite, and further reduction to ammonia which is then metabolized into central pathways.

Next to its role as a nutrient, nitrate is an important oxidant for the conservation of metabolic energy, i.e., nitrate respiration, and the dissipation of excess reducing power for redox balancing, nitrate dissimilation [4,73]. All eukaryotic and bacterial nitrate reductases are molybdenum-dependent enzymes and possess a unique cofactor, the so-called molybdopterin [74]. Nitrate-reducing archaea are also known and putative molybdopterin-binding enzymes can be identified in the genome sequence of *Archaeoglobus fulgidus* which suggests the presence of molybdopterin-dependent enzymes, such as nitrate reductase, DMSO reductase, or polysulfide reductase [69]. Four types of nitrate reductases catalyze the two-electron reduction of nitrate to nitrite [Eq. (4)].

$$NO_3^- + 2e^- + 2H^+ \rightleftharpoons NO_2^- + H_2O \quad (E_0' = +0.43\,V) \qquad (4)$$

These are the eukaryotic assimilatory nitrate reductases and three distinct bacterial enzymes, comprising the cytoplasmic assimilatory (Nas), the membrane-bound (Nar), and the periplasmic dissimilatory (Nap) nitrate reductases [11]. Nitrite oxidoreductase, a membrane-bound enzyme from nitrifying bacteria also exhibits nitrate reductase activity. This enzyme shows high sequence similarity to the membrane-bound Nar, and catalyzes the nitrite oxidation to nitrate, to allow chemoautotrophic growth [75]. Many bacteria have more than one of the four types of nitrate reductases [9]. The functional, biochemical, and structural properties of prokaryotic and eukaryotic nitrate reductases have been recently reviewed [3,4,76,77]. Protein sequence data have been used to determine phylogenetic relationships and to examine similarities in structure and function of nitrate reductases. Three distinct clades of nitrate reductase evolved: the eukaryotic assimilatory Nas, the membrane-associated prokaryotic Nar, and a clade that included both the periplasmatic Nap and prokaryotic assimilatory Nas [78].

Escherichia coli, when grown anaerobically with nitrate as respiratory oxidant, develops a respiratory chain terminated by the membrane-bound quinol-nitrate oxidoreductase, NarGHI [79]. The heterotrimeric NarGHI is composed of a Moco-containing Mo-bisMGD catalytic subunit, an Fe–S-cluster containing electron transfer subunit and a heme-containing membrane anchor subunit. The catalytic and electron transfer subunits form a membrane-extrinsic catalytic dimer anchored to the membrane. These redox-active groups are supposed to define an electron transfer chain linking a periplasmatically oriented site of quinol oxidation to a cytoplasmatically oriented site of nitrate reduction. In the respiratory chain composed of formate dehydrogenase N (FdnGHI) and NarGHI, electron transfer from formate to nitrate is coupled to the generation of a proton-motive force across the cytoplasmic membrane by a redox loop mechanism [79]. Meanwhile, high resolution X-ray structures of both FdnGHI and NarGHI have been reported which allow insights into the respiratory electron transfer pathway and proton-motive force generation [79–81].

Most recently, NarGHI has been studied by protein film voltammetry, with the enzyme adsorbed on a rotating disk pyrolytic graphite edge electrode. Nitrate binding might be influenced by the redox state of either the molybdenum site, or of the pterin moiety [82]. At the same time, the structure of the catalytic and electron-transfer subunits (NarGH) of the integral membrane protein nitrate reductase has been determined revealing the molecular architecture of the molybdenum enzyme which included a previously undetected Fe–S-cluster [83]. In this structure, an aspartate was detected as a ligand to the Mo center for the first time.

3.1.2. Nitrite Reductase

The nitrite reductase is either a cytochrome cd_1 or a copper enzyme carrying the type-1 and a type-2 Cu center [12,84–88].

3.1.2.1. Cu nitrite reductase: The Cu-dependent nitrite reductase, which transforms nitrite to nitric oxide in denitrifying bacteria, contains both the type-1 and the type-2 Cu center but lacks the binuclear type-3 site [89]. Yet, its arrangement of Cu atoms within the protein indicates a strong structural relationship with the "classical" blue multicopper oxidases [85]. It catalyzes the single-electron reduction of NO_2^- to NO and water [Eqs. (5)–(7)]:

$$NO_2^- + 2H^+ + e^- \rightleftharpoons NO + H_2O \qquad (E_0' = +0.35\ V) \quad (5)$$

$$E-Cu^{2+} + e^- + NO_2^- \rightleftharpoons E-Cu^+-NO_2^- \quad (6)$$

$$E-Cu^+-NO_2^- + 2H^+ \rightleftharpoons E-Cu^{2+} + NO + H_2O \quad (7)$$

Nitrite reductase is a homotrimer in its native state. Depending on the origin, the chain length of the respective subunits vary between 340 and 379 amino acid residues. The trimer contains three type-1 and three type-2 Cu centers. The type-1 Cu centers are localized within the subunits, whereas the type-2 Cu centers are coordinated by residues from two different subunits (Fig. 5). The

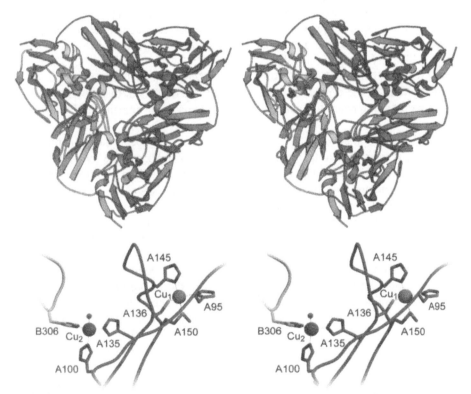

Figure 5 Stereoview of the overall structure of nitrite reductase from *A. faecalis* and the type-1 and type-2 Cu sites (PDB file 2AFN).

physiological electron donor of nitrite reductase is the type-1 Cu protein pseudoazurin, which transfers a single electron to the type-1 Cu center of nitrite reductase, from where it is transferred to the type-2 Cu [90]. The type-2 Cu is the binding-site for nitrite and where it is reduced to NO. The subunits of nitrite reductase from *Alcaligenes faecalis* contain two domains. Each domain consists of a β-barrel structure, similar to that in the small blue proteins. The type-1 Cu is embedded in one of these β-barrels and is coordinated by ligands His95 (domain 1), Cys136, His145, and Met150 (domain 2).

The type-2 copper center is coordinated by a water molecule and three histidines, it is situated at the interface of two subunits. The water molecule is displaced from the type-2 Cu ion upon nitrite binding. Although both Cu ions are coordinated by neighboring residues, they are located ~ 12.5 Å apart, which prevents a direct electron transfer. Most proteins with type-1 Cu are blue, the nitrite reductases from *Achromobacter cycloclastes*, *A. faecalis* and *Pseudomonas aureofaciens*, however, are green. The geometric distortion primarily responsible for the electronic structure changes in "green" nitrite reductase, relative to the blue type-1 Cu protein plastocyanin, was determined to involve a coupled angular movement of the Cys and Met residues toward a more flattened tetrahedral (towards square planar) structure [91].

A new class of copper-containing nitrite reductases constitutes the major anaerobically induced outer membrane protein from pathogenic *Neisseria gonorrhoeae*. The crystal structure of the soluble domain of this protein revealed a type-1 Cu with unusual visible absorption spectra, and a bidentate mode of nitrite binding to the type-2 Cu center [92]. In this organism, the expression of nitrite reductase appears to enhance the resistance against human sera [93].

3.1.2.2. Cytochrome cd_1 nitrite reductase: Cytochrome cd_1 nitrite reductase is a bifunctional multiheme enzyme catalyzing the one-electron reduction of nitrite to NO and the four-electron reduction of dioxygen to water. The enzyme has been purified from several sources as a homodimer (60 kDa/subunit), each subunit containing one heme c and one heme d_1, and is located in the bacterial periplasm. A high resolution structure has been presented for the enzyme from *Paracoccus pantotrophus* [94] and the enzyme from *Pseudomonas aeruginosa* [95]. In both enzymes, heme c is covalently linked to the N-terminal α-helical domain and heme d_1 is bound non-covalently to the C-terminal β-propeller domain. Heme c from the *P. pantotrophus* enzyme has a bis-histidine coordination, whereas the iron in heme d_1 is coordinated to Tyr25 and His200. Mutation of Tyr25 by serine does not result in decreased activity of reduction of any substate [96]. Instead, unlike the wild-type enzyme, the Tyr25Ser mutant is active as a reductase towards nitrite, oxygen and hydroxylamine without a reductive activation step. Heme c in the *P. aeruginosa* enzyme is more like the classical cytochrome c fold because it has His51 and Met88 as heme ligands. The d_1-heme domains are almost identical in both structures.

Intramolecular electron transfer between the heme c as the electron uptake site and heme d_1 as the nitrite reduction site is an essential step in the catalytic cycle and has been studied by several groups using different methods [97–101]. The rate constant of the *P. pantotrophus* enzyme [101] with $1.4 \times 10^3 \text{ s}^{-1}$ is 1000 times higher than the rate of the *P. aeruginosa* enzyme with 1 s^{-1} [97,98,100]. Recently, the electron transfer rate of the enzyme from *Pseudomonas stutzeri* has been detected with a rate of 23 s^{-1} [102].

Catalysis commences with the binding of nitrite, through its nitrogen atom, to the reduced d_1 heme [94,95]. This relatively rare example of an anion binding to a Fe(II) heme is followed by transfer of one electron and two protons to nitrite, with the result that water is released and a nitrosyl species is generated at the d_1 heme. The protons are provided by two highly conserved histidine residues on the distal side of the d_1 heme. The product of reduction and dehydration of nitrite can be regarded as $d_1\text{-Fe}^{2+}\text{-NO}^+$ or $d_1\text{-Fe}^{3+}\text{-NO}$, which are isoelectronic [103]. It is expected that the $d_1\text{-Fe}^{3+}\text{-NO}$ species should be unstable and dissociate to give NO and $\text{Fe}^{3+}\text{-}d_1$ heme. Most recently, kinetics and thermodynamics of the internal electron transfer process in the enzyme from *P. stutzeri* were found to be dominated by pronounced interactions between the c and the d_1 hemes [104]. The interactions are expressed both in significant changes in the internal electron transfer rates between these sites and in marked cooperativity in their electron affinity.

3.1.3. NO Reductase

NO reductase was the last of the terminal oxidoreductases of denitrification whose molecular properties became known [9,27]. The physiological, biochemical, and structural properties of NO reductase have been recently reviewed [24,27]. NO reductase catalyzes the reduction of NO to N_2O [Eq. (8)] by a bc heme protein [69,105,106].

$$2\text{NO} + 2\text{H}^+ + 2\text{e}^- \rightleftharpoons \text{N}_2\text{O} + \text{H}_2\text{O} \quad (E_0' = +1.175 \text{ V}) \tag{8}$$

This transmembrane enzyme is a heterodimer containing a low-spin heme c, a low-spin heme b, a high-spin heme b and a non-heme iron. The strong hydrophobicity of the cytochrome b subunit finds its explanation in 12 hydrophobic segments of the primary structure. An important development was the finding that NorB is similar to the subunit I of the heme-copper oxidase family of the terminal oxidoreductases, and from the NorB primary structure and the crystallographic data of the cytochrome c oxidase from *Paracoccus denitrificans* a structural model for NO reductase was developed. NO reductase does not contain a significant amount of Cu but does contain Fe in addition to heme iron. This iron has been shown to be a constituent of the active site coordinated by three histidines [9,27]. The other feature in the active site is the high-spin heme b. At this binuclear center the reduction of NO occurs. The stoichiometry of non-heme Fe to heme b and heme c Fe in a minimal composition of one cytochrome b and cytochrome c each is $1:2:1$ [107].

The catalytic mechanism starts with addition of one NO molecule to the non-heme iron and one NO molecule to the heme b center, both in the ferrous state. Cleavage of the bond between the proximal histidine and the heme iron makes this possible. Leaving of N_2O forms a oxo-bridged Fe(III) iron species which loses water and is reduced to the starting Fe(II) [108,109]. Kinetic studies on the proton and electron pathways in NO reductase suggest that the topologies of the redox centers and the electron-transfer pathway are similar in NO reductase and the dioxygen-reducing cytochrome c oxidases. The rate constants of electron transfer between the electron entry sites and the low-spin heme center are identical [110].

The NO reductase in gram-positive bacteria differs from the well-known enzymes in gram-negative bacteria. The cytoplasmic membrane associated enzyme from *B. azotoformans* consists of two subunits containing one non-heme iron, two copper atoms and two b-type hemes per enzyme complex [111]. The low-spin heme b center and the high-spin heme b center form a binuclear site where reduction of NO occurs instead of a non-heme iron/high-spin heme b site. The electron donor for the enzyme is menaquinol instead of cytochrome c in the gram-negative bacteria. Similarly, the hyperthermophilic archaeon *Pyrobaculum aerophilum* uses menaquinol [112,113]. This NO reductase lacks heme b, it carries stoichiometric amounts of hemes O_{p1} and O_{p2} instead, which are ethenylgeranylgeranyl and hydroxyethylgeranylgeranyl derivatives of heme b. Note that this NO reductase represents the first example which contains modified hemes reminiscent of cytochrome bo_3 and aa_3 oxidases.

Another novel candidate with NO reductase activity is a flavodiiron protein from *Moorella thermoacetica* [114]. Two genes of this strictly anaerobic, acetogenic bacterium, namely *fprA* and *hrb*, have no known function. FprA contains flavin mononucleotide (FMN) and a non-heme diiron site while Hrb contains FMN and a rubredoxin-like [Fe(SCys)$_4$] site. The combination of Hrb and FprA exhibits NADH:NO oxidoreductase activity.

3.1.4. N_2O Reductase

The reduction of N_2O to N_2 is a strictly Cu-dependent process catalyzed by nitrous oxide reductase carrying CuA and the tetranuclear μ_4-sulfide bridged cluster CuZ [25,115–117] (Fig. 6). N_2OR from *W. succinogenes* is reported to carry also a covalently bound c-type cytochrome [118,119]. The reduction of N_2O is highly exergonic ($\Delta G^{0'} = -340$ kJ mol^{-1}) [Eq. (9)]. Yet, N_2O is rather inert at room temperature and its transformation requires an effective catalyst.

$$N_2O + 2e^- + 2H^+ \rightleftharpoons N_2 + H_2O \quad (E_0' = +1.355\,V) \tag{9}$$

Much of the early biochemical and spectroscopic knowledge about nitrous oxide reductase and its several forms arose from the studies by Zumft and coworkers on the enzyme from *P. stutzeri* [25]. This enzyme had to be purified under the exclusion of dioxygen to achieve high activity [120]. It is a soluble enzyme

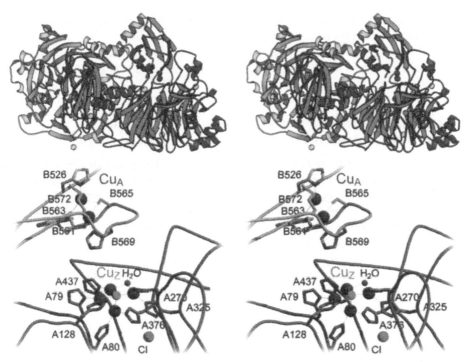

Figure 6 Stereoview of the overall structure of nitrous oxide reductase from *Pseudomonas nautica* and the CuA electron transfer center and the catalytic CuZ. (PDB file 1 FWX).

located in the periplasm as most N_2O reductases described so far [25]. Notwithstanding, a few of these enzymes may have a loose membrane association [121]. To attain its functional state, nitrous oxide reductase is subjected to a complex maturation process which involves the protein-driven synthesis of the unique CuZ catalytic center and metallation of the binuclear CuA electron transfer center in the periplasm. Using the non-denitrifier *Pseudomonas putida* the requirements for the CuA and the Cu-sulfide center assembly were determined, and a catalytically active holo-nitrous oxide reductase could be produced [122].

The crystal structure was recently reported for nitrous oxide reductase from two different microorganisms, *Pseudomonas nautica* [115,116] and *Paracoccus denitrificans* [123]. In both cases, the protein used for crystallization had been purified in the presence of dioxygen producing a form of nitrous oxide reductase with a characteristic absorption maximum $\sim$650 nm. This feature is absent in the enzyme from *P. stutzeri* obtained under the exclusion of dioxygen [25]. The enzyme is a head-to-tail homodimer stabilized by a multitude of interactions. Each monomer is composed of two domains, a C-terminal cupredoxin domain that carries the mixed-valent CuA binuclear center, and an N-terminal seven-bladed β-propeller domain which hosts the active CuZ site. In the structure of the *Pa. denitrificans* enzyme [123] the four Cu ions of CuZ are ligated by seven histidines, an oxygen (from either water or hydroxyl) and a bridging inorganic sulfide (Fig. 6).

Spectroscopic investigations combined with density functional calculations have been employed to develop a detailed bonding description of the CuZ cluster and its function during catalysis [124].

3.2. Nitrite Ammonification: Reductive Transformation of Nitrite to Ammonia

The reduction of nitrite to ammonia was described for bacteria with a fermentative rather than a respiratory metabolism [125]. However, growth of various bacteria by oxidation of non-fermentable substrates such as formate linked to the reduction of nitrite to ammonia, demonstrated that nitrite ammonification may also function as respiratory energy conserving process [126]. The enzymology and bioenergetics of respiratory nitrite ammonification have been recently reviewed [17]. In respiratory nitrite ammonification, nitrite is reduced to ammonia without the release of intermediate products, such as NO or N_2O, in a six-electron step by a cytochrome c nitrite reductase, the so-called NrfA protein [Eq. (10)] [127–131].

$$NO_2^- + 8H^+ + 6e^- \rightleftharpoons NH_4^+ + 2H_2O \quad [E_0' (NO_2^-/NH_4^+) = +0.34 \text{ V}]$$
(10)

This process can be regarded as a short circuit that bypasses denitrification and nitrogen fixation (Fig. 1). H_2 and formate are the predominant electron donors [17] but sulfide can also serve as electron donor, thus connecting the biogeochemical cycles of nitrogen and sulfur [132]. Organisms capable of nitrite ammonification usually catalyze nitrate to nitrite reduction in dissimilatory metabolism [17]. Note that the majority of sulfate-reducing *Desulfovibrio* species use nitrite but not nitrate [133]. In the sulfur- and sulfate-reducing species of the δ- and ε-proteobacteria, dissimilatory nitrite reduction to ammonia is the predominant metabolism during anaerobic growth on formate or H_2. In γ-proteobacteria like *E. coli* however, the observed expression levels of nitrite reductase were considerably lower. The presence of a highly active nitrite reduction system might serve an important purpose, namely the disposal of toxic nitrite anion.

3.2.1. From Nitrate to Nitrite

A first model for the organization of respiratory nitrite ammonification and direction of proton exchange associated with electron transfer from formate to nitrate in *E. coli* had been described earlier [126]. The reduction of nitrate to nitrite by the oxidation of formate is linked to the generation of a proton electrochemical potential (Section 3.1.1., discussion on nitrate reductase). Whether the overall architecture of the nitrate reductase complex from *E. coli* also holds for other nitrite-ammonifying bacteria, has to be investigated. *S. deleyianum* and

D. desulfuricans also showed a membrane-associated or soluble nitrate reductase catalyzing the respiratory reduction of nitrate to nitrite [134–136]. A cytochrome *c*-dependent nitrate reductase was also isolated from *Geobacter metallireducens* [137]. Obviously, nitrate uptake and nitrite extrusion into the bulk phase is dispensable for periplasmic nitrate reductases. In addition, a periplasmic location of nitrate reductases will prevent the interaction of potentially toxic compounds such as nitrite or NO with cytoplasmic proteins.

3.2.2. From Nitrite to Ammonia

In nitrite-ammonifying bacteria, especially in enterobacteria, there may exist two independently regulated dissimilatory ways of nitrite reduction to ammonia with two different physiological functions [125]. A cytoplasmic siroheme-dependent nitrite reductase activity (NADH : nitrite oxidoreductase) [138,139] confers on *E. coli* the advantage of regenerating NAD^+ during anaerobic growth to maintain glycolysis [140]. Besides the supply of ammonia for biosynthesis, this nitrite reductase leads to the detoxification of intracellularly accumulated nitrite. The other nitrite reductase is a periplasmic cytochrome *c* that was isolated from *E. coli* [141,142]. A membrane-bound formate-nitrite oxidoreductase complex with a respiratory function was first described for *E. coli* [126].

A model of the respiratory chain for the transformation of nitrite was proposed earlier for nitrite-ammonifying bacteria, such as *Sulfurospirillum deleyianum* and *Wolinella succinogenes* [126]. A periplasmically oriented membrane-bound Fdn, or a hydrogenase, was included in the model. Most likely, the primary dehydrogenases functioning in the reduction of nitrate to nitrite, and nitrite to ammonia, are identical entities [143]. The complete electron transport chain from formate to nitrite was characterized for *W. succinogenes* by reconstituting the coupled electron transfer chain into liposomes [144]. A model of the electron transport chain of *W. succinogenes* has been developed showing the three major enzyme complexes involved in electron transfer from formate (or H_2) to nitrite [17].

3.2.2.1. The cytochrome c nitrite reductase complex: The enzyme from *E. coli* was the first cytochrome *c* Nir which had been completely sequenced [145,146]. It is a heterooligomeric complex encoded by the *nrfABCDEFG* operon, with *nrfA* encoding for the periplasmic nitrite-reactive cytochrome *c* (M_r 51 kDa). Later on, sequences of cytochrome *c* Nir from other microorganisms including *S. deleyianum* and *W. succinogenes* became available in the GenBank^{TM}/EBI sequence database, and high-resolution structures of cytochrome *c* Nir (soluble part, NrfA) of *S. deleyianum*, *W. succinogenes*, *D. desulfuricans*, and *E. coli* were reported [17].

The functional nitrite reductase system of *W. succinogenes* and *S. deleyianum* was shown to be a complex of the soluble part NrfA and a membrane-anchored tetraheme *c*-type cytochrome NrfH, which acts as a quinol oxidase to receive electrons from the membraneous quinone pool [144]. The current working hypothesis thus implies that the NrfA dimer is associated with the peripheral membrane protein NrfH as depicted in the model. Preliminary crystal data for the membrane associated complex NrfAH were presented recently [147].

3.2.2.2. Reaction mechanism: Based on the observation of reaction intermediates in the crystal structure and on quantum chemical calculations Einsle et al. [148] propose an outline of the first detailed reaction mechanism of the cytochrome *c* Nir from *W. succinogenes.* Nitrite reduction starts with a heterolytic cleavage of the weak N–O bond, which is facilitated by a pronounced backbonding interaction between nitrite and the reduced active site iron. The protons come from a highly conserved histidine and tyrosine. Elimination of one of both amino acids results in a significant reduced activity. Subsequently, two rapid one-electron reductions lead to a $\{FeNO\}^8$ form and, by protonation, to a HNO adduct. A further two-electron two-proton step leads to hydroxylamine. The iron in the hydroxylamine complex is in the Fe(III) state [149], which is unusual compared to synthetic iron–hydroxylamine complexes where the iron is mainly in the Fe(II) state. Finally, it readily loses water to give the product, ammonia. This presumably dissociates from the Fe(III) form of the active site, whose re-reduction closes the reaction cycle.

3.3.   Nitrification and Nitrite Oxidation: Oxidative Transformation of Ammonia

Ammonia oxidation has been observed in many bacterial species, it is oxidized by two pathways. First, it is oxidized aerobically to hydroxylamine, which is then oxidized to nitrite in a four-electron step. Second, ammonia and nitrite are converted anaerobically to N_2 by the Anammox process (Fig. 1). Nitrifying bacteria connect the oxidized and reduced sides of the nitrogen cycle by nitrification, the conversion of ammonia to nitrogen oxyanions [3,4,150]. The most important microorganisms involved in these transformations are the lithoautotrophic ammonia- and nitrite-oxidizing bacteria placed in the family *Nitrobacteraceae* [4]. The obligate aerobic lithotrophic bacterium *Nitrosomonas europaea* derives energy from the oxidation of ammonia to nitrite. Hereby, ammonia is oxidized to hydroxylamine by ammonia monooxygenase which is converted to nitrite by hydroxylamine oxidoreductase [Eqs. (11) and (12)] [151,152].

$$NH_3 + 2e^- + 2H^+ + O_2 \rightleftharpoons NH_2OH + H_2O \tag{11}$$
$$NH_2OH + H_2O \rightleftharpoons NO_2^- + 4e^- + 5H^+ \tag{12}$$

Two of the four electrons generated from hydroxylamine oxidation are used to support the oxidation of additional ammonia molecules, while the other two electrons enter the electron transfer chain and are used for CO_2 reduction and ATP biosynthesis [153,154].

The close coupling of nitrification and denitrification has been demonstrated by measuring the overall N_2 flux from microcosms in a helium/dioxygen atmosphere [155]. Within the microenvironment of the rice rhizosphere, the activities of nitrifiers and of denitrifiers depended on the presence of dioxygen and ammonia. Carbon and nitrogen cycles are clearly interrelated. Methane monooxygenases and ammonia monooxygenases are relatively unspecific, and may also oxidize ammonia or methane, respectively [13].

The three-dimensional structure (resolution 2.8 Å) of the hydroxylamine oxidoreductase from *N. europaea* revealed a homotrimeric unit, with each monomer carrying a complex array of eight heme centers including the so-called P_{460} catalytic center [156]. Surprisingly, this enzyme revealed structural homologies to other multiheme *c*-type cytochromes, amongst them the penta-heme cytochrome *c* Nir (NrfA) from *S. deleyianum* and *W. succinogenes* [127,128]. These homologies mainly concern the arrangement of heme groups but not the surrounding protein. The five hemes of the cytochrome *c* Nir aligned to hemes 4–8 of hydroxylamine oxidoreductase with a root-mean-square deviation of 2.00 Å [128]. However, while a protoporphyrin IX heme with a novel lysine coordination forms the active site of cytochrome *c* Nir, the hydroxylamine oxidoreductase features the P_{460}-type heme with a covalent link to a tyrosine residue. The iron atom of heme P_{460} appears to be five-coordinated, with a histidine as an axial protein ligand. The sixth position of the heme iron is not occupied and hydroxylamine can bind to this position [156].

4. ASSIMILATORY PROCESSES: BUILDING THE ESSENTIAL MOLECULES OF LIFE

The bacterial assimilatory pathways exhibit quite some metabolic versatility, in contrast to the rather uniform anabolism of the autotrophic plants. Note that their assimilatory potential is based on the metabolism of only one endosymbiotic pro-karyotic group, the cyanobacteria. The prokaryotic endosymbiont enabled the host cell to grow at the expense of carbon dioxide, nitrate, phosphate, and sulfate as sole sources of carbon, nitrogen, phosphorus, and sulfur. Thus, it is not surprising that the plant assimilatory enzymes are mostly located in the chloroplast and resemble the prokaryotic systems [157]. Protein accounts for approximately half of the cell dry weight of the average prokaryotic cell, whereas nucleic acids account for approximately one fifth. Central precursor metabolites, such as acetyl-CoA, pyruvate, oxaloacetate, and other metabolites are the starting material for the synthesis of essential (close to 100) building blocks. The building blocks of other organisms can be used by prokaryotes as organic substrates. Hereby, plants constitute the most important source. These substrates are broken down into one of the central precursor metabolites, connecting degradation and biosynthesis.

The synthesis of 1 g of cell mass requires roughly 140 mg of nitrogen. The oxidation state of most forms of cellular nitrogen compounds (amino groups, N heterocyclic compounds) is equivalent to ammonia. Nitrogen substrates are usually reduced or hydrolyzed to NH_4^+ and then incorporated by means of the four central nitrogen carriers, glutamate, glutamine, and to some extent secondarily via aspartate and carbamoyl phosphate. Bacteria take up nitrate, which is subsequently reduced to nitrite and then to ammonia. This process requires a transport system for nitrate, as the negatively charged nitrate anion should not be able to cross biological membranes at fast rates [157]. For assimilation of nitrate, and nitrite, two types of uptake systems are known: ABC transporters that are driven by ATP hydrolysis, and secondary transporters reliant on a proton-motive force. Proteins homologous to the latter type of transporter are also involved in nitrate and nitrite transport in dissimilatory processes [158].

In plants, e.g., tobacco, nitrogen metabolism is a much more difficult network to study. Nitrogen moves along a complex branching and merging pathway which interacts at numerous sites with the carbon flow, pH regulation, and ion and assimilate flow at the cell and whole plant level. Superimposed on these metabolic fluxes is the strong influence of nitrate and nitrogen metabolism on plant development and plant architecture, including changes in root architecture, the timing of senescence, and flowering. Nitrate assimilation appears to be integrated with nitrate uptake, with ammonium assimilation and amino acid synthesis, and with the sugar supply in the leaves [159].

5. OUTLOOK

Our environment, as an interplay of geochemical and biochemical processes, is in a stationary phase with 20.9% O_2, 79.1% N_2, and 0.03% CO_2, and with a world ocean at pH ≈ 8, and $E_0' = +750$ mV [160]. Important elements, such as carbon, nitrogen, and sulfur, are involved in complex biogeochemical cycles.

Conversion of N_2 to NH_3, i.e., nitrogen fixation, is an essential step in the global nitrogen cycle, and it is necessary to sustain life on earth [6]. It is difficult to predict the future development of ammonia production technology. As $\sim 85\%$ of the ammonia consumption goes into the manufacture of fertilizers, it is obvious that the future of the ammonia industry is very closely bound with future fertilizer needs and the pattern of world supply. Nitrate pollution will, and has already been, become a problem in some parts of the world. A fundamental question is whether there are other options for the future than the present ammonia technology. As the major demand is in agriculture, biological processes for *in vivo* conversion of molecular nitrogen into fixed nitrogen would probably be the first choice. Alternatively abiotic *in vitro* processes with homogenous catalysis under mild reaction conditions have to be discussed. Both routes are extensively investigated [5].

The objective of atmospheric chemistry is to understand the factors that control the concentrations of chemical species in the atmosphere. Gases other than N_2, O_2, and Ar are present in the atmosphere at comparably low concentrations and are called *trace gases*, amongst them CH_4 and N_2O which are produced by the biosphere. These trace gases can be of critical importance for the greenhouse effect, the ozone layer, smog, and other environmental aspects.

An important component of aircraft exhaust is nitric oxide (NO), formed by oxidation of atmospheric N_2 at high temperatures of the aircraft engine. In the stratosphere NO rapidly reacts with ozone (O_3) to produce NO_2, which then photolyzes. Amongst the products formed in a complex set of different reactions is nitric acid (HNO_3) which will be deposited as acid rain. Acid rain falling over most of the world has little environmental effect because it will be rapidly neutralized. In some areas where the biosphere is sensitive to acid rain there has been ample evidence of its negative effects on freshwater and terrestrial ecosystems. Although ammonia in the atmosphere will neutralize acid rain by formation of NH_4^+, the acidity may be recovered in soil when NH_4^+ goes through the microbial nitrification process to form nitrate (Fig. 1). Beyond the input of acidity, deposition of NH_4^+ and NO_3^- fertilizes ecosystems by providing a source of directly assimilable nitrogen. This source has been blamed as an important contributor to the so-called *eutrophication* (excess fertilization). As a consequence algae can accumulate at the water surface, suppressing the supply of O_2 to the deep-water biosphere [161].

Unicellular microorganisms have been the most important geochemical agents by far [19]. Many of the recent advances in our understanding of microbial communities and cultured microorganisms can be attributed to the advent of genomics. Within the past few years the number of sequenced genomes of biogeochemically relevant microorganisms has increased significantly thanks to the dramatic progress in the field of genomic sequencing. A desirable goal for future sequencing projects will be to identify the major players in any given environment and bring them into culture. This will enable us to study the physiology and the biochemistry of these new isolates with the aid of genomic information. In addition, as a consequence, exciting novel metalloenzymes and metalloproteins with interesting structural, chemical and physical properties will be discovered.

ACKNOWLEDGMENTS

This work was supported by the Deutsche Forschungsgemeinschaft (PK, DFG-SPP 1071). We thank Professors S. de Vries and W. G. Zumft for providing valuable publications on NO reductase, and Professor Einsle who helped us with the figures.

ABBREVIATIONS

ADP	adenosine 5'-diphosphate
ATP	adenosine 5'-triphosphate
DMSO	dimethylsulfoxide
E_0'	midpoint redox potential for a given couple, at pH 7.0
FAD	flavin adenine dinucleotide
Fdn	formate dehydrogenase
FMN	flavin mononucleotide

ΔH_{298}	change in enthalpy, 298 K
Moco	molybdenum cofactor
MGD	bis-molybdopterin guanine dinucleotide
NAD^+	nicotinamide adenine dinucleotide (oxidized form)
NADH	nicotinamide adenine dinucleotide (reduced form)
Nap	periplasmic dissimilatory nitrate reductase
Nar	membrane bound nitrate reductase
Nas	cytoplasmic assimilatory nitrate reductase
Nir	nitrite reductase
N_2OR	N_2O reductase (nitrous oxide reductase)
Nor	NO reductase

REFERENCES

1. Sprent JI. The Ecology of the Nitrogen Cycle. Cambridge: Cambridge University Press, 1987.
2. Kroneck PMH. In: Revsbech NP, Sørensen J, eds. Denitrification in Soil and Sediment. New York and London: Plenum Press, 1990:1–20.
3. Kroneck PMH, Beuerle J, Schumacher W. In: Sigel H, Sigel A, eds. Metal Ions in Biological Systems. Vol. 28. New York, Basel, Hong Kong: Marcel Dekker Inc., 1992:455–505.
4. Kroneck PMH, Abt DJ. In: Sigel A, Sigel H, eds. Metal Ions in Biological Systems. Vol. 39. New York, Basel: Marcel Dekker Inc., 2002:369–403.
5. Appl M. Ammonia, Principles and Industrial Practice. Weinheim: Wiley-VCH Verlag, 1999.
6. Smil V. Cycles of Life: Civilization and the Biosphere. Scientific American Library. New York: W. H. Freeman and Company, 1997.
7. Kim K-R, Craig H. Science 1993; 262:1855–1857.
8. Conrad R. Microbiol Rev 1996; 60:609–640.
9. Zumft WG. Microbiol Mol Biol Rev 1997; 61:533–616.
10. Lin JT, Stewart V. Adv Microbiol Physiol 1998; 39:1–30.
11. Moreno-Vivián C, Cabello P, Martínez-Luque M, Blasco R, Castillo F. J Bacteriol 1999; 181:6573–6584.
12. Richardson DJ, Watmough NJ. Curr Opin Chem Biol 1999; 3:207–219.
13. Brune A, Frenzel P, Cypionka H. FEMS Microbiol Rev 2000; 24:691–710.
14. Ye RW, Thomas SM. Curr Opin Microbiol 2001; 4:307–312.
15. Richardson DJ. Cell Mol Life Sci 2001; 58:163–164.
16. Anbar AD, Knoll AH. Science 2002; 297:1137–1142.
17. Simon J. FEMS Microbiol Rev 2002; 26:285–309.
18. Morel FMM, Price NM. Science 2003; 300:944–947.
19. Newman DK, Banfield JF. Science 2002; 296:1071–1077.
20. Reysenbach AL, Shock E. Science 2002; 296:1077–1082.
21. Devol AH. Nature 2003; 422:575–576.
22. Strous M, Fuerst JA, Kramer EH, Logemann S, Muyzer G, van de Pas-Schoonen KT, Webb R, Kuenen JG, Jetten MS. Nature 1999; 400:446–449.
23. Kasting FJ. Science 1993; 259:902–926.

24. Wasser IM, deVries S, Moënne-Loccoz P, Schröder I, Karlin KD. Chem Rev 2002; 102:1201–1234.
25. Zumft WG, Kroneck PMH. Adv Inorg Biochem 1996; 11:193–221.
26. Hendriks J, Oubrie A, Castresana J, Urbani A, Gemeinhardt S, Saraste M. Biochim Biophys Acta 2000; 1459:266–273.
27. Zumft WG. J Inorg Biochem 2005; in press.
28. Bertini I, Sigel A, Sigel H, eds. Handbook on Metalloproteins. New York, Basel: Marcel Dekker Inc., 2001.
29. Messerschmidt A, Huber R, Poulos T, Wieghardt K, eds. Handbook of Metalloproteins. Weinheim: Wiley-VCH Verlag GmbH, 2001.
30. Holm RH, Solomon EI. Chem Rev 2004; 104:347–1200.
31. Butler A. Science 1998; 281:207–209.
32. Lee SC, Holm RH. Proc Natl Acad Sci USA 2003; 100:3595–3600.
33. Seefeldt LC, Dance IG, Dean DR. Biochemistry 2004; 43:1401–1409.
34. Holleman AF, Wiberg N. Lehrbuch der Anorganischen Chemie. Berlin, New York: Walter de Gruyter Verlag, 1995.
35. Shriver DF, Atkins PW, Langford CH. In: Heck J, Kaim W, Weidenbruch M, eds. Inorganic Chemistry. Weinheim, New York, Chichester, Brisbane, Singapore, Toronto: Wiley-VCH, 1997.
36. Sellmann D, Hantsch B, Rösler A, Heinemann FW. Angew Chem Int Ed Engl 2001; 40:1505–1507.
37. Zhang Y, Zuo JL, Zhou HC, Holm RH. J Am Chem Soc 2002; 124:14292–14293.
38. Halbleib CM, Ludden PW. Am Soc Nutr Sci 2000; 130:1081–1084.
39. Burgess BK, Lowe DJ. Chem Rev 1996; 96:2983–3011.
40. Howard JB, Rees DC. Chem Rev 1996; 96:2965–2982.
41. Smith BE. Adv Inorg Chem 1999; 47:159–218.
42. Christiansen J, Dean DR, Seefeldt LC. Annu Rev Plant Physiol Plant Mol Biol 2001; 52:269–295.
43. Kim J, Rees DC. Science 1992; 257:1677–1682.
44. Kim J, Rees DC. Nature 1992; 360:553–560.
45. Chan MK, Kim J, Rees DC. Science 1993; 260:792–794.
46. Peters JW, Stowell MHB, Soltis SM, Finnegan MG, Johnson MK, Rees DC. Biochemistry 1997; 36:1181–1187.
47. Schmid B, Einsle O, Chiu HJ, Willing A, Yoshida M, Howard JB, Rees DC. Biochemistry 2002; 41:15557–15565.
48. Einsle O, Tezcan FA, Andrade SLA, Schmid B, Yoshida M, Howard JB, Rees DC. Science 2002; 297:1696–1700.
49. Kim J, Woo D, Rees DC. Biochemistry 1993; 32:7104–7115.
50. Bolin JT, Campobasso N, Muchmore SW, Morgan TV, Mortenson LE. In: Stiefel EI, Coucouvanis D, Newton WE, eds. Molybdenum Enzymes, Cofactors and Model Systems. ACS Symposium Series 535. Washington, DC: American Chemical Society, 1993:186–195.
51. Mayer SM, Lawson DM, Gormal CA, Roe SM, Smith BE. J Mol Biol 1999; 292:871–891.
52. Dean DR, Jacobsen MR. In: Stacey G, Burris RH, Evans HJ, eds. Biological Nitrogen Fixation. New York: Chapman and Hall, 1992:763–834.
53. Georgiadis MM, Komiya H, Chakrabarti P, Woo D, Kornuc JJ, Rees DC. Science 1992; 257:1653–1659.

54. Schindelin H, Kisker C, Schlessman JL, Howard JB, Rees DC. Nature 1997; 387:370–376.
55. Allen AD, Senoff CV. Chem Commun 1965; 621–622.
56. Leigh GJ. Science 2003; 301:55–56.
57. Yandulov DV, Schrock RR. Science 2003; 301:76–78.
58. Thorneley RNF, Lowe DJ. Biochem J 1983; 215:393–403.
59. Leigh GJ. In: Leigh GJ, ed. Nitrogen Fixation at the Millennium. Amsterdam: Elsevier, 2003:299–322.
60. Henderson RA, Leigh GJ, Pickett CJ. Adv Inorg Radiochem 1983; 27:197–292.
61. Dance I. Chem Commun 2003; 7:324–325.
62. Hinnemann B, Nørskov JK. J Am Chem Soc 2003; 125:1466–1467.
63. Rees DC, Howard JB. Science 2003; 300:929–931.
64. Grönberg KLC, Gormal CA, Durrant MC, Smith BE, Henderson RA. J Am Chem Soc 1998; 120:10613–10621.
65. Durrant MC. Biochemistry 2002; 41:13934–13945.
66. Masepohl B, Schneider K, Drepper T, Müller A, Klipp W. In: Leigh GJ, ed. Nitrogen Fixation at the Millennium. Amsterdam: Elsevier, 2002:191–222.
67. Ribbe M, Gadkari D, Meyer O. J Biol Chem 1997; 272:26627–26633.
68. Ferguson SJ. Curr Opin Chem Biol 1998; 2:182–193.
69. Richardson DJ. Microbiol 2000; 146:551–571.
70. Moura I, Moura JJG. Curr Opin Chem Biol 2001; 5:168–175.
71. Thauer RK, Jungermann K, Decker K. Bacteriol Rev 1977; 41:100–180.
72. Oremland RS, Stolz JF. Science 2003; 300:939–944.
73. Richardson DJ, Berks BC, Russell DA, Spiro S, Taylor CJ. Cell Mol Life Sci 2001; 58:165–178.
74. Kisker C, Schindelin H, Baas D, Rétey J, Meckenstock RU, Kroneck PMH. FEMS Microbiol Rev 1999; 22:503–521.
75. Sundermeyer-Klinger H, Meyer W, Warninghoff B, Bock E. Arch Microbiol 1984; 149:153–158.
76. Blasco F, Guigliarelli B, Magalon A, Asso M, Giordano G, Rothery RA. Cell Mol Life Sci 2001; 58:179–193.
77. Campbell WH. Cell Mol Life Sci 2001; 58:194–204.
78. Stolz JF, Basu P. Chem Biochem 2002; 3:198–206.
79. Jormakka M, Törnroth S, Byrne B, Iwata S. Science 2002; 295:1863–1868.
80. Jormakka M, Byrne B, Iwata S. FEBS Lett 2003; 545:25–30.
81. Bertero MG, Rothery RA, Palak M, Hou C, Lim D, Blasco F, Weiner JH, Strynadka NCJ. Nat Struct Biol 2003; 10:681–687.
82. Elliot SJ, Hoke KR, Heffron K, Palak M, Rothery RA, Weiner JH, Armstrong FA. Biochemistry 2004; 43:799–807.
83. Jormakka M, Richardson D, Byrne B, Iwata S. Structure 2004; 12:95–104.
84. Adman ET, Godden JW, Turley S. J Biol Chem 1995; 270:27458–27474.
85. Murphy MEP, Lindley PF, Adman ET. Prot Sci 1997; 6:761–770.
86. Williams PA, Fülop V, Garman EF, Saunders NF, Ferguson SJ, Hajdu J. Nature 1997; 389:406–412.
87. Dodd FE, Hasnain SS, Abraham ZHL, Eady RR, Smith BE. Acta Cryst 1997; D53:406–418.
88. Gray HB, Malmström BG, Williams RJP. J Biol Inorg Chem 2000; 5:551–559.
89. Suzuki S, Kataoka K, Yamaguchi K. Acc Chem Res 2000; 33, 10:728–735.

90. Murphy LM, Dodd FE, Yousafzai FK, Eady RR, Hasnain SS. J Mol Biol 2002; 315:859–871.
91. LaCroix L, Shadle SE, Wang Y, Averill BA, Hedman B, Hodgson KO, Solomon EI. J Am Chem Soc 1996; 118:7755–7768.
92. Boulanger MJ, Murphy MEP. J Mol Biol 2002; 315:1111–1127.
93. Cardinale JA, Clark VL. Infect Immun 2000; 68:4368–4369.
94. Fülöp V, Moir JWB, Ferguson SJ, Hajdu J. Cell 1995; 81:369–377.
95. Nurizzo D, Silvestrini MC, Mathieu M, Cutruzzola F, Bougeois D, Fülöp V, Hajdu J, Brunori M, Tegoni M, Cambillau C. Structure 1997; 5:1157–1171.
96. Gordon EHJ, Sjögren T, Löfqvist M, Richter CD, Allen JWA, Higham CW, Hajdu J, Fülöp V, Ferguson SJ. J Biol Chem 2003; 278:11773–11781.
97. Silvestrini MC, Tordi MG, Musci G, Brunori M. J Biol Chem 1990; 265:11783–11787.
98. Parr SR, Barber D, Greenwood C. J Biochem 1977; 167:447–455.
99. Blatt Y, Pecht I. Biochemistry 1979; 18:2917–2922.
100. Schichman SA, Gray HB. J Am Chem Soc 1981; 103:7794–7795.
101. Kobayashi K, Koppenhöfer A, Ferguson SJ, Tagawa S. Biochemistry 1997; 36:13611–13616.
102. Farver O, Kroneck PMH, Zumft WG, Pecht I. Biophys Chem 2002; 98:27–34.
103. Williams PA, Fülöp V, Leung YC, Chan C, Moir JWB, Howlett G, Ferguson SJ, Radford SE, Hajdu J. Nature Struct Biol 1995; 2:975–982.
104. Farver O, Kroneck PMH, Zumft WG, Pecht I. Proc Natl Acad Sci USA 2003; 100:7622–7625.
105. Hendriks JHM, Gohlke U, Saraste M. J Bioenerg Biomembr 1998; 30:15–24.
106. Cheesman MR, Ferguson SJ, Moir JWB, Richardson DJ, Zumft WG, Thomson AJ. Biochemistry 1997; 36:16267–16276.
107. Girsch P, deVries S. Biochem Biophys Acta 1997; 1318:202–216.
108. Moënne-Loccoz P, deVries S. J Am Chem Soc 1998; 120:5147–5152.
109. Moënne-Loccoz P, Richter OMH, Huang HW, Wasser IM, Ghiladi RA, Karlin KD, deVries S. J Am Chem Soc 2000; 122:9344–9345.
110. Hendriks JHM, Jasaitis A, Saraste M, Verkhovsky MI. Biochemistry 2002; 41:2331–2340.
111. Suharti, Strampraad MJF, Schröder I, deVries S. Biochemistry 2001; 40:2632–2639.
112. deVries S, Schröder I. Biochem Soc Trans 2002; 30:662–667.
113. deVries S, Strampraad MJF, Lu S, Moënne-Loccoz P, Schröder I. J Biol Chem 2003; 278:35861–35868.
114. Silaghi-Dumitrescu R, Coulter ED, Das A, Ljungdahl LG, Jameson GNL, Huynh BH, Kurtz DM Jr. Biochemistry 2003; 42:2806–2815.
115. Brown K, Tegoni M, Prudêncio M, Pereira AS, Besson S, Moura JJG, Moura I, Cambillau C. Nat Struct Biol 2000; 7:191–195.
116. Brown K, Djinovic-Craugo K, Haltia T, Cabrito I, Saraste M, Moura JJG, Moura I, Tegoni M, Cambillau C. J Biol Chem 2000; 275:41133–41136.
117. Rasmussen T, Berks BC, Sanders-Loehr J, Dooley DM, Zumft WG, Thomson AJ. Biochemistry 2000; 39:12753–12756.
118. Teraguchi S, Hollocher SC. J Biol Chem 1989; 264:1972–1979.
119. Zhang C, Hollocher TC, Kolodziej AF, Orme-Johnson WH. J Biol Chem 1991; 266:2199–2202.

120. Coyle CL, Zumft WG, Kroneck PMH, Körner H, Jakob W. Eur J Biochem 1985; 153:459–467.
121. Hole UH, Vollack KU, Zumft WG, Eisenmann E, Siddiqui RA, Friedrich B, Kroneck PMH. Arch Microbiol 1995; 165:55–61.
122. Wunsch P, Herb M, Wieland H, Schiek UM, Zumft WG. J Bacteriol 2003; 185:887–896.
123. Haltia T, Brown K, Tegoni M, Cambillau C, Saraste M, Mattila K, Djinovic-Carugo K. Biochem J 2003; 369:77–88.
124. Cheng P, Cabrito I, Moura JJG, Moura I, Solomon EI. J Am Chem Soc 2002; 124:10497–10507.
125. Cole JA. FEMS Microbiol Lett 1996; 136:1–11.
126. Schumacher W, Hole U, Kroneck PMH. In: Winkelmann G, Carrano CJ, eds. Transition Metals in Biology. Amsterdam: Harwood Academic, 1997:329–356.
127. Einsle O, Messerschmidt A, Stach P, Bourenkov GP, Bartunik HD, Huber R, Kroneck PMH. Nature 1999; 400:476–480.
128. Einsle O, Stach P, Messerschmidt A, Simon J, Kröger A, Huber R, Kroneck PMH. J Biol Chem 2000; 275:39608–39616.
129. Almeida MG, Macieira S, Gonçalves LL, Huber R, Cunha CA, Romão MJ, Costa C, Lampreia J, Moura JJG, Moura I. Eur J Biochem 2003; 270:3904–3915.
130. Dias JM, Cunha CA, Teixeira S, Almeida G, Costa C, Lampreda J, Moura JJG, Moura I, Romão MJ. Acta Cryst 2000; D56:215–217.
131. Bamford VA, Angove HC, Seward HE, Thomson AJ, Cole JA, Butt JN, Hemmings AM, Richardson DJ. Biochemistry 2002; 41:2921–2931.
132. Eisenmann E, Beuerle J, Sulger K, Kroneck PMH, Schumacher W. Arch Microbiol 1995; 164:180–185.
133. Mitchell GJ, Jones JG, Cole JA. Arch Microbiol 1986; 144:35–40.
134. Schumacher W. Ph.D. Thesis, Universität Konstanz, Konstanz, Germany, 1993.
135. Costa C, Moura JJG, Moura I, Liu M-Y, Peck HD Jr, LeGall J, Wang Y, Huynh BH. J Biol Chem 1990; 265:14382–14388.
136. Pereira IAC, LeGall J, Xavier AV, Teixeira M. Biochim Biophys Acta 2000; 1481:119–130.
137. Naik RR, Murillo FM, Stolz JF. FEMS Microbiol Lett 1993; 106:53–58.
138. Jackson RH, Cornish-Bowden A, Cole JA. J Biochem 1981; 193:861–867.
139. Harborne NR, Griffiths L, Busby SJ, Cole JA. Mol Microbiol 1992; 6:2805–2813.
140. Cole JA, Brown CM. FEMS Microbiol Lett 1980; 7:65–72.
141. Cole JA. Biochim Biophys Acta 1968; 162:356–368.
142. Fujita T. J Biochem (Tokyo) 1966; 60:204–215.
143. Berks BC, Ferguson SJ, Moir JWB, Richardson DJ. Biochim Biophys Acta 1995; 1232:97–173.
144. Simon J, Gross R, Einsle O, Kroneck PMH, Kröger A, Klimmek O. Mol Microbiol 2000; 35:686–696.
145. Darwin A, Hussain H, Griffiths L, Grove J, Sambongi Y, Busby S, Cole JA. Mol Microbiol 1993; 9:1255–1265.
146. Hussain H, Grove J, Griffiths L, Busby S, Cole JA. Mol Microbiol 1994; 12:153–163.
147. Einsle O, Stach P, Messerschmidt A, Klimmek O, Simon J, Kröger A, Kroneck PMH. Acta Cryst 2002; D58:341–342.
148. Einsle O, Messerschmidt A, Huber R, Kroneck PMH, Neese F. J Am Chem Soc 2002; 124:11737–11745.

149. Rudolf M, Einsle O, Neese F, Kroneck PMH. Biochem Soc Trans 2002; 30:649–653.
150. Hooper AB, Vannelli T, Bergmann DJ, Arciero DM. Antonie van Leeuwenhoek 1996; 71:59–67.
151. Hooper AB, Logan M, Arciero DM, McTavish H. Biochim Biophys Acta 1991; 1058:13–16.
152. Whittaker M, Bergmann D, Arciero DM, Hooper AB. Biochim Biophys Acta 2000; 1459:346–355.
153. Ensign SA, Hyman MR, Arp DJ. J Bacteriol 1993; 175:1971–1980.
154. Frijlink MJ, Abee T, Laanbroek HJ, de Boer W, Konings WN. Arch Microbiol 1992; 157:194–199.
155. Arth I, Frenzel P, Conrad R. Soil Biol Biochem 1998; 30:509–515.
156. Igarashi N, Moriyama H, Fujiwara T, Fukumori Y, Tanaka N. Nature Struct Biol 1997; 4:276–284.
157. Lengler JW, Drews G, Schlegel HG. Biology of the Prokaryotes. Stuttgart, New York: Thieme Verlag, 1999.
158. Moir JWB, Wood NJ. Cell Mol Life Sci 2001; 58:215–224.
159. Stitt M, Müller C, Matt P, Gibon Y, Carillo P, Morcuende R, Scheibe W-R, Krapp A. J Exp Bot 2002; 53:959–970.
160. Zehnder AJB, Stumm W. In: Zehnder AJB, ed. Biology of Anaerobic Microorganisms. New York: John Wiley and Sons, 1988:1–38.
161. Jacob DJ. Introduction to Atmospheric Chemistry. Princeton, New Jersey: Princeton University Press, 1999.

5

The Biological Cycle of Sulfur

Oliver Klimmek

Institut für Mikrobiologie, Johann Wolfgang Goethe-Universität, Marie-Curie-Strausse 9, D-60439 Frankfurt am Main, Germany

1.	Sulfur in Biology	106
2.	Chemistry of Elemental Sulfur	107
	2.1. Polysulfide Sulfur as Soluble Sulfur Compound	107
3.	Polysulfide Sulfur as an Intermediate in Sulfur Respiration	108
4.	Polysulfide Sulfur Respiration of Bacteria	108
	4.1. Bacterial Sulfur Reducers	108
	4.2. Polysulfide Sulfur Respiration of *W. succinogenes*	109
	4.2.1. Electron Transport Chain of Polysulfide Sulfur Respiration	109
	4.2.2. Mechanism of Electron Transfer and Proton Translocation	119
5.	Sulfur Respiration of Archaea	123
	5.1. Archaeal Sulfur Reducers	123
	5.2. Sulfur Respiration of *Acidianus ambivalens*	124
	5.3. Sulfur Respiration of *Pyrodictium abyssi*	124
6.	Polysulfide Sulfur Transferases	124
	6.1. Polysulfide Sulfur Transferase of *W. succinogenes*	125
7.	Conclusions and Outlook	127
	Acknowledgment	127
	Abbreviations	127
	References	128

1. SULFUR IN BIOLOGY

It is well established that most of the known anaerobic prokaryotes perform "oxidative phosphorylation" without O_2. Depending on the species and the metabolic conditions, these bacteria may use a large variety of inorganic (e.g., nitrate, nitrite, sulfate, thiosulfate, elemental sulfur, polysulfide sulfur) or organic compounds (e.g., fumarate, dimethylsulfoxide, trimethylamine-*N*-oxide, vinyl- or arylchlorides) as terminal electron acceptor instead of oxygen. The redox reactions with these acceptors are catalyzed by membrane-integrated electron transport chains and are coupled to the generation of an electrochemical proton potential (Δp) across the membrane. Oxidative phosphorylation in the absence of O_2 is also termed "anaerobic respiration". Oxidative phosphorylation with elemental sulfur is called "sulfur respiration". Oxidative phosphorylation with polysulfide sulfur is called "polysulfide respiration".

Certain anaerobic *Archaea* (e.g., members of the genera *Acidianus, Thermoproteus, Pyrodictium*) and *Bacteria* (e.g., *Wolinella succinogenes*, members of the genera *Desulfuromonas* and *Sulfurospirillum*) use H_2 as the electron donor of sulfur respiration [Reaction (1)]. The organisms grow with H_2 and

$$H_2 + 1/8\,S_8 \xrightarrow{\hspace{1.5cm}} HS^- + H^+ \qquad (1)$$
$$1H_i^+ \quad 1H_o^+$$

$$H_2 + 1/2\,O_2 \xrightarrow{\hspace{1.5cm}} H_2O \qquad (2)$$
$$8H_i^+ \quad 8H_o^+$$

elemental sulfur as the sole substrates of energy metabolism [1,2]. Therefore it is likely that Reaction (1) is coupled to the consumption of protons from the inside (H_i^+) and to proton release on the outside (H_o^+) of the membrane, and that the Δp generated in this way drives ATP synthesis.

Aerobic bacteria (e.g., *Paracoccus denitrificans* and *Bacillus subtilis*) grow by oxidative phosphorylation driven by Reaction (2) where the sulfur in Reaction (1) is replaced by O_2 [3]. The two reactions differ by the amount of protons apparently translocated per electron transported from H_2 to the acceptor substrate (H^+/e ratio, n_{H+}/n_e). The H^+/e ratio is estimated to be 0.5 for Reaction (1) and 4 for Reaction (2). These numbers were calculated from the redox potentials of the substrates (Table 1 [4–6]) according to Eq. (3) with $\Delta p = 0.17$ V and assuming

$$n_{H+}/n_e = q \cdot \Delta E/\Delta p \qquad (3)$$

that 50% of the free energy of the redox reactions is conserved ($q = 0.5$). The relatively low H^+/e ratio of Reaction (1) is due to the much lower redox potential of the couple $1/8\,S_8/HS^-$ as compared to that of O_2/H_2O. If three protons have to be translocated for the synthesis of a molecule of ATP, the H^+/e ratio of

Table 1　Comparison of Several Redox Potentials

Reaction[a]	E_o' (mV)	Ref.
$H_2 \rightarrow 2H^+ + 2e^-$	-420	
$HCO_2^- + H_2O \rightarrow HCO_3^- + 2H^+ + 2e^-$	-413	[4]
$MQH_2 \rightarrow MQ + 2H^+ + 2e^-$	-90	[5]
$MMH_2 \rightarrow MM + 2H^+ + 2e^-$	-220	[O. Klimmek and F. MacMillan, unpublished results]
$HS^- \rightarrow 1/8\,S_8 + H^+ + 2e^-$	-275	[4]
$4HS^- \rightarrow S_4^{2-} + 4H^+ + 6e^-$	-260	[6]
Succinate $\rightarrow$ Fumarate $+ 2H^+ + 2e^-$	$+30$	[4]
$H_2O \rightarrow 1/2\,O_2 + 2H^+ + 2e^-$	$+815$	[4]

[a]See list of abbreviations.

Reaction (1) means that six electrons have to be transported from H_2 to sulfur for the synthesis of one molecule of ATP.

The earlier is a striking example of the advantage of oxidative phosphorylation as compared to substrate phosphorylation. In contrast to substrate phosphorylation, oxidative phosphorylation allows ATP synthesis to be driven by reactions which do not allow the synthesis of 1 mol ATP per mol of substrate. In oxidative phosphorylation, the amount of protons translocated can be adjusted to the free energy of the driving redox reaction. This probably explains the abundance of oxidative phosphorylation especially in anaerobic organisms, where the free energy change of the catabolic reactions is usually much lower than in aerobic organisms.

In this chapter, the polysulfide sulfur respiration of *W. succinogenes* and other microorganisms will be described and the coupling mechanism of apparent proton translocation to polysulfide sulfur reduction will be discussed. Sulfur respiration involving also sulfate and sulfite [1–3,7] as well as the anaerobic respiration [3,7–10] with other acceptors have been reviewed previously.

2.　CHEMISTRY OF ELEMENTAL SULFUR

2.1.　Polysulfide Sulfur as Soluble Sulfur Compound

As sulfur flower is nearly insoluble in water (5 μg L^{-1} at 25°C) [11], it is unlikely that elemental sulfur is the substrate of sulfur respiration for mesophilic bacteria and for thermophilic archaea. Sulfur flower consists of S_8 rings, the most stable form of elemental sulfur, and readily dissolves in aqueous solutions containing sulfide (HS^-) which is the product of sulfur respiration. Thus, polysulfide sulfur is formed abiotically according to Reaction (4). The amount of sulfur dissolved increases with the concentration of HS^-, with pH and with the temperature of the solution [12].

$$n/8\,S_8 + HS^- \rightleftharpoons S_{n+1}^{2-} + H^+ \tag{4}$$

In a solution containing 10 mM HS^-, the equilibrium concentration of polysulfide sulfur is 0.1 mM at pH 6.7 and 30°C. At 90°C 0.1 mM sulfur was dissolved already at pH 5.5.

Tetrasulfide (S_4^{2-}) and pentasulfide (S_5^{2-}) are the predominant species of polysulfide sulfur at pH values >6 [13,14]. The pK_a values the of acid dissociations of HS_4^- and HS_5^- were measured to be 6.3 and 5.7, respectively. The two species dismutate rapidly according to Reaction (5). The equilibrium constant of Reaction (5) was determined to be 4×10^{-9} M at 20°C [14]. From

$$3S_5^{2-} + HS^- \rightleftharpoons 4S_4^{2-} + H^+ \tag{5}$$

this equilibrium constant it can be calculated that the concentration of S_4^{2-} is twice that of S_5^{2-} in a solution (pH 8.5 and 37°C) containing 1 mM polysulfide sulfur and 2 mM HS^- [15]. The redox potential of polysulfide sulfur can be calculated from that of $1/8\,S_8/HS^-$ (Table 1) and the equilibrium constant of Reaction (4) (3.6×10^{-9} M at 37°C) with the simplification that only S_4^{2-} is formed upon dissolution of sulfur. The redox potential of polysulfide sulfur so obtained is only slightly more positive than that of $1/8\,S_8/HS^-$ (Table 1) [6].

3. POLYSULFIDE SULFUR AS AN INTERMEDIATE IN SULFUR RESPIRATION

Polysulfide sulfur may be the intermediate of sulfur reduction to organisms growing under condition which allow the dissolution of at least 0.1 mM sulfur [1]. Comparison reveals that all eubacterial strains and most of the archaeal strains grow best under conditions where polysulfide sulfur is stable up to at least 0.1 mM dissolved polysulfide sulfur. Therefore, it is likely that these organisms use polysulfide sulfur as the actual substrate of sulfur respiration. Polysulfide sulfur is not available to the extremely acidophilic archaea growing at pH 3 or below (e.g., members of the genera *Acidianus*, *Stygioglobus* and *Thermoplasma*) [2]. Under the growth conditions of these sulfur reducers elemental sulfur has to be mobilized in a different way which is not yet known.

4. POLYSULFIDE SULFUR RESPIRATION OF BACTERIA

4.1. Bacterial Sulfur Reducers

The ability to reduce polysulfide sulfur using H_2 or organic substrates as electron donors is widespread among bacteria and archaea (a list of archaeal and bacterial genera which are able to reduce elemental sulfur is reported in a review [2]). Mesophilic and thermophilic polysulfide sulfur reducers, mostly from the bacterial domain [16], have been isolated from environments such as anoxic marine or brackish sediments, fresh water sediments, hot water pools from solfataric fields, volcanic hot springs and in the case of *W. succinogenes* from bovine rumen [17]. Polysulfide sulfur reducing bacteria, e.g., members of the genera *Ammonifex, Desulfurobacterium, Desulferomonas, Geobacter, Pelobacter, Sulfospirillum*, and *Wolinella*, are able to gain ATP by lithotrophic polysulfide sulfur respiration.

4.2. Polysulfide Sulfur Respiration of *W. succinogenes*

The mechanism of polysulfide sulfur respiration in *W. succinogenes* has been investigated most thoroughly. *W. succinogenes* belongs to the ε-group of proteobacteria and has been isolated from bovine rumen [17,18]. It does not ferment sugars and grows solely by anaerobic respiration with sulfur, polysulfide sulfur, fumarate, dimethylsulfoxide, nitrate, nitrite, or N_2O as electron acceptor. H_2 or formate serve as electron donors of anaerobic respiration. Polysulfide sulfur was identified to be the intermediate of sulfur reduction. Disulfides ($R-S-S-R$), thiosulfate ($S_2O_3^{2-}$) or tetrathionate ($S_4O_6^{2-}$) do not serve as electron acceptors.

Wolinella succinogenes grows by polysulfide sulfur respiration with either H_2 [Reaction (6)] or formate [Reaction (7)] as electron donor [2,19]. Acetate and

$$H_2 + S_n^{2-} \longrightarrow HS^- + S_{n-1}^{2-} + H^+ \tag{6}$$

$$HCO_2^- + S_n^{2-} + H_2O \longrightarrow HCO_3^- + HS^- + S_{n-1}^{2-} + H^+ \tag{7}$$

glutamate serve as sources of carbon. The cell density in the culture increased proportional to the amounts of polysulfide sulfur reduced, and equimolar amounts of formate and polysulfide sulfur were consumed. The specific activity of polysulfide sulfur reduction by formate or H_2 of cells of *W. succinogenes* grown at the expense of Reactions (6) or (7) was 20% higher than the activity calculated from the specific growth rate and the cell yield of the growing bacteria [1,15]. This confirms the view that Reactions (6) and (7) are responsible for growth.

4.2.1. Electron Transport Chain of Polysulfide Sulfur Respiration

Cells of *W. succinogenes* catalyzing Reactions (6) or (7) were found to take up tetraphenylphosphonium (TPP^+) form the external medium [20,21]. TPP^+ uptake was prevented by the presence of a protonophore. From the amount of TPP^+ taken up, the $\Delta\Psi$ (0.17 V) was calculated to be approximately the same as that generated by fumarate respiration with formate [Reaction (8)] (Table 2).

$$HCO_2^- + fumarate + H_2O \longrightarrow HCO_3^- + succinate \tag{8}$$

The values of $\Delta\Psi$ and Δp were nearly the same, since the ΔpH across the membrane was negligible. The H^+/e^- and the ATP/e^- ratios were determined for fumarate respiration but not for polysulfide sulfur respiration.

Table 2   Bioenergetic Data of the Polysulfide Sulfur Respiration with Formate of
W. succinogenes [2]. The Data are Compared to Those of Fumarate Respiration

Electron acceptor	pH	Y (g cells/ mol formate)	$-\Delta E$ (V)	Δp (V)	H^+/e	ATP/e	$(\Delta E \cdot F)/(ATP/e)$ (kJ mol^{-1} ATP)
Polysulfide sulfur	8.4	3.2	0.20	0.17	(1/2)	(1/6)	116
Fumarate	7.9	7.0	0.44	0.18	1	1/3	127

Note: The values of pH, Y (growth yield) and ΔE refer to the middle of the exponential growth phase at 37°C. ΔE was calculated from the E'_o given in Table 1 with the given values of pH and equal concentrations of HCO_2^- and HCO_3^-, polysulfide sulfur and HS^-, and fumarate and succinate. The numbers in parentheses were estimated as described in the text.

However, the values for polysulfide respiration can be estimated from the growth yields (Y) of *W. succinogenes* growing at the expense of Reactions (7) and (8). The growth yield of polysulfide sulfur respiration was measured to be approximately half that of fumarate respiration, suggesting that the H^+/e^- and ATP/e^- ratio of polysulfide sulfur respiration was also half that of fumarate respiration [1]. This view is confirmed by the redox potential differences (ΔE) between formate and each of the two electron acceptors under the growth conditions of the bacteria. The ΔE of polysulfide sulfur respiration is approximately half that of fumarate respiration. The free energy required for ATP synthesis calculated from ΔE and the ATP/e^- ratios are 116 and 127 kJ mol^{-1} in anaerobic respiration with polysulfide sulfur and fumarate, respectively. Both values are consistent with the general observation that phosphorylation requires about 100 kJ mol^{-1} ATP in growing bacteria in most instances.

The membrane fraction isolated from *W. succinogenes* cells grown with formate and either polysulfide sulfur or fumarate catalyzes polysulfide sulfur reduction by H_2 or formate [1,22]. The corresponding electron transport chain consists of polysulfide reductase, 8-methyl-menaquinone$_6$ (Fig. 1), and either hydrogenase or formate dehydrogenase (Fig. 2) [2]. This was shown by reconstituting a functional electron transport chain form the isolated enzymes (Table 3). Proteoliposomes containing polysulfide reductase, 8-methyl-menaquinone and either hydrogenase or formate dehydrogenase catalyzed polysulfide sulfur reduction by H_2 [Reaction (6)] or formate [Reaction (7)]. The turnover number of polysulfide reductase in electron transport was commensurate with that measured in the membrane fraction of *W. succinogenes*. When prepared according to the method designed by Rigaud et al. [24], proteoliposomes containing polysulfide reductase, 8-methyl-menaquinone, and formate dehydrogenase catalyzed polysulfide sulfur reduction by formate which was coupled to the uptake of TPP^+ from the external medium (Fig. 3). TPP^+ uptake was prevented by a protonophore. The $\Delta \Psi$ (0.14 V) calculated from the

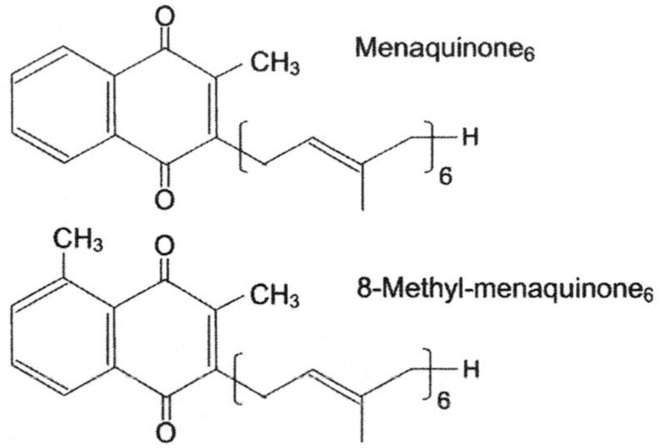

Figure 1 Structures of the quinones of *W. succinogenes* [23]. The second methyl group of methyl-menaquinone is bound in position 8 [O. Klimmek and F. MacMillan, unpublished results].

amount of TPP$^+$ taken up in the steady state of electron transport was not much lower than that generated across the membrane of cells by the same reaction (Table 3). These experiments demonstrate that the isolated enzymes contain all the components required for electron transport and for Δp generation by polysulfide sulfur respiration.

Each of the three enzymes involved in polysulfide sulfur respiration consists of two hydrophilic and a hydrophobic subunit (Fig. 2). The latter subunits are integrated in the membrane. The hydrophobic subunit of polysulfide reductase probably carries 8-methyl-menaquinone (MM$_b$). The hydrophobic subunits of hydrogenase and formate dehydrogenase are related di-heme cytochromes *b* [25]. The larger hydrophilic catalytic subunits of the enzymes carry the substrate sites and are oriented towards the periplasmic side of the membrane. The smaller hydrophilic subunits carry three (HydA) or four iron–sulfur centers (PsrB and FdhB) which probably mediate the electron transfer from the catalytic to the membrane-integrated subunit or vice versa (PsrB). The genes corresponding to the subunits of each enzyme are located in operons, the sequences of which were determined [26,27]. Deletion mutants were constructed [27] which lack one or more genes of each operon. The mutants can be complemented by genomic integration of the respective genes. This allows the construction of variants of the enzymes by site-directed mutagenesis.

4.2.1.1.   Polysulfide reductase: The isolated polysulfide reductase consists of the three subunits predicted by the *psrABC* operon, and contains a molybdenum ion coordinated by two molecules of molybdopterin guanine dinucleotide (MGD) [26,28] (Fig. 2). These cofactors are likely to be bound to the catalytic subunit PsrA together with a [4Fe–4S] iron–sulfur center which is predicted by the sequence of PsrA. Four additional [4Fe–4S] iron–sulfur centers are predicted by the sequence of PsrB. The isolated enzyme contains $\sim$20 mol of free iron and

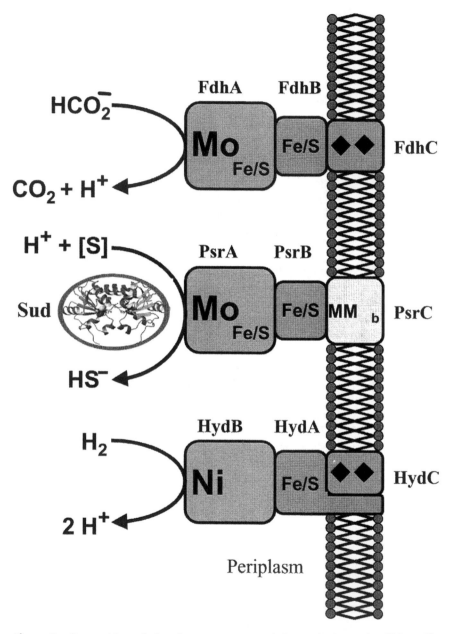

Figure 2 Composition of the electron transport chain catalyzing polysulfide sulfur respiration with H_2 or formate in *W. succinogenes*. FdhA/B/C, formate dehydrogenase; PsrA/B/C, polysulfide reductase; HydA/B/C, hydrogenase; Ni, nickel-iron center; Fe/S, iron–sulfur centers; Mo, molybdenum ion bound to MGD; the dark squares designate heme *b* groups; Sud, polysulfide sulfur transferase.

Table 3 Activities of Polysulfide Sulfur Respiration in Proteoliposomes Containing Different Naphthoquinones [6]

Quinone	$H_2 \to [S]$	$HCO_2^- \to [S]$ (s^{-1})	$HCO_2^- \to$ Fumarate
Without added	25	5	17
8-Methyl-menaquinone$_6$	370	175	35
Menaquinone$_6$	27	5	1490
Menaquinone$_4$	34	7	1455
Vitamin K_1	25	5	1180

Note: Polysulfide sulfur respiration with H_2 ($H_2 \to [S]$) or with formate ($HCO_2^- \to [S]$) was measured in proteoliposomes containing polysulfide reductase and either hydrogenase or formate dehydrogenase isolated from *W. succinogenes*. Fumarate respiration with formate ($HCO_2^- \to$ fumarate) was measured in proteoliposomes containing fumarate reductase and formate dehydrogenase. The activities are given as substrate turnovers of polysulfide reductase or fumarate reductase at 37°C.

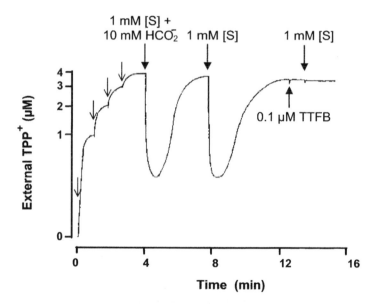

Figure 3 Recording of the external TPP^+ concentration in a suspension of proteoliposomes catalyzing polysulfide sulfur reduction by formate. Proteoliposomes (3.0 g phospholipide L^{-1}) containing formate dehydrogenase (62 nmol g^{-1} phospholipid), polysulfide reductase (30 nmol g^{-1} phospholipid) and 8-methyl-menaquinone (10 μmol g^{-1} phospholipid) were suspended in an anoxic Tris chloride buffer (pH 8.0, 37°C). Calibration of the TPP^+ electrode was done by four additions of 1 μM TPP^+ (thin arrows). Electron transport was started by the addition of polysulfide sulfur ([S]) and formate.

sulfide, in agreement with the presence of five [4Fe–4S] iron–sulfur centers [2]. The enzyme does not contain heme, flavin, or heavy metal ions in addition to molybdenum. About 1 mol of a mixture of methyl-menaquinone and menaquinone was consistently found to be bound to the isolated enzyme after correction for the amount of quinone which is associated with the phospholipid present in the preparation. Because of its lipophilic nature, the quinone is likely to be bound to the membrane anchor PsrC of the enzyme.

The isolated enzyme catalyzes the reduction of polysulfide by BH_4^- [Reaction (9)] and the oxidation of sulfide by dimethylnaphthoquinone (DMN) [Reaction (10)] at commensurate turnover numbers (1.7×10^3 s^{-1} and

$$S_n^{2-} + BH_4^- \longrightarrow S_{n-1}^{2-} + HS^- + BH_3 \tag{9}$$

$$(n + 1)HS^- + nDMN + (n - 1)H^+ \longrightarrow S_{n+1}^{2-} + nDMNH_2 \tag{10}$$

1.1×10^3 s^{-1}) [2]. The apparent K_M for polysulfide sulfur [Reaction (9)] is 50 μM and that for sulfide [Reaction (10)] is 25 mM. PsrC is not required for either catalytic activity. A mutant lacking *psrC* still catalyzed both reactions. In the absence of the membrane anchor PsrC the enzyme was located in the periplasmic cell fraction [6,27]. Furthermore, 19 mutants with a residue of PsrC replaced had wild-type activities with respect to Reactions (9) and (10), and PsrA was bound to the membrane [6]. However, six of these mutants did not catalyze electron transport from H_2 or formate to polysulfide sulfur, suggesting that the oxidation of reduced 8-methyl-menaquinone by polysulfide sulfur was impaired in these mutants (Fig. 7). Therefore, the site of DMN reduction [Reaction (10)] appears to be different from that of 8-methyl-menaquinone interaction. Each of the four iron–sulfur centers of PsrB appears to be required for the activities of Reactions (9) and (10) as well as for electron transport [Reactions (6) and (7)]. This is suggested by mutants, in which each cysteine residue of the four cysteine clusters coordinating the four iron–sulfur centers was replaced by alanine (O. Klimmek and W. Dietrich, unpublished results). PsrA of the mutants was bound to the membrane, suggesting that the structure of the enzyme was not drastically impaired. However, the mutants did not catalyze Reactions (9) and (10) or electron transport from H_2 or formate to polysulfide sulfur, suggesting that each of the four iron–sulfur centers of PsrB was required for enzymic activity.

As seen from its amino acid sequence, the catalytic subunit PsrA of polysulfide reductase belongs to the DMSO reductase family of molybdooxidoreductases [26,29]. The structure of several single-subunit enzymes of this family is known. As a rule, the molybdenum ion coordinated by two MGD molecules appears to be the electron donor or acceptor to the respective substrate in these enzymes. A cavity extending from the surface of the protein to the molybdenum close to its center is seen in all structures. The substrates probably reach the active site near the molybdenum through this cavity, and the products are released at the surface through the cavity. With the assumption that PsrA resembles these enzymes, it is likely that the substrates, polysulfide sulfur and protons, reach the molybdenum through the cavity. Here a sulfur atom is reductively cleaved from the end of the polysulfide sulfur chain of polysulfide according to Reaction (11) (Fig. 4), and the products are released through the

$$S_n^{2-} + 2e^- + H^+ \longrightarrow S_{n-1}^{2-} + HS^- \tag{11}$$

cavity. Since PsrA is oriented towards the periplasmic side of the membrane, the substrates are taken up from the periplasm and the products are released to the same side. The electrons [Reaction (11)] are thought to be provided by the molybdenum ion in the reduced state (+4). In the oxidized state, molybdenum (+6) ion is thought to be reduced via the iron–sulfur centers of PsrA and PsrB by reduced 8-methyl-menaquinone which is bound to PsrC.

4.2.1.2. Formate dehydrogenase: The two operons of the formate dehydrogenase in the genome of *W. succinogenes* differ in their promoter regions, but are nearly identical in their gene sequences [30–32]. Deletion mutants lacking one of the operons still grow by anaerobic respiration with polysulfide sulfur or fumarate using formate as electron donor. The formate

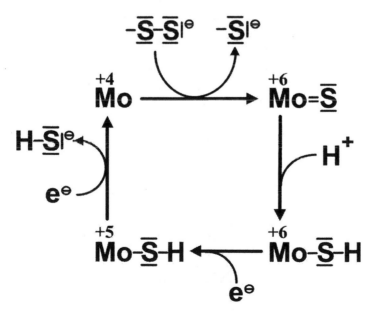

Figure 4 Hypothetical mechanism of polysulfide sulfur reduction at the substrate site of polysulfide reductase. A sulfur atom is cleaved from the end of the polysulfide sulfur chain $(-\bar{S} - \bar{S}|^{\ominus})$ and bound to the molybdenum ion (Mo) which is thereby oxidized. After the uptake of a proton and two electrons, HS^- is released and molybdenum ion is in the reduced state again.

dehydrogenase serving in polysulfide sulfur respiration is also involved in fumarate respiration where it catalyzes the reduction by formate of menaquinone which is dissolved in the membrane [9,25]. Like *E. coli* formate dehydrogenase-N, the enzyme consists of three different subunits, whose sequences are similar to those of the *E. coli* enzyme. The structure of *E. coli* formate dehydrogenase-N was determined recently, and that of the *W. succinogenes* enzyme is thought to be similar [33]. The two enzymes are anchored in the membrane by related di-heme cytochromes *b* which carry the site of quinone reduction. The two heme groups of the *E. coli* enzyme are nearly on top of each other when viewed along the membrane normal. The iron–sulfur subunit is situated between the cytochrome *b* and the catalytic subunit which carries a molybdenum ion coordinated by two MGD molecules and an [4Fe–4S] iron–sulfur center (Fig. 2). The prosthetic groups of the enzyme form a pathway for rapid electron transfer from the molybdenum ion to the distal heme group which is thought to be close to the site of quinone reduction (Fig. 5). Similar as in other molybdenum enzymes, the substrate formate reaches the active site near the molybdenum

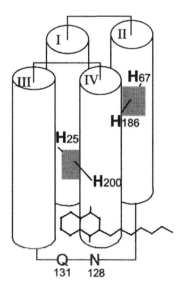

Cytoplasm

Figure 5 Hypothetical arrangement of the four predicted membrane-spanning helices of *W. succinogenes* HydC [25,40]. The scheme is based on the crystal structure of *E. coli* formate dehydrogenase-N [33]. The shaded squares represent the heme *b* groups. A molecule of HQNO bound to the site of quinone reduction which is confined by the axial ligand H200 of the distal heme group and by N128 and Q131 in the stretch connecting the hydrophobic parts of helices II and III.

through a cavity extending from the molybdenum to the surface of the subunit. After oxidation of formate according to Reaction (12), the products (CO_2 and H^+)

$$HCO_2^- \longrightarrow CO_2 + H^+ + 2e^- \tag{12}$$

leave the substrate site through the cavity. Since the catalytic subunit FdhA of formate dehydrogenase is exposed to the periplasmic side of the membrane, formate is taken up from the periplasm and the products are released to the same side of the membrane [34]. The electrons [Reaction (12)] are thought to be passed to the cytochrome b subunit via the iron–sulfur centers of FdhA and FdhB.

4.2.1.3. Hydrogenase: Hydrogenase consists of the three subunits encoded by the *hydABC* operon [35]. The enzyme is identical with hydrogenase involved in fumarate respiration with H_2, where it catalyzes the reduction of menaquinone by H_2 [9,25]. The two hydrophilic subunits HydA and B of the enzyme (Fig. 2) are similar to those making up the periplasmic Ni-hydrogenases of two *Desulfovibrio* species, the structures of which are known [36,37]. At the active site of these enzymes, H_2 is split to yield protons and electrons. The protons are released via a proton pathway extending from the active site to the surface of the larger catalytic subunit. The electrons are guided by three consecutive iron–sulfur centers to the binding site of the electron acceptor, a cytochrome c, on the surface of the smaller hydrophilic subunit. Since all the relevant residues are also conserved in HydA and B of *W. succinogenes*, it is likely that the catalytic mechanism also applies here. In the *W. succinogenes* enzyme the membrane-integrated di-heme cytochrome b subunit HydC serves as the acceptor of electrons from HydA to which HydC is tightly bound (Fig. 2). The midpoint potentials of the two heme groups of HydC were determined as -240 and -100 mV relative to the standard hydrogen electrode [38]. In contrast to the periplasmic hydrogenases, HydA of *W. succinogenes* carries a C-terminal hydrophobic helix which is integrated in the membrane [39]. The protons generated by H_2 oxidation are released into the periplasm, since the catalytic subunit HydB is oriented to the periplasmic side of the membrane.

The sequences of the di-heme cytochrome b subunits of hydrogenase (HydC) and of formate dehydrogenase (FdhC) of *W. succinogenes* are similar to that of formate dehydrogenase-N of *E. coli* [33]. Especially the axial ligands of the two heme groups in subunit FdnI of the *E. coli* enzyme are conserved also in HydC and FdhC [25,38,40]. Therefore, the structures of the di-heme cytochrome b subunits of the *W. succinogenes* enzymes are thought to be similar to the known structure of *E. coli* FdnI. In Fig. 5, *W. succinogenes* HydC is drawn according to the structure of *E. coli* FdnI. The proximal heme group of FdnI is in electron transfer distance to an iron–sulfur center of FdnH and to the distal heme group. The proximal and the distal heme groups represent the upper and the lower shaded squares in Fig. 5. The site of quinone reduction on FdnI seems to be occupied by a molecule of HQNO which is located close to the distal heme group. HQNO is in close proximity to the axial heme ligand on helix IV and to three residues (N110, G112, and Q113 of FdnI) within the hydrophilic stretch connecting helices II and III. Another two residues (M172 and A173 of FdnI) at the cytoplasmic end of helix IV make contact to HQNO.

HQNO is known to inhibit menaquinone reduction by formate or H_2 in *W. succinogenes* [41–43]. The midpoint potential of at least one heme *b* group in the membrane fraction of *W. succinogenes* was shifted from -190 to -230 mM upon the addition of HQNO [44]. Therefore, HQNO probably interacts at the site of quinone reduction of hydrogenase and of formate dehydrogenase of *W. succinogenes*.

To find out whether the site of quinone reduction on HydC is situated similarly as in *E. coli* FdnI, mutants were constructed in which each of the three residues (H122, N128, and Q131) of *W. succinogenes* HydC which are predicted to be located in the loop connecting helix II and III were replaced [40]. Residues N128 and Q131 of HydC correspond to N110 and Q113 of FdnI which are in contact with HQNO, and are conserved also in *W. succinogenes* FdhC. Mutants N128D and Q131L did not grow by anaerobic respiration with fumarate or polysulfide sulfur and H_2 as electron donor. The membrane fraction of mutants N128D and Q131L grown with formate and fumarate catalyzed benzyl viologen reduction by H_2 with wild-type specific activities, whereas the activities of DMN or polysulfide reduction by H_2 were drastically inhibited [40]. Replacement of H122, which is not conserved, did not affect the enzymic activities. The heme groups of HydC were present in the mutants, and their reduction by H_2 was not impaired by the mutations. This is seen from the amount of heme *b* reduced by H_2 after the addition of Triton X-100 and fumarate to the membrane fraction of the mutants.

M203 and A204 at the cytoplasmic end of helix IV of HydC are conserved in FdhC and in *E. coli* FdnI where they make contact with HQNO [33]. The corresponding *W. succinogenes* mutants M203I and A204F were found to have wild-type properties. In contrast, mutant Y202F showed negligible activities of DMN or polysulfide reduction together with wild-type activity of benzyl viologen reduction by H_2. However, heme *b* was not reduced by H_2 in this mutant, although the HydC protein was detected by ELISA. This suggests that the electron transfer from HydB to HydC was interrupted in mutant Y202F. This view is supported by the structure of *E. coli* formate dehydrogenase-N. The tyrosine residue corresponding to Y202 is seen to be bound to a conserved histidine residue in the C-terminal membrane helix of the iron–sulfur subunit. A mutant (H305M) with the corresponding histidine residue of *W. succinogenes* HydA replaced was found to have the same properties as Y202F [38]. This result indicates that Y202 of HydC and H305 of HydA serve the same function as the corresponding residues in *E. coli* formate dehydrogenase-N. Furthermore, tight binding of Y202 to H305 appears to be required for electron transfer from HydA to HydC which in turn is a prerequisite for quinone and for polysulfide sulfur reduction.

The results suggest that the site of quinone reduction on HydC is located close to the cytoplasmic membrane surface as on *E. coli* FdhI. Furthermore, HydC and its site of quinone reduction appear to be involved in the electron transfer from hydrogenase to polysulfide reductase. The function of FdhC of *W. succinogenes* formate dehydrogenase in quinone reduction and in electron transfer to polysulfide reductase is thought to be equivalent to that of HydC.

4.2.2. Mechanism of Electron Transfer and Proton Translocation

Proteoliposomes containing polysulfide reductase and either hydrogenase or formate dehydrogenase isolated from *W. succinogenes* do not catalyze polysulfide sulfur respiration unless 8-methyl-menaquinone is present (Table 3). Menaquinone with a side chain consisting of six or four isoprene units, or vitamin K_1 served in reconstituting fumarate respiration, but did not replace 8-methyl-menaquinone in polysulfide sulfur respiration. The low activities of polysulfide sulfur respiration observed without added 8-methyl-menaquinone were probably due to the small amounts of this quinone associated with the enzyme preparations used. Maximum activity of polysulfide sulfur respiration required ~ 10 μmol 8-methyl-menaquinone per gram phospholipid [O. Klimmek and W. Dietrich, unpublished results].

Most of the 8-methyl-menaquinone of *W. succinogenes* is probably dissolved in the lipid phase of the bacterial membrane. The redox potential of menaquinone dissolved in the membrane was estimated to be 170 mV more positive than that of polysulfide sulfur (Table 1) [6]. Therefore, it is unlikely that the electron transfer from the dehydrogenases to polysulfide reductase is mediated by diffusion of menaquinone in the membrane. The 8-methyl-menaquinone involved in polysulfide sulfur respiration is thought to be bound to PsrC. In the reduced state, the bound quinone (MM_b) is thought to form the hydroquinone anion (MM_bH^-). This assumption is consistent with the finding that an arginine residue (R305) in one of the membrane helices of PsrC is absolutely required for polysulfide sulfur respiration (see Fig. 7 below). Mutants with the arginine residue replaced by phenylalanine or lysine had negligible activities of polysulfide sulfur respiration with H_2 or formate, whereas the activities of polysulfide reductase [Reactions (9) and (10)] were close to those of the wild-type strain, and PsrA was bound to the membrane of the mutants [6]. The hydroquinone anion MM_bH^- is probably bound and stabilized by the positive charge of arginine. The redox potential of the MM_b/MM_bH^- couple is expected to be close to that of polysulfide or more negative. In ethanol the redox potential of 8-methyl-menaquinone was found to be -220 mV (Table 1) [O. Klimmek and F. MacMillan, unpublished results].

Since electron transfer from the dehydrogenases to polysulfide reductase by diffusion of 8-methyl-menaquinone is unlikely, two alternative mechanism have to be considered. The dehydrogenases may either form a stable electron transport complex with polysulfide reductase, or the electron transfer is achieved by diffusion and collision of the enzymes within the membrane. The latter mechanism is supported by experimental evidence (Fig. 6). When equimolar amounts of a dehydrogenase and polysulfide reductase are incorporated into liposomes containing 8-methyl-menaquinone, the activity of electron transport increases proportional to the protein/phospholipid ratio. The same linear relation was observed, when the membrane fraction of *W. succinogenes* was fused with increasing amounts of liposomes containing 8-methyl-menaquinone [2]. The linear relation argues against the existence of stable electron transport complexes within the membrane, and cannot be explained by complex dissociation upon dilution with phospholipid. Complex dissociation would result in a hyperbolic rather than a linear relation of electron transport activity to the protein/phospholipid ratio according to Ostwalds law of dilution. The linear relation suggests that the activity of electron transport is limited by the diffusion of the enzymes within the membrane. The diffusion coefficient (D) evaluated from Fig. 6 according to Eq. (13) is $\sim 10^{-8}$ cm^2 s^{-1}, in

$$D = d^2/t \tag{13}$$

agreement with direct measurements of the diffusion coefficients of membrane proteins of similar sizes [45]. The average membrane surface occupied by an enzyme molecule (d^2) was calculated from the protein/phospholipid ratio, and

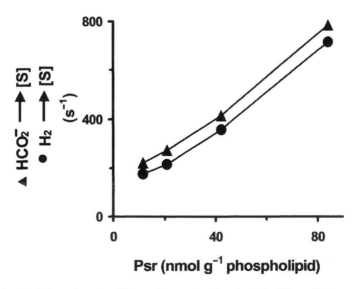

Figure 6 Activity of polysulfide sulfur reduction by H$_2$ (H$_2$ → [S]) or formate (HCO$_2^-$ → [S]) in proteoliposomes at 37°C. The proteoliposomes contained 8-methyl-menaquinone (10 µmol g^{-1} phospholipid) and approximately the same molar amount of hydrogenase or formate dehydrogenase as polysulfide reductase. The activities were determined as described [6].

the turnover numbers of the enzyme in electron transport was used as $1/t$. Thus the experiment of Fig. 6 suggests that the electron transfer from the dehydrogenases to polysulfide reductase is achieved by diffusion and collision of the enzyme molecules within the membrane. This view is supported by the finding that the activity of fumarate respiration is not decreased by enzyme dilution with phospholipid. In this case, the electron transfer from the dehydrogenases to fumarate reductase is mediated by diffusion of menaquinone. The diffusion of menaquinone is two orders of magnitude faster than that of the enzymes and, therefore, does not limit electron transport.

PsrC is predicted to form eight membrane helices. The 19 residues of PsrC which were replaced by others are indicated in Fig. 7. Active polysulfide reductase was bound to the membrane in all mutants, but six of them had negligible electron transport activity with H_2 or formate and polysulfide sulfur. The corresponding residues (in bold type in Fig. 7) are either charged at neutral pH or are tyrosines, the phenolic hydroxyl groups of which appear to be essential for polysulfide sulfur respiration. Y23, Y159, and R305 may serve in binding MM_b and MM_bH^- to PsrC (Fig. 8). D218 and E225 are well suited for release of the proton formed by MM_bH^- oxidation into the periplasmic aqueous phase.

The activity of polysulfide sulfur respiration in cells of *W. succinogenes* or in proteoliposomes is decreased to ~30% upon the addition of a protonophore [6]. This effect can be explained if it is assumed that the activity is limited by the amount of MM_bH^- in the absence of a Δp, and that MM_bH^- dissociates from

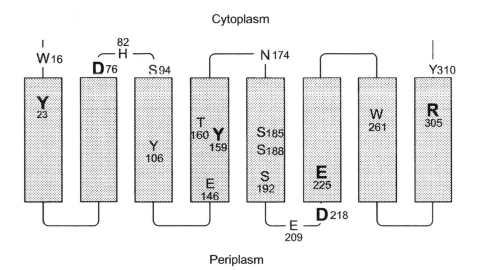

Figure 7 Hypothetical topology of PsrC [6]. Residues replaced in PsrC by site-directed mutagenesis are indicated. Residues in bold letters correspond to mutants with 5% or less of the wild-type specific activities of polysulfide sulfur respiration with H_2 or formate. Reproduced from Ref. [6] with the permission of *Eur. J. Biochem.*

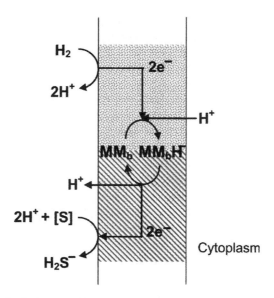

Figure 8   Hypothetical mechanism of polysulfide sulfur respiration with H_2 [6]. The reduction of 8-methyl-menaquinone bound to PsrC (MM_b) is coupled to the uptake of a proton from the cytoplasmic side of the membrane. The oxidation of the hydroquinone anion of MM_b (MM_bH^-) by polysulfide is coupled to the release of a proton to the periplasmic side. The dotted and the striped areas designate HydC and PsrC, respectively. Reproduced from Ref. [6] with the permission of *Eur. J. Biochem.*

PsrC (C) according to Reaction (14), where H_i^+ designates a proton taken up from

$$MM_bH^- + H_i^+ \rightleftharpoons MMH_2 + C \tag{14}$$

the cytoplasmic side of the membrane and MMH_2 reduced methyl-menaquinone dissolved in the lipid phase of the membrane. In the presence of a Δp (negative inside), MM_bH^- formation from MMH_2 is favored according to Reaction (14). It is feasible that D76 which is located near the cytoplasmic end of a PsrC membrane helix (Fig. 7) is involved in the uptake and release of H_i^+. Replacement of D76 by asparagine or leucine resulted in nearly inactive mutants, possibly because the formation of MM_bH^- from MMH_2 is prevented. The view that the electron transport is limited by MM_bH^- in the absence of a Δp is supported by the finding that the activity of polysulfide sulfur reduction by H_2 is stimulated twofold by the incorporation of additional 8-methyl-menaquinone into the membrane fraction of *W. succinogenes* [6]. Considering its redox potential (Table 1), the 8-methyl-menaquinone dissolved in the membrane should be fully reduced in the steady state of polysulfide sulfur respiration. Therefore, the amount of MM_bH^- is predicted to increase with the quinone content of the membrane according to Reaction (14).

The coupling mechanism during the electron transfer from HydC of hydrogenase to 8-methyl-menaquinone bound to PsrC (MM_b) at the moment of collision of the two enzymes is illustrated in Fig. 8. The Δp is drawn to be generated by proton translocation across the membrane which is coupled to the redox reactions of MM_b. The same mechanism is thought to apply with formate dehydrogenase replacing hydrogenase. The proton consumed in MM_b reduction is taken up from the cytoplasmic side of the membrane, and the proton liberated by MM_bH^- oxidation is released on the periplasmic side. Proton uptake is envisaged to be accomplished by HydC and proton release by PsrC. D218 and E225 of PsrC (Fig. 7) possibly serve in the release of the proton formed by MM_bH^- oxidation to the periplasmic side. Mutants D218N and E225Q did not catalyze polysulfide sulfur respiration with H_2 or formate, whereas mutant E225D showed 25% of the wild-type specific activity. The H^+/e^- ratio of polysulfide sulfur respiration with H_2 (or formate) predicted by the mechanism (0.5) is consistent with the value given in Table 2.

According to this mechanism, the Δp is exclusively generated by MM_b reduction with H_2 (Fig. 8). In this process, two protons are released on the periplasmic side from H_2 oxidation, and one proton disappears from the cytoplasmic side for MM_b reduction. Consistently, the site of quinone reduction on HydC is probably located close to the cytoplasmic membrane surface (Fig. 5). Furthermore, menaquinone reduction by H_2 catalyzed by hydrogenase in the membrane of *W. succinogenes* or in proteoliposomes was found to generate a Δp [9,25].

MM_bH^- oxidation by polysulfide sulfur dissipates part of the Δp generated by MM_b reduction with H_2 according to the mechanism (Fig. 8), since one proton per two electrons disappears from the periplasmic (positive) side in this process. Hence, MM_bH^- oxidation by polysulfide sulfur is predicted to be driven by a Δp across the membrane. The view that Δp generation is coupled solely to MM_b reduction is consistent with the energetics of the two reactions [6]. Only MM_b reduction is considered to be sufficiently exergonic to drive the translocation of a proton against a Δp of 0.17 V. In contrast, the oxidation of MM_bH^- might be slightly endergonic under standard conditions.

5. SULFUR RESPIRATION OF ARCHAEA

5.1. Archaeal Sulfur Reducers

Most of the archaeal polysulfide sulfur reducers are hyperthermophilic and grow optimally above 80°C [2]. They have been isolated from environments such as water containing volcanic areas, e.g., terrestrial solfataric fields and hot springs, or from shallow and abyssal submarine hydrothermal systems. Polysulfide sulfur reducing archaea, e.g., members of the genera *Acidianus*, *Pyrobaculum*, *Pyrodictium*, *Stygiolobus*, and *Thermoproteus*, are able to gain ATP by lithotrophic polysulfide sulfur respiration.

5.2. Sulfur Respiration of *Acidianus ambivalens*

The hyperthermophilic and acidophilic crenarchaeote *Acidianus ambivalens* grows at 80°C and pH 2.5 with H_2 and elemental sulfur. Under anaerobic conditions the reduction of elemental sulfur with H_2 as electron donor is the only energy-yielding reaction [46,47]. The electron transport chain consists of a sulfur reductase and a hydrogenase which form a multienzyme complex within the membrane [48]. The sulfur reductase of this complex is similar to polysulfide reductase of *W. succinogenes*. The enzyme probably consists of three subunits similar to the Psr of *W. succinogenes* and the amino acid sequence of the catalytic subunit of this sulfur reductase revealed that the protein belongs to the DMSO reductase family of molybdoenzymes. The hydrogenase of this multienzyme complex belongs to the typical NiFe hydrogenases. It is likely that the electron transfer between hydrogenase and sulfur reductase in this complex is mediated by quinones, preferably sulfolobusquinone [49].

5.3. Sulfur Respiration of *Pyrodictium abyssi*

The hyperthermophilic and chemolithoautotrophic Archaea *Pyrodictium abyssi* grows at 105°C and pH 6.0 with H_2 and elemental sulfur [50]. The membrane-integrated electron transport chain catalyzing polysulfide sulfur reduction by H_2 in *P. abyssi* differs from that of *W. succinogenes* and *A. ambivalens*, respectively. A H_2 : sulfur oxidoreductase complex has been purified from membranes of *P. abyssi*. This complex is composed of nine different subunits containing a NiFe hydrogenase, a sulfur reductase and cytochromes *b* and *c*. No quinone has been detected in the isolated complex or in the membrane fraction of *P. abyssi* [51]. The catalytic properties of the enzyme complex suggest that it represents the entire respiratory chain of this organism, with hydrogenase, electron transport components, and sulfur reductase arranged in one stable multienzyme complex.

6. POLYSULFIDE SULFUR TRANSFERASES

Wolinella succinogenes grows via polysulfide sulfur respiration with hydrogen or formate as electron donors. Growth and survival by polysulfide sulfur respiration may be critical for bacteria living at low pH and/or low sulfide concentration, since the concentration of polysulfide sulfur may become very low under these conditions [Reaction (4)]. Therefore, it is not surprising that sulfur reducers synthesize binding proteins for polysulfide sulfur which permit rapid polysulfide sulfur respiration at low polysulfide sulfur concentration. The periplasmic polysulfide sulfur transferases Sud and Str of *W. succinogenes* appear to serve as such binding proteins [52,53]. Sud and Str are formed by the bacteria when grown with elemental sulfur or polysulfide. Both proteins are nearly absent upon growth with fumarate or other terminal electron acceptors. The apparent K_M for polysulfide sulfur is 70 μM in polysulfide sulfur reduction by H_2 catalyzed by bacteria grown with fumarate, and is 10 μM with polysulfide sulfur-grown bacteria. A similar decrease of K_M was observed upon the addition of Sud protein to the bacterial membrane fraction or proteoliposomes catalyzing polysulfide sulfur reduction.

6.1. Polysulfide Sulfur Transferase of *W. succinogenes*

The polysulfide sulfur transferase (Sud) of *W. succinogenes* consists of two identical subunits (14.3 kDa) and does not contain prosthetic groups or heavy metal ions. The protein was originally isolated as a sulfide dehydrogenase [54]. Sud catalyzes the transfer of sulfur from polysulfide sulfur to cyanide [Reaction (15)] at a high

$$S_n^{2-} + CN^- \longrightarrow SCN^- + S_{n-1}^{2-} \tag{15}$$

turnover number (10^4 s^{-1} at 37°C). Sud binds up to 10 mol sulfur per mol subunit when incubated in a polysulfide sulfur solution. In solutions containing <0.1 mM polysulfide sulfur the dissociation constant of sulfur bound to Sud was 10 μM or less. Under these conditions, the decrease of K_M by Sud was observed in polysulfide sulfur reduction. At higher concentrations of polysulfide sulfur, the dissociation constant was 0.2 mM and no effect of Sud on the K_M for polysulfide sulfur was observed.

Sud contains a single cysteine residue which is essential for polysulfide sulfur transferase activity [Reaction (15)], for sulfur binding, and for the decrease of K_M for polysulfide sulfur in polysulfide sulfur reduction [53]. Sulfur appears to be covalently bound to the cysteine residue. When subjected to MALDI mass spectrometry, Sud incubated with polysulfide sulfur was found to carry one or two sulfur atoms per monomer. No sulfur was bound to the monomer after treatment of Sud with cyanide. A Sud variant carrying a serine instead of the cysteine residue did not carry sulfur even upon incubation with polysulfide sulfur. To explain the effect of Sud on the apparent K_M for polysulfide sulfur in polysulfide sulfur reduction, it is assumed that sulfur is transferred from Sud to the active site of polysulfide reductase. From the concentration of Sud required for K_M decrease and from its concentration in the bacterial periplasm it is likely that Sud is bound to polysulfide reductase during polysulfide sulfur transfer. The two enzymes occur in about equimolar amounts in cells of *W. succinogenes* grown with polysulfide sulfur.

The structure of the Sud dimer was determined by NMR spectroscopy [55–57] (Fig. 9). The two cysteine residues (C89), one above (left side) and one below the paper surface (right side), are highlighted [Fig. 9(A)]. The monomer has an α/β topology with six α-helices packing against a central core of five parallel β-strands. The dimer interface consists of a four helix bundle around the rotational symmetry axis. There is no indication of cooperative interaction of the two cysteine residues which are 25 Å apart. The interaction of Sud with polysulfide reductase was investigated comparing the [^{1}H,^{15}N] TROSY spectra of Sud and of an equimolar mixture of Sud dimer and polysulfide reductase in the presence of polysulfide sulfur at low concentration (0.1 mM). The resonances of eight residues (D47, D49, M54, S66, R67, G115, D118, and K119)

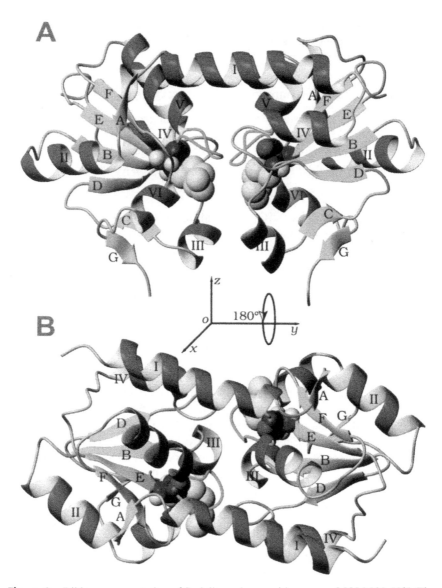

Figure 9   Ribbon representation of Sud dimer drawn with program MOLMOL [58]. The A and B views are related by a 90° rotation around the symmetry axis. The catalytic cysteines (with a five-atoms long polysulfide sulfur chain attached) are depicted using a CPK model (spheres with WdV radius and standard colors: red for O, yellow for S, blue for N, black for C and gray for H). The active site loop (residues 89–94) is colored in green [57]. Reproduced from Ref. [57] with the permission of *Biochemistry*.

of Sud were found to be shifted by the presence of polysulfide reductase. These residues which form a ring around the cysteine residue, are thought to make contact with polysulfide reductase. In the functional complex consisting of Sud and polysulfide reductase, the end of the polysulfide sulfur chain bound to the cysteine residue of Sud may come close to the molybdenum where a sulfur atom is reductively cleaved off according to the mechanism depicted in Fig. 4. This view was supported by the response of the EPR signal of the molybdenum (+5) ion in polysulfide reductase which indicated a slight change in the geometry of ligand sphere of molybdenum when Sud was added in the presence of polysulfide sulfur at low concentration [59]. The effect of Sud was not observed in the absence of polysulfide sulfur or with polysulfide sulfur present at high concentration.

7. CONCLUSIONS AND OUTLOOK

Energy conservation via polysulfide sulfur respiration in Bacteria and Archaea appears to be similar. Membrane bound respiratory chains generate a chemiosmotic potential, which is utilized by membrane bound ATP synthases to form ATP. Yet, due to their extreme habitats, thermophiles like *A. ambivalens* and hyperthermophiles like *P. abyssi* may have adapted their electron transport systems to high temperatures by forming stable electron transport complexes in contrast to the mesophilic bacteria *W. succinogenes*.

The compositions of the described electron transport chains show participations of similar NiFe hydrogenases and similar polysulfide sulfur reductases in the case of *W. succinogenes* and *A. ambivalens*, whereas the sulfur reductase of *P. abyssi* may be different. In this microorganism the electron transfer between the hydrogenases and the polysulfide sulfur reductases mediated by 8-methylmenaquinone, sulfolobusquinone or cytochromes *b* and *c*, respectively, seems to be adapted to the synthesis capabilites of the species.

This chapter has focused only on biological dissolubility of elemental sulfur by sulfur transferases and on biological reduction of sulfur and polysulfide sulfur by terminale reductases. Other sulfur containing species (sulfite, sulfate, and others) which play an important role in the biogeochemical cycle of sulfur compounds have been reviewed previously [2,7,60–63].

ACKNOWLEDGMENT

The work was supported by grants from the Deutsche Forschungsgemeinschaft (SFB 472) and the Fonds der Chemischen Industrie.

ABBREVIATIONS

Δp	electrochemical proton potential across a membrane
$\Delta \Psi$	electrical proton potential across a membrane
DMN	2,3-dimethyl-1,4-naphthoquinone

ELISA	enzyme-linked immunosorbent assay
FdhA/B/C	formate dehydrogenase
HQNO	2-(n-heptyl)-4-hydroxyquinoline-N-oxide
HydA/B/C	hydrogenase
MALDI	matrix assisted laser desorption/ionization
MGD	molybdopterin guanine dinucleotide
MM	8-methyl-menaquinone$_6$ (Fig. 1)
MMH_2	hydroquinone of MM
MM_b	8-methyl-menaquinone bound to PsrC
MM_bH^-	hydroquinone anion of MM_b
MQ	menaquinone$_6$ (Fig. 1)
MQH_2	hydroquinone of MQ
PsrA/B/C	polysulfide reductase
[S]	polysulfide sulfur
Sud	polysulfide sulfur transferase (<u>su</u>lfide <u>d</u>ehydrogenase)
TPP^+	tetraphenylphosphonium ion
TROSY	transverse relaxation-optimized spectroscopy
TTFB	4,5,6,7-tetrachloro-2-trifluoromethylbenzimidazol

REFERENCES

1. Schauder R, Kröger A. Bacterial sulphur respiration. Arch Microbiol 1993; 159:491–497.
2. Hedderich R, Klimmek O, Kröger A, Dirmeier R, Keller M, Stetter KO. Anaerobic respiration with sulfur and with organic disulfides. FEMS Microbiol Rev 1999; 22:353–381.
3. Richardson DJ. Bacterial respiration, a flexible process for a changing environment. Microbiology 2000; 146:551–571.
4. Thauer R, Jungermann K, Decker K. Energy conservation in chemotrophic anaerobic bacteria. Bacteriol Rev 1977; 41:100–180.
5. Clark WM. Oxidation-Reduction Potentials of Organic Systems. Baltimore: Willims & Wilkins, 1960.
6. Dietrich W, Klimmek O. Eur J Biochem 2002; 296:1086–1095.
7. Kroneck PMH, Beuerle J, Schumacher W. Metal-dependent conversion of inorganic nitrogen and sulfur compounds; see Sections 1.2 (The sulfur cycle) and 3 (Transformation of inorganic sulfur compounds). Met Ions Biol Syst 1992; 28:455–505.
8. Unden G, Bongaerts J. Alternative respiratory pathways of *Escherichia coli*: energetics and transcriptional regulation in response to electron acceptors. Biochim Biophys Acta 1997; 1320:217–234.
9. Kröger A, Biel S, Simon J, Gross R, Unden G, Lancaster CRD. Fumarate respiration of *Wolinella succinogenes*, enzymology, energetics, and coupling mechanism. Biochim Biophys Acta 2002; 1553:23–38.
10. Simon J. Enzymology and bioenergetics of respiratory nitrite ammonification. FEMS Microbiol Rev 2002; 26:285–309.
11. Boulègue J. Phosph Sulf 1978; 5:127–128.
12. Schauder R, Müller E. Arch Microbiol 1993; 160:377–382.

13. Schwarzenbach G, Fischer A. Helv Chim Acta 1960; 43:1365–1388.
14. Giggenbach W. Inorg Chem 1972; 11:1201–1207.
15. Klimmek O, Kröger A, Steudel R, Holdt G. Arch Microbiol 1991; 155:177–182.
16. Widdel F, Pfennig N. The genus *Desulfuromonas* and other Gram-negative sulfur-reducing eubacteria. In: Balows A, Trüper HG, Dwarkin M, Harder W, Schleifer K-H, eds. The Prokaryotes. 1991:3379–3389.
17. Wolin MJ, Wolin EA, Jacobs NJ. J Bacteriol 1961; 81:911–917.
18. Simon J, Gross R, Klimmek O, Kröger A. The Genus *Wolinella*. In: Dworkin M, ed. The Prokaryotes, an Evolving Electronic Resource for the Microbiology Community. 3d ed. New York: Springer-Verlag, 2000.
19. Macy JM, Schöder I, Thauer RK, Kröger A. Arch Microbiol 1986; 144:147–150.
20. Wloczyk C, Kröger A, Göbel T, Holdt G, Steudel R. Arch Microbiol 1989; 152:600–605.
21. Fauque G, Klimmek O, Kröger A. Meth Enzymol 1994; 243:367–383.
22. Klimmek O. Aufbau und Funktion der Polysulfid-Reduktase von *Wolinella succinogenes*, Doctoral (PhD) Thesis, Johann-Wolfgang-Goethe-Universität Frankfurt, FB Biologie 1996.
23. Collins MD, Fernandez F. Federation Eur Microbiol Soc 1984; 22:273–276.
24. Rigaud JL, Pitard B, Levy D. Biochim Biophys Acta 1995; 1231:223–246.
25. Biel S, Simon J, Gross R, Ruiz T, Ruitenberg M, Kröger A. Eur J Biochem 2002; 269:1974–1983.
26. Krafft T, Bokranz M, Klimmek O, Schröder I, Fahrenholz F, Kojro E, Kröger A. Eur J Biochem 1992; 206:503–510.
27. Krafft T, Gross R, Kröger A. Eur J Biochem 1995; 230:601–606.
28. Jankielewicz A, Schmitz RA, Klimmek O, Kröger A. Arch Microbiol 1994; 162:238–242.
29. Kisker C, Schindelin H, Rees DC. Annu Rev Biochem 1997; 66:233–267.
30. Bokranz M, Gutmann M, Körtner C, Kojro E, Fahrenholz F, Lauterbach F, Kröger A. Arch Microbiol 1991; 156:119–128.
31. Lenger R, Herrmann U, Gross R, Simon J, Kröger A. Eur J Biochem 1997; 246:646–651.
32. Baar C, Eppinger M, Raddatz G, Simon J, Lanz C, Klimmek O, Nandakumar R, Gross R, Rosinus A, Keller H, Jagtap P, Linke B, Meyer F, Lederer H, Schuster SC. Proc Natl Acad Sci USA 2003; 100:11690–11695.
33. Jormakka M, Tornroth S, Byrne B, Iwata S. Science 2002; 295:1863–1868.
34. Kröger A, Dorrer E, Winkler E. Biochim Biophys Acta 1980; 589:119–136.
35. Dross F, Geisler V, Lenger R, Theis F, Krafft T, Fahrenholz F, Kojro E, Duchêne A, Tripier D, Juvenal K, Kröger A. Eur J Biochem 1992; 206:93–102.
36. Volbeda A, Charon MH, Piras C, Hatchikian EC, Frey M, Fontecilla-Camps JC. Nature 1995; 373:580–587.
37. Higuchi Y, Tatsuhiko Y, Yasuoka N. Structure 1997; 5:1671–1680.
38. Gross R, Simon J, Lancaster CRD, Kröger A. Mol Microbiol 1998; 30:639–646.
39. Gross R, Simon J, Theis F, Kröger A. Arch Microbiol 1998; 170:50–58.
40. Gross R, Pisa R, Sanger M, Lancaster CR, Simon J. J Biol Chem 2004; 279:274–281.
41. Kröger A, Innerhofer A. Eur J Biochem 1976; 69:487–495.
42. Unden G, Kröger A. Biochim Biophys Acta 1982; 682:258–263.
43. Kröger A, Winkler E, Innerhofer A, Hackenberg H, Schägger H. Eur J Biochem 1979; 94:465–475.

44. Schröder I, Roberton AM, Bokranz M, Unden G, Böcher R, Kröger A. Arch Microbiol 1985; 140:380–386.
45. Chazotte B, Hackenbrock CR. J Biol Chem 1988; 28:14359–14367.
46. Fischer F, Zillig W, Stetter KO, Schreiber G. Nature 1983; 301:511–513.
47. Zillig W, Yeats S, Holz I, Böck A, Rettenberger M, Gropp F, Simon G. Syst Appl Microbiol 1986; 8:197–203.
48. Laska S, Kletzin A. J Chromatogr B, Biomed Sci Appl 2000; 737:151–160.
49. Laska S, Lottspeich F, Kletzin A. Microbiology 2003; 149:2357–2371.
50. Stetter KO, König H, Stackebrandt E. Syst Appl Microbiol 1983; 4:535–551.
51. Dirmeier R, Keller M, Frey G, Huber H, Stetter KO. Eur J Biochem 1998; 252:486–491.
52. Klimmek O, Kreis V, Klein C, Simon J, Wittershagen A, Kröger A. Eur J Biochem 1998; 253:263–269.
53. Klimmek O, Stein T, Pisa R, Simon J, Kröger A. Eur J Biochem 1999; 263:79–84.
54. Kreis-Kleinschmidt V, Fahrenholz F, Kojro E, Kröger A. Eur J Biochem 1995; 227:137–142.
55. Lin J-Y, Pfeiffer S, Löhr F, Klimmek O, Rüterjans H. J Biomol NMR 2000; 18:285–286.
56. Löhr F, Pfeiffer S, Lin J-Y, Hartleib J, Klimmek O, Rüterjans H. J Biomol NMR 2000; 18:337–346.
57. Lin J-Y, Dancea F, Löhr F, Klimmek O, Pfeiffer-Marek S, Nilges M, Wienk H, Kröger A, Rüterjans H. Biochemistry 2004; 43:1418–1424.
58. Koradi R, Billeter M, Wüthrich K. J Mol Graphics 1996; 14:51–55.
59. Prisner T, Lubenova S, Atabay Y, MacMillan F, Klimmek O. J Biol Inorg Chem 2003; 8:419–426.
60. Fauque G, LeGall J, Barton LL. Sulfate-reducing and sulfur-reducing bacteria. In: Shively JM, Barton LL, eds. Variations in Autotrophic Life. London: Academic Press, 1991:271–337.
61. Friedrich CG. Physiology and genetics of sulfur-oxidizing bacteria. Adv Microbial Physiol 1998; 39:235–289.
62. Cook AM, Laue H, Junker F. Microbial desulfonation. FEMS Microbiol Rev 1999; 22:399–419.
63. Kertesz MA. Riding the sulfur cycle—metabolism of sulfonates and sulfate esters in Gram-negative bacteria. FEMS Microbiol Rev 1999; 24:135–175.

6

Biological Cycling of Phosphorus

Bernhard Schink

*Fakultät für Biologie, Universität Konstanz, Postfach 5560 ⟨M654⟩,
D-78457 Konstanz, Germany*

1.	Introduction: Chemistry of Phosphorus Minerals	131
2.	Phosphates in Biology	133
	2.1. Phosphate Uptake by Algae, Bacteria, and Higher Plants	134
	2.2. Polyphosphate Synthesis and Degradation	137
3.	Metabolism of Phosphorus Compounds with C–P Linkages	138
	3.1. Synthesis of C–P and C–P–C Compounds	138
	3.2. Degradation of C–P and C–P–C Compounds	140
4.	Metabolism of Reduced Inorganic Phosphorus Compounds	143
	4.1. Assimilation of Phosphite and Hypophosphite	143
	4.2. Dissimilation of Phosphite	144
5.	Formation of Phosphine	145
6.	General Conclusions	147
	Acknowledgments	148
	Abbreviations	148
	References	148

1. INTRODUCTION: CHEMISTRY OF PHOSPHORUS MINERALS

Phosphorus is a very reactive element and does not occur in a free state in nature. It is typically found in combinations with oxygen, as phosphate, at the redox state

+5. Phosphorus makes up 0.09% by mass of the Earth's crust and, with this, it is the 14th most abundant element on Earth. The most important natural phosphate minerals are calcium phosphates such as apatite $Ca_5(PO_4)_3(OH,F,Cl)$ and, as a sedimentary derivative of apatites and of organic matter of biological origin, phosphorite $Ca_3(PO_4)_2$. To a minor degree, also iron and aluminum phosphates are found, e.g., vivianite, $Fe_3(PO_4)_2 \cdot 8\ H_2O$ and wavellite $Al_3(PO_4)_2(OH,F)_3 \cdot 5\ H_2O$ [1]. Phosphorus makes up $\sim 2-3\%$ of dry biomass and is especially enriched in animal bones (up to 10%). Actually, the first preparation of pure phosphorus by the alchemist Hennig Brandt in Hamburg in 1669 started from biological material, i.e., urine, which was condensed and heated, leading to a reductive transformation of $NaNH_4HPO_4$ by organic matter to white phosphorus which glows in the dark and gave this element its name (*Greek* phosphoros = light carrier). The history of phosphorus discovery and its utilization by mankind, among others in the production of matches and of firebombs (attributing to it its surname "the Devil's element"), has recently been described in detail in a shocking novel [2].

Preparation of elemental phosphorus by reduction of phosphates is technically achieved by heating fluoroapatite with coke in the presence of SiO_2 (quartzite) in an electrical light-arc furnace at $1400-1500°C$. In aqueous solution at pH 7.0, reduction of phosphate via phosphite, hypophosphite to elemental phosphorus and further to phosphine requires reduction at very low redox potentials, on average at -0.713 V (Table 1). It is not surprising, therefore, that reduced phosphorus compounds are very rare in nature, and that the dominant phosphorus form in nature is the phosphoric acid or phosphate form.

Phosphorus is an important constituent of biomass, and is metabolized by all forms of life. Through their activities, among others, phosphate is exchanged between the various pools of living and non-living systems in the global cycle. The most important pools of phosphates on Earth and the estimated annual rates of exchange between these pools is depicted in Fig. 1. As is obvious from this figure, the main reservoir of phosphate on Earth are marine and

Table 1 Standard Potentials of Redox Transitions of Phosphorus Compounds

	E_0' (mV)[a]
HPO_4^{2-}/HPO_3^{2-}	-690
$HPO_3^{2-}/H_2PO_2^-$	-913
$H_2PO_2^-/P_{white}$	-922
P_{white}/PH_3	-525
Average	-713

[a]Calculated for 1 M concentrations at pH 7.0 (after Ref. [3]).

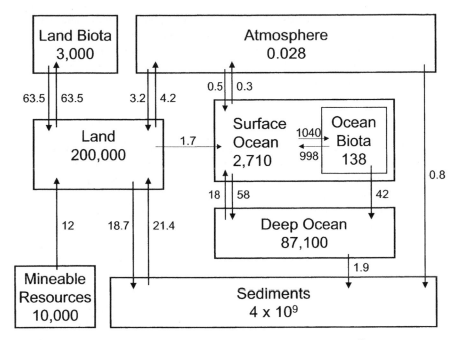

Figure 1 The global phosphorus cycle. Values shown are in Tg (10^{12} g or millions of metric tons) phosphorus. Redrawn after Ref. [4].

freshwater sediments, terrestrial resources in soil, and the Deep Ocean. The terrestrial and the aquatic phosphate budgets are internally well balanced between biotic phosphate uptake and mineralization, and there is a minor phosphate transfer from land to the oceans through riverine transport. A continuous import by mining from the pool of mineable phosphate is used for synthetic inorganic and organic chemistry and for fertilization, and has contributed significantly to eutrophication of surface waters and the oceans (see below). Since the only gaseous phosphorus derivative is phosphine (PH_3) which is unstable in the presence of oxygen, the phosphorus content of the atmosphere is very small, comprising mainly phosphates in dust particles and aerosols. Details on the global phosphorus cycle and the background for the calculation of the pools and fluxes in Fig. 1 can be looked up in Ref. [4].

In the following sections, the interaction of phosphorus compounds with living systems will be discussed in detail.

2. PHOSPHATES IN BIOLOGY

Since phosphate does not change its redox state in living cells (except in a few instances, see below), it is present in the cell only as free inorganic phosphate, as esters or as acid anhydrides, e.g., as constituent of nucleotides and their derivatives, of nucleic acids, sugar phosphates, phospholipids, and various co-enzymes. Most often, phosphoryl groups act as activators of sugars, acids, and numerous other compounds; phosphoric acid anhydride linkages are called "energy-rich" because their hydrolysis in water releases a substantial amount of energy, in the range of $20-40$ kJ mol^{-1} at pH 7.0.

2.1. Phosphate Uptake by Algae, Bacteria, and Higher Plants

Because the water solubility of phosphate minerals (including bones and teeth!) is very low, the concentration of phosphate in surface waters is very low as well. Freshwater lakes and rivers carry contents of total phosphate in the range of $0.1-5$ μM, and the same is true for seawater (2 μM [5]). Ultra-oligotrophic lakes have total phosphorus contents in the range of less than 5 μg L^{-1}, oligotrophic lakes in the range of $5-10$ μg L^{-1}, and eutrophic lakes in the range of $30-100$ μg L^{-1} [6]. These very low concentrations (below 3 μM!) have a pronounced effect on the productivity of lakes because phosphate is in most cases the limiting nutrient for photosynthetic biomass production. Indirectly, the phosphate availability also determines the type of primary producers active in the respective water: in oligotrophic and mesotrophic lakes, green algae including diatoms dominate primary production as long as phosphate and not the nitrogen availability limits the primary productivity. In eutrophic lakes often cyanobacteria proliferate which profit from the enhanced phosphorus availability and cover their nitrogen needs by nitrogen fixation. Mass development of cyanobacterial blooms is a very undesired secondary effect of lake eutrophication because these organisms not only develop ill-smelling scums on the water surface but also often produce cyanotoxins which may kill fish and pasture lifestock. For this reason and for securing a sufficient supply of clean surface water for mankind, efforts were taken in the last 40 years to decrease the phosphate delivery into surface waters (see below).

Of the "total phosphate" pools in lake water mentioned earlier, $<1\%$ is free orthophosphate. The major part ($>90\%$) of the total phosphorus content is particulate organic phosphate which is bound in all types of living organisms and dead biomass. The second largest pool ($<10\%$) is particulate inorganic phosphate, mainly calcium apatites, iron phosphates, and aluminum phosphate, which precipitate either free or associated with humic water constituents or calcite, or with imported clay and silt particles, and are transported down to the sediment. A further minor pool ($1-2\%$) of phosphates in the water are dissolved organic phosphorus compounds, such as free nucleotides and nucleic acids released from lyzing cells, as well as complexes of phosphate with humic compounds. Phosphates accumulate in the sediment up to $0.2-2$ mg g^{-1} sediment, i.e., up to $10,000-100,000$ times higher than in the overlying water. Phosphate can be released from the sediment mainly by reductive activities, e.g., via microbial iron and sulfate reduction, which transform, e.g., ferric phosphates to ferrous derivatives, causing a redissolution of phosphate. This reductive redissolution supplies the water column again with phosphate and increases the productivity, a "vicious cycle" which hampers efforts for restoration of eutrophic lakes, especially of shallow lakes [6].

Whereas animals cover their phosphate needs with their organic food, primary producers such as algae and cyanobacteria depend on phosphate supply from their inorganic environment. However, also heterotrophic bacteria require an additional phosphate supply because their relative phosphate content is higher than that of their organic food. Despite the obvious importance of phosphate supply for the trophic status of surface waters, rather little is known about the mechanisms how phosphate is being taken up by algae. Whereas at first sight diatoms appear to have lower K_S values (half-saturation concentration, analogous to the Michaelis constant K_M) and with this, higher affinities for phosphate than bacteria, the latter compensate this shortcoming by their higher surface-to-volume ratio and higher maximum uptake rates (Table 2). The relative importance of algae, cyanobacteria, and heterotrophic bacteria in phosphate uptake and phosphate exchange between them in mixed communities in lakes is still a matter of discussion. In any case, these communities maintain their phosphate supply sufficiently at total phosphate concentrations of less than 1 μM phosphate, thus proving their high affinity for this essential nutrient.

Only little is known so far about the biochemical basis of phosphorus uptake by algae. With the green alga *Chlamydomonas reinhardtii*, active uptake of inorganic phosphate has been observed, and the uptake rates are increased under conditions of phosphate deprivation [7]. In *Escherichia coli*, there are two different phosphate uptake systems, one for low and one for high affinity uptake. At high phosphate supply, phosphate enters the periplasm by diffusion through a porine in the outer membrane, and crosses the inner membrane through a constitutively expressed low-affinity transport protein. Under

Table 2 Phosphate Uptake Kinetics of Two Diatoms and Three Bacterial Species

	K_S (μM P)	v_{max} (10^{-9} μmol $\approx$ μm^{-2} $\approx$ d)	$v_S^{0\,a}$ (mmol P $\approx$ m^{-2} $\approx$ d^{-1} $\approx$ μM^{-1})
Diatoms			
Cyclotella nana	0.6	2.0	3.3
Thalassiosira fluviatilis	1.7	7.0	4.1
Bacteria			
Corynebacterium bovis	6.7	7.7	1.15
P. aeruginosa	12.2	17.9	1.47
B. subtilis	11.3	12.5	1.11

[a]This term gives the initial slope of the saturation curve at [S] = 0 and is derived by dividing v_{max} by K_S.
Source: Data for K_S and v_{max} taken from Ref. [6].

conditions of phosphate limitation, orthophosphate is bound in the periplasmic space by a specific phosphate-binding protein which transfers the phosphate to an ATP-dependent, phosphate-specific transport system in the cytoplasmic membrane. In addition, also an alkaline phosphatase, a nucleotidase, and a cyclic phosphodiesterase are induced which are localized in the periplasmic space and thus allow to tap further phosphate resources. The whole system is under control of a two-component regulatory system that governs 20 promoters and more than 100 proteins which all respond to the phosphate levels of the environment [8]. Similar regulatory systems have been identified in *Bacillus subtilis* [9], *Corynebacterium glutamicum* [10] and the cyanobacterium *Synechococcus* sp. [11]. The phosphate-binding protein of *E. coli* was suggested to be used as the recognition site in biosensors for phosphate, due to its unusually high affinity to phosphate [12].

Since phosphate is available in aquatic environments only at very low concentration and has to be accumulated in the cell by energy-dependent transport systems to intracellular concentrations of > 10 mM [13], exposure of such organisms to growth media of high phosphate content as commonly used in microbiology will cause severe disorders in cell energetics if such transport systems cannot immediately be switched off. It is not surprising, therefore, that many environmental isolates of bacteria from freshwater ecosystems are sensitive to elevated concentrations of phosphate, e.g., many phototrophic bacteria [14]. Sensitivity towards phosphate in growth media may be one of the reasons that have so far prevented cultivation of the majority of microbes from natural environments.

Also in soil, the majority of phosphate is bound in water-insoluble minerals, and the soil water phase contains bioavailable phosphate only at very low concentrations (<10 μM [15]). Thus, acquisition of phosphate by plant roots is one of the key problems in plant nutrition. Experiments with various vascular plants provided evidence of an active transport system in the fine roots which takes up primary orthophosphate ($H_2PO_4^-$) in a cotransport with protons [16]. Most plants enhance their nutrient uptake capacity (mainly phosphate and nitrogen compounds) by association with fungi in various forms of Mycorrhiza. In these symbiotic associations, the fungus extends the surface area of the plant roots, thus increasing the nutrient transport capacity also for phosphate acquisition [16–18]. Alternatively, also bacteria in the rhizosphere can improve the phosphate supply to plants. Several types of "phosphate solubilizing bacteria" have been described, such as *Pseudomonas cepacia, Serratia marcescens, Erwinia herbicola, Phizobium* spp. and *Bacillus* spp. [19–22] which are supposed to increase the phosphorus supply in the rhizosphere. The phosphate-mobilizing effect of these bacteria is mainly due to the release of gluconic acid which is formed from glucose via specifically induced, pyrroloquinoline quinone-dependent glucose dehydrogenases; the produced gluconic acid helps to solubilize mineral phosphates especially in alkaline, carbonate-rich soils. Other bacteria may contribute to orthophosphate supply by excretion of extracellular phosphatases to tap the pool of organic phosphates.

2.2. Polyphosphate Synthesis and Degradation

Numerous algae, cyanobacteria and heterotrophic bacteria are known to take up phosphate at sufficient supply far beyond their physiological needs ("luxurious uptake"), and to store it inside in the form of polyphosphate ("volutin granula") which can make up up to 15% of total bacterial cell mass. In these polyphosphates, the phosphoryl units are linked by acid anhydride linkages which are synthesized through polyphosphate kinase, according to Reaction (1)

$$ATP + poly\text{-}P_n \longrightarrow ADP + poly\text{-}P_{n+1} \tag{1}$$

Polyphosphate can replace ATP in the phosphorylation of glucose in numerous microorganisms [23,24]. Since the equilibrium of Reaction (1) is on the side of polyphosphate synthesis, this enzyme is not employed in polyphosphate degradation. Instead, a poly-P:AMP phosphotransferase activity was found in cell-free extracts of the phosphate-accumulating bacterium *Acinetobacter johnsonii* which catalyzes Reaction (2)

$$poly\text{-}P_n + AMP \longrightarrow poly\text{-}P_{n-1} + ADP \tag{2}$$

The ADP thus formed can be converted to ATP by adenylate kinase which catalyzes the reaction $2ADP \rightleftharpoons ATP + AMP$ and thus simultaneously regenerates the acceptor molecule AMP for the earlier reaction [25–28]. Thus, polyphosphate is not only a phosphate storage system but can also act as an energy resource, although only for short terms of energy deprivation.

The ability of several aerobic and facultatively nitrate-reducing bacteria to accumulate phosphate at ample supply and to release it again under conditions of energy limitation is being exploited today in advanced wastewater treatment systems [28,29]. The eutrophication of surface waters in the recent four decades by excess phosphate input, mainly through polyphosphate-containing laundry detergents and surface run-off from over-fertilized agricultural soils, has necessitated strict control of phosphate discharge into surface waters. In conventional phosphate containment in waste water treatment systems, phosphate is chemically precipitated with aluminum or ferric chlorides or sulfates, either before or after the activated sludge step. This chemical treatment follows the mass action law of chemistry: to remove small amounts of phosphate (in the range of a few milligrams per liter), an excess of precipitating agent is needed. Instead, the biological accumulation of polyphosphates by certain bacteria, including *Acinetobacter* spp. and several Gram-positive bacteria, can be used to accumulate phosphate from the purified wastewater inside the bacterial cells. Cycling of the activated sludge between oxic and anoxic incubation steps selects for such phosphate-accumulating bacteria [28–33] and allows a much more efficient phosphate removal either with the sludge biomass itself or by precipitation in the anoxic incubation tank in which the phosphate freight has been significantly concentrated (Fig. 2). This strategy has been developed in the recent

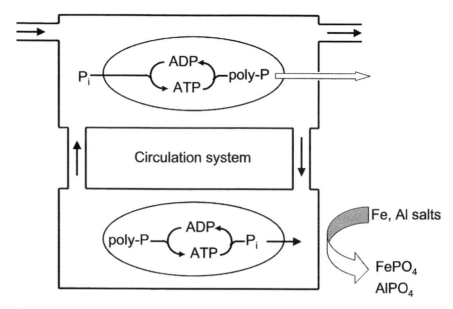

Figure 2 Flow scheme of phosphate accumulation in the biological phosphate elimination process. Phosphate is taken up by bacteria as a phosphorus and energy reserve in the presence of oxygen, and released again under anoxic conditions. Redrawn after Ref. [78].

past to large-scale application in waste water treatment, especially in the Netherlands and in South Africa.

3. METABOLISM OF PHOSPHORUS COMPOUNDS WITH C–P LINKAGES

Whereas in nearly all bioorganic compounds the phosphate is bound to the next carbon atom via an oxygen bridge, there are very few exceptions with C–P or C–P–C linkages in which the phosphorus assumes the redox state +III or +I, accordingly. Contrary to phosphate esters and phosphoric acid anhydride linkages, such C–P linkages are extremely stable and resist chemical hydrolysis, thermal decomposition, or photolysis [34]. Therefore, the biological formation of such compounds and the biochemical cleavage of these linkages represent special challenges to biochemistry.

3.1. Synthesis of C–P and C–P–C Compounds

Several protozoa and other lower animals such as the sea anemone *Anthopleura elegantissima* contain phosphonolipids in their membranes which carry 2-aminoethylphosphonate as a polar group, analogous to ethanolamine phosphate in other organisms. Other naturally occurring phosphonates are the antibiotic phosphonomycin (1,2-*cis*-epoxypropylphosphonate, $CH_3-CH-(-O-)CH-PO_3^{2-}$), produced by *Streptomyces fradiae*, or the antibiotically active compound L-alanyl-L-alanyl-phosphinothricin which is produced by *Streptomyces viridochromogenes*. The latter contains phosphinothricin, $^-OOC-CH(NH_3^+)-(CH_2)_2-PO-(OH)-CH_3$, which acts as an antimetabolite of glutamate in glutamine synthetase. Phosphinothricin is the phytotoxic part of bialaphos, and is widely used as a non-selective herbicide known under the name of BASTA.

The key reaction in the synthesis of biological organophosphates is the formation of a C–P linkage by the phosphoenolpyruvate : phosphonopyruvate mutase reaction [Fig. 3(A)]. The enzyme has been purified from *Tetrahymena pyriformis* [35]. The equilibrium of the reaction is far on the side of phosphoenolpyruvate, and is shifted towards phosphonopyruvate formation by subsequent decarboxylation of phosphonopyruvate to phosphonoacetaldehyde [36]. Further phosphoenolpyruvate mutase enzymes have been characterized in *Trypanosoma cruzi* [37] and *Mytilus edulis* [38]. Phosphonoacetaldehyde is the key compound from which the synthesis of phosphonoethanolamine and phosphonomycin proceeds. Biosynthesis of phosphinothricin starts with an analogous reaction

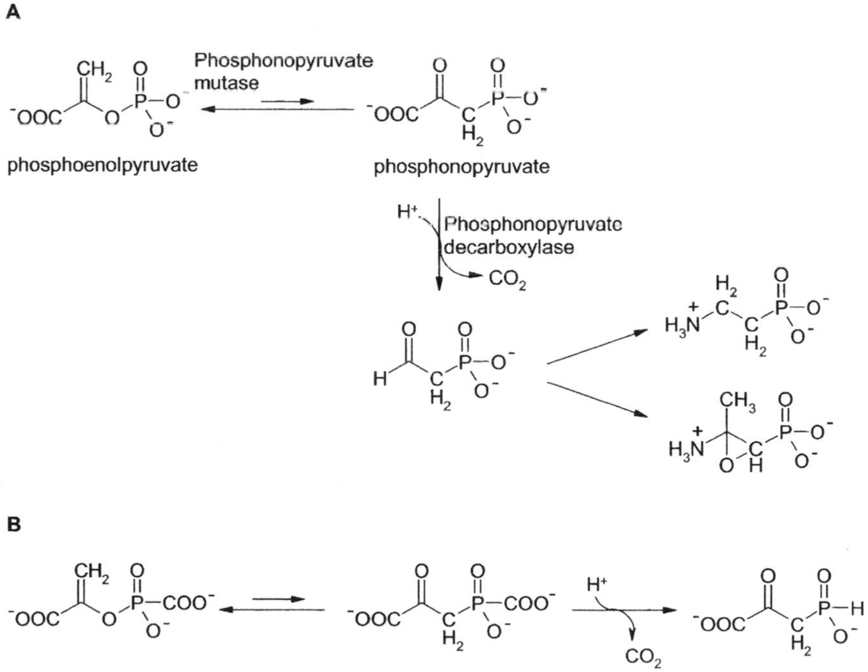

A

B

Figure 3 Reactions leading to the formation of carbon–phosphorus linkages. Modified and redrawn after Ref. [34].

catalyzed by carboxyphosphoenolpyruvate mutase; the produced carboxyphos-phonopyruvate is decarboxylated to form phosphinylpyruvate [Fig. 3(B)].

A further enzyme, the P-methylation enzyme, adds the further methyl group that results in the formation of the second C–P bond in phosphinothricin. The methyl donor for this reaction is methyl cobalamine [39]. The addition of two alanyl residues to phosphinothricin to form bialaphos in *Streptomyces virido-chromogenes* is catalyzed by synthetase enzymes, and proceeds independently of ribosomes [40].

3.2.   Degradation of C–P and C–P–C Compounds

Different from the biosynthesis of phosphonates which follows basically similar strategies in all cases studied (see above), the microbial degradation of phosphonates and, more specifically, the cleavage of C–P linkages, is carried out by a broad variety of different organisms employing enzymes with rather different chemical strategies. The matter has been reviewed twice in the recent past [34,41].

A broad variety of further organophosphonates is produced by synthetic chemistry, mainly in the form of detergent additives, and our knowledge of their fate in the environment is still limited [41,42].

So far, four different pathways for cleavage of C–P linkages have been described in bacteria, the phosphonatase, phosphonoacetate hydrolase, phospho-nopyruvate hydrolase, and C–P lyase. All these enzymes differ with respect to their substrate specificity and the cleavage mechanism (Fig. 4 [34]).

Other organophosphonates, often also carrying sulfur substituents at the phosphorus atom, are used as pesticides [43]. Their degradation by white-rot fungi has been studied in some detail, but little is known about the chemistry of cleavage of C–P linkages in these compounds [43].

Phosphonate cleavage by phosphonatase [Fig. 4(A)] requires an activated group at the carbon atom vicinal to the phosphonate-carrying carbon. The reaction is initiated by formation of a Schiff base with a lysine amino residue to the carbonyl group of phosphonoacetaldehyde which activates it for C–P bond cleavage. A nucleophilic substituent is supposed to accept the phosphonyl residue from which it is released as orthophosphate by hydrolysis [44]. All organophosphonates that can easily be converted to phosphonoacetaldehyde, such as phosphonoethanolamine, can be degraded by phosphonatase. Also phosphono-mycin degradation by *Rhizobium huakuii* is likely to enter the phosphonatase degradation pathway [45]. Phosphonatase of *Bacillus cereus* resembles alkaline phosphatase in many properties but does not hydrolyze a wide range of phosphate monoesters; it is activated by magnesium ions [46]. Similar enzymes have been isolated and described from *Pseudomonas aeruginosa, Salmonella typhimurium, Pseudomonas* sp., *Arthrobacter arthrocyaneus* (they are involved in glyphosate decomposition [44]), and *Salmonella typhimurium.* Phosphonatase appears to be regulated in most cases by the *pho*-regulon via exogenous orthophosphate availability.

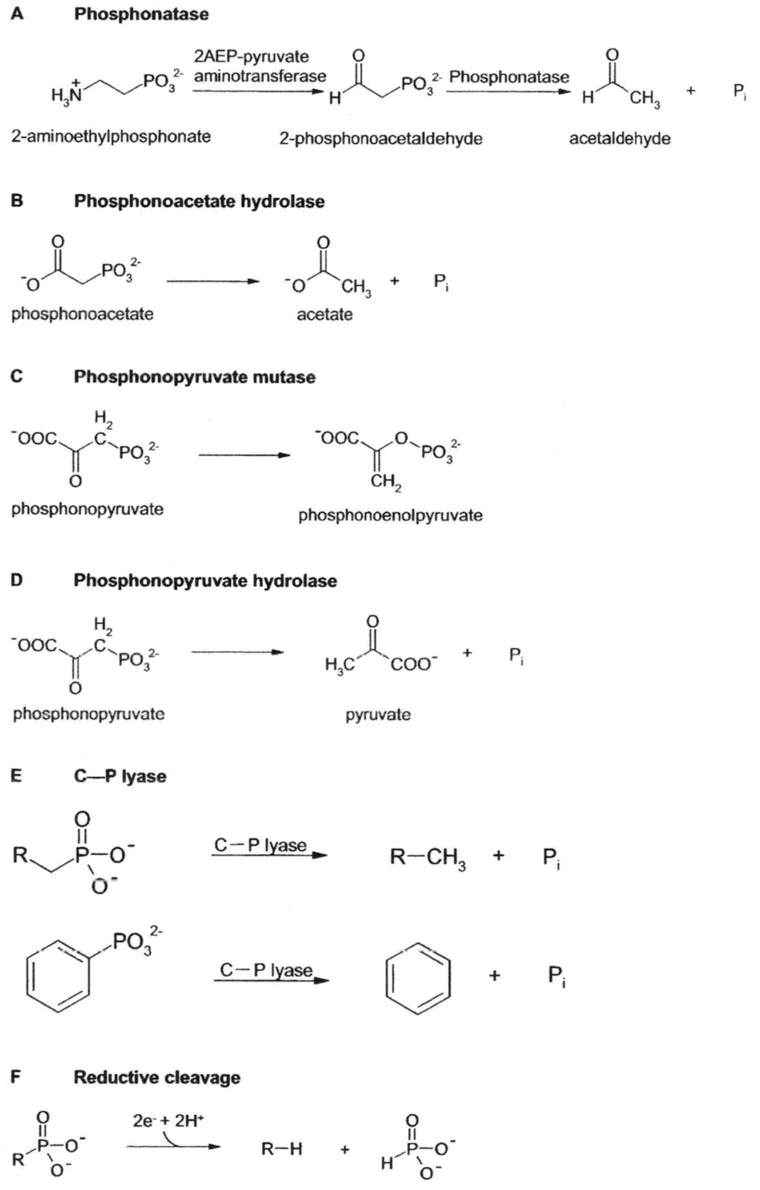

A   **Phosphonatase**

2-aminoethylphosphonate   2-phosphonoacetaldehyde   acetaldehyde

B   **Phosphonoacetate hydrolase**

phosphonoacetate   acetate

C   **Phosphonopyruvate mutase**

phosphonopyruvate   phosphonoenolpyruvate

D   **Phosphonopyruvate hydrolase**

phosphonopyruvate   pyruvate

E   **C—P lyase**

F   **Reductive cleavage**

Figure 4 Microbially catalyzed reactions involved in the cleavage of carbon–phosphorus linkages. Modified and redrawn after Ref. [34].

Contrary to phosphonatase, phosphonoacetate hydrolase [Fig. 4(B)] is induced by phosphonoacetate and is independent of orthophosphatase supply. This enzyme was detected in *Pseudomonas fluorescens* and is metal ion-dependent [47–49]. These bacteria express this enzyme in addition to phosphonatase, but phosphonoacetate hydrolase is induced only by its substrate. The biochemistry of the cleavage reaction is not clear yet, but may be comparable to that of phosphonatase. Whether the phosphate starvation-independent degradation of aminoethylphosphonic acid by *Pseudomonas putida* [50] and the degradation of 4-aminobutylphosphonate by the yeast *Kluyveromyces fragilis* [51] proceed through the same pathway has still to be elucidated. An alternative, metal-independent phosphonoacetate hydrolase activity was found in *Penicillium oxalicum* [52].

Degradation of phosphonoalanine and phosphonopyruvate can proceed through a pathway basically analogous to phosphonopyruvate synthesis, i.e., a reversal of the phosphoenolpyruvate mutase reaction [Fig. 4(C)] which converts phosphonopyruvate to phosphoenolpyruvate, a central intermediate of metabolism [53]. A further phosphonopyruvate hydrolase type [Fig. 4(D)] was described recently with a *Variovorax* strain. It is induced in the presence of phosphonoalanine or phosphonopyruvate and is independent of the phosphate supply of the cells. The enzyme uses only phosphonopyruvate as substrate and depends on cobalt ions for activity. Based on the sequence of the coding gene, this enzyme is related to phosphoenolpyruvate phosphomutase and entirely different from phosphonoacetaldehyde hydrolase and phosphonoacetate hydrolase, which belong to the haloacetate dehalogenase and the alkaline phosphatase superfamily, respectively [54–56].

Different from all the C–P cleaving enzymes mentioned so far, cleavage of C–P linkages by C–P lyase [Fig. 4(E)] does not require an activated function at the β-carbon atom of the phosphonate substrate, and proceeds by a reaction mechanism involving radicals. This enzyme is induced again in response to phosphorus limitation and has been measured so far only with intact cells, never in cell-free extracts [57–60]. In accordance with the assumed radicalic nature of the reaction mechanism, C–P lyase appears to react rather unspecifically with a broad variety of different organic phosphonates, including alkyl and aryl derivatives. Experiments with simple model substrates, i.e., alkyl phosphonates and the C–P lyase system of *Rhizobium* sp., the reaction produced orthophosphate and the corresponding alkanes and alkenes [61]. Accordingly, phenylphosphonic acid yielded benzene and biphenyl. Participation of hydroxyl radicals could be ruled out by control assays with hydroxyl radical scavengers. Rather, the participation of a heavy metal, e.g., iron, was suggested, leading to a reaction mechanism as depicted in Fig. 5. According to this mechanism, the metal replaces a proton at the acidic phosphonyl oxygen and leaves after valency change, thus forming a phosphonyl radical which subsequently yields an organic carbon radical and a free metaphosphate residue which, upon water addition, forms orthophosphate. Addition of a single electron and a proton forms the reduced organic derivative [34,62]. Although not all details of this reaction have been understood sufficiently, the radicalic nature of this cleavage reaction appears to be well established. As a common feature of all reactions cleaving carbon phosphonate linkages, there is no doubt that in all these cases

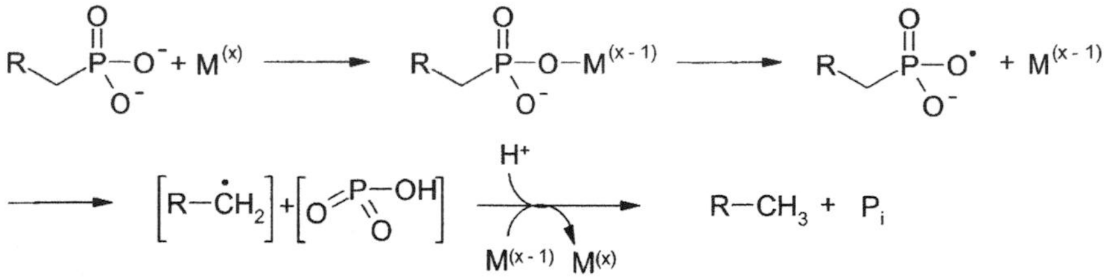

Figure 5 Possible reaction mechanism of carbon–phosphorus lyase involving radicals. Modified and redrawn after Ref. [34].

the phosphonyl residue is liberated as orthophosphate. The same enzyme system is involved also in the degradation of many phosphorus-containing herbicides, e.g., glyphosate [63,64].

An alternative mechanism of a C–P lyase reaction [Fig. 4(F)] was suggested on the basis of genetic experiments with an *E. coli* strain assimilating organophosphonates as phosphorus sources. Mutants of these bacteria always lost the capacity to assimilate organophosphonates together with the capacity to assimilate phosphite. Complementation always restored both capacities together. Based on these findings, it was suggested that the cleavage reaction proceeds by a reductive step, forming phosphite as phosphorous product [65,66]. Although the formed phosphite was never detected and the chemistry of the assumed reductive C–P cleavage reaction is still enigmatic, the indirect genetic evidence of this type of reaction appears convincing and is of specific interest because such a reaction could be a source of phosphite in anoxic environments (see below).

4. METABOLISM OF REDUCED INORGANIC PHOSPHORUS COMPOUNDS

4.1. Assimilation of Phosphite and Hypophosphite

The utilization of reduced inorganic phosphorus sources such as phosphite or hypophosphite instead of orthophosphate appears to be widespread among many different groups of bacteria, and may also be associated with the utilization of organophosphonates as assimilatory phosphorus sources. Whereas phosphorus supply from organophosphonates through the phosphorus-regulated enzyme systems phosphonatase and C–P lyase always produce orthophosphate, C–P lyase of an *E. coli* strain is likely to produce phosphite from organophosphonates [65] which is subsequently assimilated as phosphorus source. Utilization of reduced inorganic phosphorus sources had been reported in the past for a *Bacillus caldolyticus* strain. This activity was attributed to a "hypophosphite oxidase" which oxidized both hypophosphite and phosphite to orthophosphate [67]. Unfortunately, this *Bacillus* strain is not available anymore and the experiments cannot be reproduced.

The only organisms which have been studied in detail with respect to assimilation of reduced phosphorus compounds are *Klebsiella aerogenes* [68], *E. coli* [65], and *Pseudomonas stutzeri* strain WM 88 [69]. With the latter bacterium, it was shown by genetic analysis that hypophosphite is oxidized via phosphite to orthophosphate [69]. Whereas the step from hypophosphite to phosphite is probably carried out by a dioxygenase-type enzyme, oxidation of phosphite to phosphate is coupled to reduction of NAD^+. This NAD-dependent phosphite dehydrogenase was heterologously expressed in *E. coli*, purified, and characterized in detail [70]. Phosphite oxidation with NAD^+ by this enzyme appears to be irreversible which is not surprising if one compares the redox transition potential of phosphite oxidation (-630 mV; Table 1) with that of the $NAD^+/NADH$ couple (-320 mV). The enzyme is obviously very specific for phosphite as electron donor and is related to D-hydroxyacid dehydrogenases. Three different reaction mechanisms have been discussed so far, two of which involve an initial nucleophilic attack of NAD^+ on the phosphorus atom and subsequent hydride transfer, and the third one includes a primary hydride transfer from phosphite to NAD^+; the latter mechanism appears to be substantiated by kinetic data [71].

4.2. Dissimilation of Phosphite

Enrichment cultures have been tried repeatedly in the past to search for bacteria that can utilize reduced phosphorus compounds as electron donors in their dissimilatory metabolism. Both hypophosphite and phosphite could be excellent electron donors for a microbial energy metabolism because their oxidation releases electrons at very low redox potentials (Table 1). So far, only one bacterium has been isolated that can run its energy metabolism on the basis of phosphite oxidation to phosphate.

This bacterium, *Desulfotignum phosphitoxidans*, couples phosphite oxidation either to the reduction of sulfate to sulfide [72] or with the homoacetogenic reduction of CO_2 to acetate [73]. There is indirect evidence from growth yield data that the bacteria exploit all the energy available in phosphite-dependent sulfate reduction or homoacetogenesis. Although the phosphite-oxidizing enzyme activity could not yet be studied in cell-free extracts, growth yield data suggest that phosphite oxidation itself leads to formation of an energy-rich linkage which can be transformed directly or indirectly into an ATP equivalent. Such a reaction could start with a primary binding of phosphite to an acid or thiol residue which, after oxidation, forms an energy-rich acid anhydride or thioester linkage from which ATP can be formed by a subsequent phosphoryl group transfer [74].

Failure to demonstrate the activity of the phosphite-oxidizing enzyme *in vitro* so far may be due to the fact that we did not find the right cosubstrate for this enzyme yet. Our sulfate-reducing bacterium cannot oxidize hypophosphite or any other reduced phosphorus compounds. Interestingly, there are no reports of any other metabolic groups of organisms with phosphite as electron donor, e.g., aerobic, or nitrate-reducing bacteria, and we did not succeed either in enriching for such organisms. This failure may indicate either that the phosphite-oxidizing enzyme system requires a low redox potential as this is typically associated with the metabolic activity of sulfate-reducing bacteria, or that the phosphite-oxidizing enzyme system may have derived from the sulfate-reducing enzyme apparatus, e.g., the adenosine phosphosulfate reductase enzyme which carries out a similar redox reaction between sulfite and a sulfuryl residue. Phosphite oxidation in our bacterium is associated with the expression of a major protein band with a molecular mass of 42 kDa which is substantially different from the NAD^+-reducing phosphite dehydrogenase of *Pseudomonas stutzeri* (36.2 kDa [70]). Different from the *Desulfotignum* enzyme, the NAD^+-dependent enzyme of *Pseudomonas stutzeri* appears not to involve a substrate level phosphorylation step in this oxidation, indicating that these two enzymes may have very little in common.

5. FORMATION OF PHOSPHINE

Formation of phosphine from phosphate has been claimed repeatedly through the older literature to be catalyzed by the action of microorganisms [75,76]. These older papers should be looked at today with serious scepticism. They did not include any reliable analytical proof of phosphine formation; typically, a "garlic-like odor" was taken as a proof of phosphine formation, but could have been caused as well by mercaptanes which form as side products of aldehydes and ketones with hydrogen sulfide. In an assessment of the phosphorus economy of a sewage treatment plant in Hungary, Devai and coworkers detected a gap in the phosphorus balance of 30–45% which they attributed to microbial reduction of phosphate to phosphine [77]. One can only hope that the authors were wrong with their assumption, otherwise all higher life in the area would have been killed! As listed in Table 1, phosphate reduction via phosphite, hypophosphite, and phosphorus to phosphine proceeds through steps of extremely low redox potentials, on average at -713 mV, and can hardly be coupled to biomass oxidation which releases electrons on average at about -430 mV at pH 7.0.

Despite these physicochemical limitations, phosphate reduction has often been postulated in the literature in analogy to sulfate reduction or nitrate reduction, although the average redox potentials of these reduction processes are substantially more positive, at -218 or $+363$ mV, respectively [78]. It is obvious from this comparison that phosphate reduction to phosphine can never proceed as an energy-conserving respiratory process [79].

Nonetheless, traces of phosphine (in the nanomolar range) have been detected in certain anoxic environments such as sediments [80,81], paddy fields [82,83], and manure samples [84], but it was never proven that its formation was caused by biological phosphate reduction. Some of these phosphine findings are interpreted today as the result of a biologically enhanced hydrolysis of phosphorus-rich metallic iron or steel [85]. Traces of phosphine (in the pico- to nanomolar range) were also found in undefined anaerobic cultures obtained from sewage sludge or from animal feces [86,87], but again the origin of this phosphine remains obscure. One cannot entirely rule out that a biological conversion of a phosphoryl group could lead via phosphoenolpyruvate mutase to phosphonopyruvate, and this compound through a C–P cleavage reaction analogous to the *E. coli*-type enzyme [65] to phosphite, but such a (very hypothetical!) pathway would run at a very low rate and would never produce substantial amounts of phosphite. Even if traces of phosphite could be formed in this way (at the expense of the energy-rich linkage in phosphoenolpyruvate!), further reduction of phosphite to phosphine would require even more energy. Also a dismutation of phosphite, according to Eq. (3)

$$4HPO_3^{2-} + 2H^+ \longrightarrow 3HPO_4^{2-} + PH_3 \tag{3}$$

would still be endergonic at pH 7.0 ($\Delta G_0' = +17.6$ kJ mol^{-1} reaction) and would become exergonic only at pH <5.0. Things would become easier after formation of hypophosphite; its dismutation according to Eq. (4)

$$2H_2PO_2^- H_2PO_2^- \longrightarrow HPO_4^{2-} + PH_3 \tag{4}$$

is exergonic at pH 7.0 ($\Delta G_0' = -68.4$ kJ mol^{-1}) and could thus even support a microbial energy metabolism. Perhaps partly reduced phosphorus derivatives such as phosphite or hypophosphite entering natural environments at trace levels could be the sources for phosphine formation via these dismutation reactions, or could be a source for the trace amounts of phosphine found in certain anoxic environments, beyond the already described hydrolytic release of phosphine from phosphide steels. Phosphine formation in nature remains enigmatic and a challenging problem since it is still considered the only solution to the phenomenon of methane self-ignition over swamps called "will-o'-the-wisp". However, the extremely low phosphine concentrations detected in natural environments can hardly give rise to a substantial methane ignition.

The ecological role of dissimilatory phosphite oxidation and assimilatory phosphite utilization in nature poses another problem to be discussed. On today's oxygen-exposed earth, phosphite and other reduced inorganic phosphorus compounds are not stable as minerals. In prebiotic and archaean times, reduced phosphorus compounds might have been more important. Phosphonic acids have been suggested as evolutionary precursors of biological phosphate compounds [88,89]. An involvement of phosphite radicals in the synthesis of phosphonoacetaldehyde, an analogue of glycolaldehyde phosphate, has been reported [90]. Reduced phosphorus compounds might have been introduced by meteorites [91] or might have been produced during core formation of the early Earth [88]. Recently, electrical discharges associated with volcanic eruptions were proposed as possible sources of phosphite formation [92]. Thus, it appears plausible to speculate that the (oxygen-free) early Earth might have contained reduced phosphorus compounds. Dissimilatory phosphite oxidation and assimilatory phosphite utilization may therefore be regarded as an ancient evolutionary trait.

6. GENERAL CONCLUSIONS

Phosphorus is an important element in biochemistry and plays a key role as a constituent of essential cell components and of the central energy carrier ATP. Numerous new biochemical reactions including also redox changes of phosphorus have been discovered in the recent past which indicates that the biochemistry of this element still provides challenges and surprises.

Eutrophication of surface waters due to excess discharge of phosphates is no longer a problem in Central Europe after the banning of polyphosphate-containing laundry detergents and a better control of phosphate distribution in agricultural fertilization. The exploitation of phosphate accumulation by bacteria in advanced sewage treatment technology may further help to control and minimize the phosphate discharge into rivers and lakes, also in threshold countries which are just at the state of establishing efficient sewage treatment devices.

Natural compounds containing carbon-phosphorus linkages make up only a very small fraction of biomass, especially in bacterial and fungal secondary metabolites, and the biosynthesis of these compounds has been studied in detail. A broad variety of bacterial reactions has been described which can degrade natural and synthetic compounds containing C–P linkages. Some of these enzyme systems underly control by phosphate limitation; others are independent of phosphate supply and allow the degradation of the organic moieties of organophosphonates as energy sources. Especially the C–P lyase enzymes react with a broad variety of substrates and allow microbial degradation also of unusual man-made C–P containing herbicides and biocides. New C–P-cleaving enzymes are likely to be discovered in the future when new types of C–P substrates are released into the environment.

Redox changes of phosphorus between the $+V$, $+III$ and $+I$ state are catalyzed by microbes not only with phosphorus linked to organic residues but also with inorganic phosphorus compounds. Only one enzyme has been described in detail, an NAD^+-dependent phosphite dehydrogenase which supplies *Pseudomonas stutzeri* with phosphite as phosphorus source. Phosphite can also serve as electron donor for the energy metabolism of a sulfate-reducing bacterium which couples phosphite oxidation with sulfate reduction. The phosphite-oxidizing enzyme in this organism has not been characterized yet but is likely to form an energy-rich phosphate linkage in the dehydrogenation step.

The formation of phosphine as a product of microbial metabolism is still enigmatic and cannot be associated so far with any defined microbial culture. Phosphine cannot be formed from phosphate by microbial activities to any significant amounts, but traces of phosphine may be formed through metabolic side reactions that have not been understood yet. One of the key sources of phosphine in anoxic sediments is probably the chemical hydrolysis of phosphides in steels which are released with scrap metals, etc.

ACKNOWLEDGMENTS

Work on phosphorus redox reactions was initiated in my lab on the basis of some enrichment cultures set up by an enthusiastic undergraduate student, Michael Friedrich. He as well as Volker Thiemann and Heike Laue have contributed to the experimental work in our lab on this subject. Markus Göbel helped with the drawings for this contribution. My interest in phosphorus biochemistry was aroused by Dr. Ralph S. Wolfe who introduced me to the phenomenon of will-o'-the-wisp. He and his colleague, Dr. William Metcalf, maintained my interest in this matter and stimulated my further activities in this field.

ABBREVIATIONS

ADP	adenosine 5'-diphosphate
AMP	adenosine 5'-monophosphate
ATP	adenosine 5'-triphosphate
K_S	half-saturation concentration, analogous to Michaelis constant K_M
NAD^+	nicotinamide adenine dinucleotide
NADH	nicotinamide adenine dinucleotide (reduced)
P_i	inorganic phosphate
poly-P_n	polyphosphate

REFERENCES

1. Wiberg N. Holleman-Wiberg's Lehrbuch der anorganischen Chemie, 33rd ed. Berlin: Walter de Gruyter, 1985:1307.
2. Emsley J. The Shocking History of Phosphorus: A Biography of the Devil's Element. Macmillan, 2000:1–326.
3. Weast RC, Astle MJ, Beyer WH, eds. CRC Handbook of Chemistry and Physics. Boca Raton, Florida: CRC Press, 1988.
4. Jahnke RA. In: Butcher SS, Charlson RJ, Orians GH, Wolfe GV, eds. Global Biogeochemical Cycles. London: Academic Press, 1992:301–315.
5. Stumm W, Morgan JJ. Aquatic Chemistry, 2d ed. New York: Wiley & Sons, 1981:568.
6. Wetzel RG. Limnology, 3rd ed. San Diego: Academic Press, 2001:1–841.
7. Shimogawara K, Wykoff DD, Usuda H, Grossman AR. Plant Physiol 1999; 120:685–694.
8. Wanner BL. J Cell Biochem 1983; 51:47–54.
9. Hulett FM. Mol Microbiol 1996; 19:933–939.
10. Ishige T, Krause M, Bott M, Wendisch VF, Sahm H. J Bacteriol 2003; 185:4519–4529.
11. Aiba H, Nagaya M, Mizuno T. Mol Microbiol 1993; 8:81–91.

12. Salins LL, Deo SK, Daunert S. Sens Actuators B Chem 2004; 1:81–89.
13. Thauer RK, Jungermann K, Decker K. Bacteriol Rev 1977; 41:100–180.
14. Stanier RY, Pfennig N, Trüper HG. In: Starr MP, Stolp H, Trüper HG, Balows A, Schlegel HG, eds. The Prokaryotes. Chapter 7, Berlin, Heidelberg, New York: Springer Verlag, 1981:197–211.
15. Marschner H. Mineral Nutrition of Higher Plants. London: Academic Press, 1995:1–273.
16. Rausch C, Bucher M. Planta 2002; 216:23–37.
17. Maldonado-Mendoza IE, Dewbre GR, Harrison MJ. Mol Plant Microbe Interact 2001; 14:1140–1148.
18. Harrison MJ, Dewbre GR, Liu J. Plant Cell 2002; 14:2413–2429.
19. Liu ST, Lee LY, Tai CY, Hung CH, Chang YS, Wolfram JH, Rogers R, Goldstein AH. J Bacteriol 1992; 174:5814–5819.
20. Babu-Khan S, Yeo TC, Martin WL, Duron MR, Rogers RD, Goldstein AH. Appl Environ Microbiol 1995; 61:972–978.
21. Rodriguez H, Fraga R. Biotechnol Adv 1999; 17:319–339.
22. Krishnaraj PU, Goldstein AH. FEMS Microbiol Lett 2001; 205:215–220.
23. Wood HG, Clark JE. Annu Rev Biochem 1988; 57:235–260.
24. Kortstee GJJ, Appeldoorn KJ, Bonting CFC, van Niel EWJ, van Veen HW. FEMS Microbiol Rev 1994; 15:137–153.
25. van Groenestijn JW, Deinema MH, Zehnder AJB. Arch Microbiol 1987; 148:14–19.
26. Bonting CFC, Kortstee GJJ, Zehnder AJB. J Bacteriol 1991; 173:6484–6488.
27. Bonting CFC, Kortstee GJJ, Zehnder AJB. Arch Microbiol 1992; 158:139–144.
28. Kortstee GJJ, Appeldoorn KJ, Bonting CFC, van Niel EWJ, van Veen HJ. In: Schink B, ed. Advances in Microbial Ecology. Vol. 16. New York: Kluwer Academic/Plenum Publishers, 2000:169–200.
29. McGrath JW, Quinn JP. Adv Appl Microbiol 2003; 52:75–100.
30. Mullan A, Quinn JP, McGrath JW. Microb Ecol 2002; 44:69–77.
31. Hu ZR, Wentzel MC, Ekama GA. Water Res 2002; 36:4927–4937.
32. Lin YM, Liu Y, Tay JH. Appl Microbiol Biotechnol 2003; 62:430–435.
33. Shoji T, Satoh H, Mino T. Water Sci Technol 2003; 47:23–29.
34. Kononova SV, Nesmeyanova MA. Biochemistry (Moscow) 2002; 67:184–195.
35. Seidel HM, Freeman S, Seto H, Knowles JR. Nature 1988; 335:457–458.
36. Bowman E, McQueney M, Barry RJ, Dunaway-Mariano D. J Am Chem Soc 1988; 110:5575–5576.
37. Sarkar M, Hamilton CJ, Fairlamb AH. J Biol Chem 2003; 278:22703–22708.
38. Kim A, Kim J, Martin BM, Dunaway-Mariano D. J Biol Chem 1998; 273:4443–4448.
39. Kamigiri K, Hidaki T, Imai S, Murakami T, Seto H. J Antibiot 1992; 45:781–787.
40. Schwartz D, Alijah R, Nussbaumer B, Pelzer S, Wohlleben W. Appl Environ Microbiol 1996; 62:570–577.
41. Ternan NG, McGrath JW, McMullan G, Quinn JP. World J Microbiol Biotechnol 1998; 14:635–647.
42. Ghisalba O, Küenzi M, Ramos Tombo GM, Schär H-P. Chimia 1987; 41:206–215.
43. Jauregui J, Valderrama B, Albores A, Vazquez-Duhalt R. Biodegradation 2003; 14:397–406.
44. Baker AS, Ciocci MJ, Metcalf WW, Kim J, Babbitt PC, Wanner BL, Martin BM, Dunaway-Mariano D. Biochemistry 1998; 37:9305–9315.

45. McGrath JW, Hammerschmidt F, Quinn JP. Appl Environ Microbiol 1998; 64:356–358.
46. La Nauze JM, Rosenberg H, Shaw DC. Biochim Biophys Acta 1970; 212:332–350.
47. McMullan G, Quinn JP. Biochem Biophys Res Commun 1992; 184:1022–1027.
48. McMullan G, Quinn JP. J Bacteriol 1994; 176:320–324.
49. McGrath JW, Wisdom GB, McMullan G, Larkin MJ, Quinn JP. Eur J Biochem 1995; 234:225–230.
50. Ternan NG, Quinn JP. Syst Appl Microbiol 1998; 21:346–352.
51. Ternan NG, McMullan G. FEMS Microbiol Lett 2000; 184:237–240.
52. Klimek-Ochab M, Lejczak B, Forlani G. FEMS Microbiol Lett 2003; 222:205–209.
53. Ternan NG, McGrath JW, Quinn JP. Appl Environ Microbiol 1998; 64:2291–2294.
54. Ternan NG, Quinn JP. Biochem Biophys Res Commun 1998; 248:378–381.
55. Ternan NG, Hamilton JT, Quinn JP. Arch Microbiol 2000; 173:35–41.
56. Kulakova AN, Wisdom GB, Kulakov LA, Quinn JP. J Biol Chem 2003; 278:23426–23431.
57. Wackett LP, Shames SL, Venditti CP, Walsh CT. J Bacteriol 1987; 169:710–717.
58. Murata K, Higaki N, Kimura A. Biochem Biophys Res Commun 1988; 157:190–195.
59. Murata K, Higaki N, Kimura A. Agric Biol Chem 1989; 53:1225–1229.
60. McMullan G, Watkins R, Harper DB, Quinn JP. Biochem Intern 1991; 25:271–279.
61. Vandepitte V, Verstraete W. Biodegradation 1995; 6:157–165.
62. Frost JW, Loo S, Cordeiro ML, Li D. J Am Chem Soc 1987; 109:2166–2171.
63. Lerbs W, Stock M, Parthier B. Arch Microbiol 1990; 153:146–150.
64. Jacob GS, Garbow JR, Hallas LE, Kimack NM, Kishore GM, Schaefer J. Appl Environ Microbiol 1988; 54:2953–2958.
65. Metcalf WW, Wanner BI. J Bacteriol 1991; 173:587–600.
66. Wanner BL, Metcalf WW. FEMS Microbiol Lett 1992; 100:133–140.
67. Heinen W, Lauwers AM. Arch Microbiol 1974; 95:267–274.
68. Imazu K, Tanaka S, Kuroda A, Anbe Y, Kato J, Ohtake H. Appl Environ Microbiol 1998; 64:3754–3758.
69. Metcalf WW, Wolfe RS. J Bacteriol 1998; 180:5547–5558.
70. Garcia Costas AM, White AK, Metcalf WW. J Biol Chem 2001; 276:17429–17436.
71. Vrtis JM, White AK, Metcalf WW, van der Donk WA. J Am Chem Soc 2001; 123:2672–2673.
72. Schink B, Friedrich M. Nature 2000; 406:37.
73. Schink B, Thiemann V, Laue H, Friedrich MW. Arch Microbiol 2002; 177:381–391.
74. Buckel W. Angew Chem 2001; 113:1463–1464.
75. Barrenscheen HK, Beckh-Widmanstetter HA. Biochem Z 1923; 140:279–283.
76. Rudakow KJ. Centralbl Bakteriol Infektionskr 1929; 29:229–245.
77. Devai I, Felföldy L, Wittner I, Plosz S. Nature 1988; 333:343–345.
78. Schink B. In: Lengeler JW, Drews G, Schlegel HG, eds. Biology of the Prokaryotes, Chapter 30, Ecophysiology and Ecological Niches of Prokaryotes. Stuttgart: Thieme, 1999:723–762.
79. Roels J, Verstraete W. Bioresource Technol 2001; 79:243–250.
80. Gassmann G, Schorn F. Naturwissenschaften 1993; 80:78–80.
81. Devai I, DeLaune RD. Org Geochem 1995; 23:277–279.
82. Tsubota G. I Soil Plant Food 1959; 5:10–15.
83. Han S-H, Zhuang Y-H, Liu J-A, Glindemann D. Sci Tot Environ 2000; 258:195–203.

84. Devai I, DeLaune RD, Devai G, Patrick WH, Czegeny I. Anal Lett 1999; 32:1447–1457.
85. Glindemann D, Eismann F, Bergmann A, Kuschk P, Stottmeister U. Environ Sci Pollut Res 1998; 5:71–74.
86. Rutishauser BV, Bachofen R. Anaerobe 1999; 5:525–531.
87. Jenkins RO, Morris T-A, Craig PJ, Ritchie AW, Ostah N. Sci Tot Environ 2000; 250:73–81.
88. Schwartz AW. Origins Life Evol Biosphere 1997; 27:505–512.
89. Schwartz AW. J Theor Biol 1997; 187:523–527.
90. De Graaf RM, Visscher J, Schwartz AW. Nature 1995; 378:474–477.
91. Cooper GW, Onwo WM, Cronin JR. Geochim Cosmochim Acta 1992; 56:4109–4115.
92. Glindemann D, De Graaf RM, Schwartz AW. Origins Life Evol Biosphere 1999; 29:555–561.

Iron, Phytoplankton Growth, and the Carbon Cycle

Joseph H. Street and Adina Paytan

*Department of Geological and Environmental Sciences, Braun Hall,
Stanford University, Stanford, California 94305-2115, USA*

1.	Iron, an Essential Nutrient for Marine Organisms		154
	1.1.	Iron Function in Cells	154
	1.2.	Cellular Uptake Mechanisms	155
	1.3.	Chemical and Physical Limits on Uptake Rates	156
	1.4.	Expression of Stress	157
2.	Iron Chemistry in Seawater		158
	2.1.	Chemical Forms	158
	2.2.	Speciation and Redox Chemistry	159
	2.3.	Interaction with Organic Compounds	159
3.	Iron Distribution and Cycling in the Ocean		160
	3.1.	External Iron Sources	160
		3.1.1. Atmospheric Dust Deposition	160
		3.1.2. River Input	162
		3.1.3. Hydrothermal Sources	163
		3.1.4. Mobilization in Sediments	163
	3.2.	Iron Cycling in the Ocean	164
		3.2.1. Upwelling and Advection	164
		3.2.2. Biological Cycling, Export, and Regeneration	165
		3.2.3. Particle Scavenging	166
	3.3.	Patterns in the Distribution of Iron in the Ocean	166
		3.3.1. Dissolved Iron in the Open Ocean	167

3.3.2.	Dissolved Iron in the Coastal Ocean, Shallow Seas, and Semi-enclosed Basins	169
3.3.3.	Particulate Iron	169
3.3.4.	Suboxic/Anoxic Zones	169

4. Iron Limitation of Marine Primary Productivity and Control on Ecosystem Structure   170
  4.1. Evidence for the Role of Iron in Regulating Productivity in High Nutrient Low Chlorophyll Regions   170
    4.1.1. Bottle Incubation Experiments   171
    4.1.2. Mesoscale Enrichment Experiments   173
    4.1.3. Evidence of Iron Limitation from Other Regions   176
  4.2. Interaction of Iron with Other Limiting Factors   177
  4.3. Carbon Export and Iron Fertilization   180
5. The Role of Iron in Regulating Atmospheric CO_2 180
  5.1. The Iron Hypothesis   180
  5.2. Changes in Dust Input and Productivity in Glacial Periods   181
  5.3. Consequences for Atmospheric CO_2 and Global Climate   182
6. Summary and Conclusions   183
Acknowledgments   184
Abbreviations   184
References   185

1. IRON, AN ESSENTIAL NUTRIENT FOR MARINE ORGANISMS

1.1. Iron Function in Cells

Iron is an essential nutrient supporting the growth and metabolism of marine organisms. Iron-bearing molecules are involved in photosynthetic and respiratory electron transport, nitrate and nitrite reduction, N_2 fixation, sulfate reduction, and, as enzyme components, in the detoxification of reactive oxygen species such as O_2^- and H_2O_2 [1].

Relative to other nutrients, iron is disproportionately utilized in cell metabolism over cell structure. As a result, cellular Fe : C ratios tend to vary with Fe availability (and, by extension, metabolic rate), in contrast to the more or less fixed ratios of C : N : P measured in marine phytoplankton [2,3]. Fe : C ratios measured in pelagic phytoplankton and marine particulate matter range from 1 : 3000 to 1 : 500,000 [3–7]. This variation in cellular Fe content is intrinsic to different classes of phytoplankton as has been shown in culture experiments [8]. In particular, because iron is present in the catalytic centers of nitrate- and nitrite-reductase enzymes, in reducing equivalent molecules (NADPH, ferredoxin) and in the N_2-fixing enzyme nitrogenase, organisms assimilating nitrate as a nitrogen source or fixing N_2 from the atmosphere require much more cellular iron to support growth than those utilizing reduced nitrogen species [1,9–12].

1.2. Cellular Uptake Mechanisms

By almost any measure, iron is vanishingly scarce in seawater, and much of the iron that is present exists in forms not directly accessible to organisms (Sections 2 and 3). Marine phytoplankton are for the most part restricted to the uptake of dissolved Fe species, although Fe can be solubilized from particles or colloids. Marine phytoplankton have evolved several strategies for efficiently acquiring iron and transporting it across the hydrophobic, semi-impermeable lipid bilayer membrane surrounding most cells. The general mechanism involves the use of cell-regulated membrane transport proteins ("transporters"), but prokaryotic and eukaryotic cells have each developed distinct variations on this approach.

Prokaryotic microorganisms, including marine phototrophic and heterotrophic bacteria, are known to acquire Fc(III) by means of sidcrophorc-based transport systems. In these systems, cells under Fe stress synthesize and release low-mass, high-affinity Fe(III)-chelating molecules ("siderophores") into the surrounding seawater, where they capture and solubilize otherwise unavailable Fe species (e.g., oxide minerals, particle-bound, organic complexes) [13–16]. The Fe-siderophore chelates are then taken up via chelate-specific transporters, which move the complex across the membrane and release Fe inside the cell via reduction to Fe(II) or degradation of the siderophore itself. Seawater surveys have indicated that a large fraction of the bacteria were capable of producing strong Fe-binding chelators (probably siderophores) [17–20]. Siderophore transport systems, sometimes utilizing novel chelating agents, have also been described in many cyanobacteria [11,21,22] (Fig. 1).

In contrast to marine bacteria, there is little evidence for the production of siderophores by eukaryotic marine phytoplankton [23,24]. Rather, eukaryotes appear to utilize transporters that directly capture dissolved inorganic iron at the cell membrane. Experiments with two diatoms and a coccolithophore found that for these organisms Fe uptake depended on the concentration of labile inorganic Fe and was independent of the concentration of total iron [5,23,25]. Tracer experiments have shown that phytoplankton will make use of iron bound to siderophores or other organic ligands only after photo- or biologically-mediated dissociation from the ligand [15]. It is likely that the transporters involved in Fe uptake are transmembrane proteins, powered by ATP or cross-membrane electrochemical gradients, and controlled by a DNA switch that represses gene transcription and hence synthesis of transport proteins when intracellular Fe concentrations become elevated [1,25,26].

Certain classes of photosynthetic protozoan marine algae are able to ingest cells or abiotic particles directly as supplementary sources of fixed carbon and essential nutrients, including the bulk of the iron required to support photosynthesis and respiration [27]. These "mixotrophic" organisms

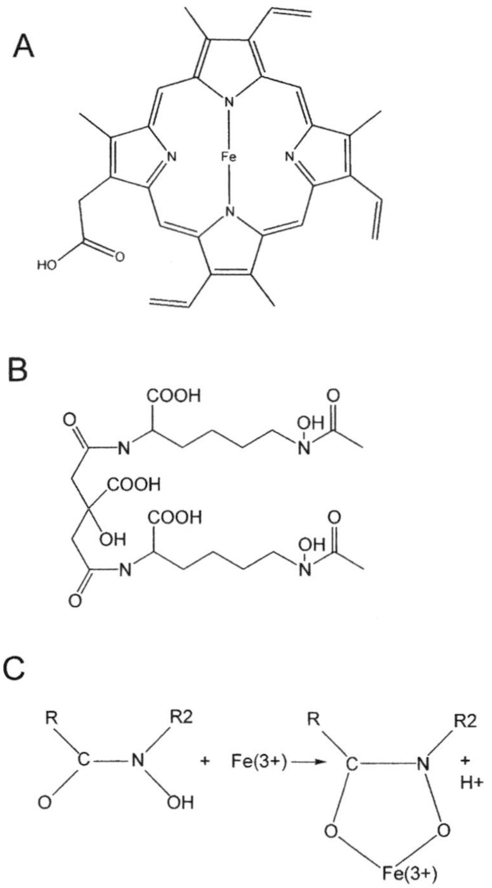

Figure 1 (A) Heme, one of several iron-bearing porphyrin compounds, occurring in cytochromes and other metabolic proteins in marine microorganisms; (B) Alterobactin, a siderophore produced by marine bacteria. Reproduced from Ref. [283] by courtesy of the American Society of Limnology and Oceanography; (C) Chelation of Fe^{3+} by a hydroxamate siderophore. Adapted from Ref. [284] by courtesy of Marcel Dekker, Inc.

have been shown to be able to sustain moderate growth rates (35–70% of maximum) using only ingested mineral colloids as Fe sources in seawater that has been artificially depleted in dissolved iron [28].

1.3. Chemical and Physical Limits on Uptake Rates

Iron uptake rates for most phytoplankton species appear to be limited by the rate of cross-membrane transport, itself a function of ligand exchange kinetics and the amount of space available on the outer cell membrane for transporter proteins

[1,23,29]. This conclusion is born out by experiments with coastal diatoms and dinoflagellates of a range of cell sizes (3.5–32 μm), which show that cell surface- and volume-normalized Fe uptake rates, as well as specific growth rates, flatten out with increasing dissolved Fe concentrations. Although part of this effect may have resulted from Fe-hydroxide precipitation at high Fe concentrations, cell surface normalized Fe uptake was similar for all species, regardless of size, consistent with the notion of uptake rates near their physical and chemical limits [30]. This finding was replicated in a comparison of an oceanic diatom (*Thalassiosira oceanica*) with a related coastal diatom (*T. pseudonana*) [1]. However, in spite of identical surface-normalized uptake rates, the oceanic diatom grew much faster than the coastal species at low Fe concentrations. These results suggest that oceanic species have adapted to the low iron conditions of the open ocean by increasing their Fe-use efficiency and reducing their growth requirement for iron by up to sixfold compared to coastal phytoplankton, which have adapted to higher ambient Fe concentrations [4,5,10].

Large cells with low surface : volume ratios are at a disadvantage with regard to iron uptake, especially in iron-poor environments. In the experiments discussed earlier, the dinoflagellate *Prorocentrum micans* (30 μm) achieved surface-normalized uptake rates similar to those of smaller species, but lower volume-normalized uptake rates because of its larger size. This translated into lower intracellular Fe : C ratios and lower growth rates relative to the smaller species, especially at low Fe concentrations [30]. Moreover, for large cells (>60 μm), diffusion of bioavailable Fe species to the cell surface becomes a significant factor limiting uptake rates [29]. The cumulative effect of these physical limits is perhaps most clearly illustrated by the widespread observation that small-celled phytoplankton communities dominate the iron-limited, high nutrient low chlorophyll (HNLC) regions of the ocean [31–35].

Models of carrier-mediated Fe uptake indicate that the transport process is also limited by the rate of reaction of labile Fe binding to the membrane transporter [1,23]. Observed rates of this reaction closely match those for the reaction of dissolved Fe(III), abbreviated here as Fe(III)′, with microbial siderophores [36,37] and with the naturally-occurring strong organic ligands in the surface ocean [36,38–40]. Because Fe-chelates and colloids have slow ligand exchange kinetics, it follows that these forms would be essentially unavailable for direct uptake by phytoplankton [1]. Experiments indicate that Fe(III)′ are the species actually taken up by phytoplankton, hinting at the existence of an auto-oxidation step following the binding of Fe(II) to the membrane transporter [23].

1.4. Expression of Stress

Cells growing under suboptimal iron conditions undergo a number of physiological and functional changes, some of which can be measured as indicators of iron stress. Iron-stressed phytoplankton are less photosynthetically efficient than their iron-replete counterparts [41,42], a response that has been clearly demonstrated in the many Fe enrichment studies conducted in low iron regions of the ocean (Section 4). In iron-stressed communities, average cell volumes decline, and cellular ratios of major elements (C, N, P, Si) to Fe and to one another will change [5,10,43]. Not coincidentally, chronically iron-limited ecosystems are dominated by small species with high Fe-use efficiencies (Section 4).

Many iron-stressed microorganisms, including diatoms, replace the Fe-containing redox protein ferredoxin with flavodoxin, a redox protein of similar function containing riboflavin $5'$-phosphate as a cofactor instead of Fe [44–46]. This substitution of proteins lowers the intracellular iron requirements of diatoms [47], and provides oceanographers with a potentially powerful probe for identifying iron-limited ocean regions and teasing apart differences in iron stress among different phytoplankton taxa. Flavodoxin abundances (relative to total protein) in diatom extracts have been used to trace a gradient of increasing iron stress moving from coastal to open ocean areas in the eastern subarctic Pacific, providing further confirmation of iron limitation in this region [47].

2. IRON CHEMISTRY IN SEAWATER

Iron is the fourth most abundant element in the Earth's crust, yet is scarce as a dissolved element in seawater to the point of limiting primary productivity in many parts of the ocean. Although biological uptake greatly intensifies this scarcity, iron availability in seawater is ultimately limited by the inorganic chemistry of Fe in the oxic, slightly basic (pH $\sim$ 8) conditions that dominate the modern ocean.

2.1. Chemical Forms

Iron occurs in seawater in multiple forms, particulate and dissolved, organic and inorganic, over a wide range of physical size classes. Particulate Fe, commonly defined as that fraction removed by filtration through 0.2 or 0.4 μm pore size filters, is contained in particles of alumino-silicate clays, the intact cells of marine microorganisms, and a variety of biogenic detritus, ranging in size from submicron to >100 μm in diameter [48]. Dissolved Fe (<0.2–0.4 μm) consists largely of iron complexed with organic molecules, including several classes of organic chelators, cell lysis products, and humic compounds [36,38,49–51]. Inorganic iron hydroxides comprise a small but important fraction of the dissolved Fe pool, while free iron (Fe^{3+}, Fe^{2+}) is all but absent.

A significant portion of operationally defined "dissolved Fe" actually consists of small (0.02–0.45 μm) colloidal particles, ranging from $<10\%$ of total dissolved Fe in open ocean samples to 90% in coastal waters [52–54]. It has traditionally been assumed that dissolved Fe is equivalent to bioavailable Fe; this is complicated by ligand binding, which keeps Fe in solution but not in a directly accessible form, and by the large colloidal fraction of the dissolved pool, which is not directly bioavailable [54].

2.2. Speciation and Redox Chemistry

Iron can exist in two oxidation states, Fe(III) or Fe(II), free or complexed with inorganic or organic ligands. In oxic seawater, Fe(III) is the dominant state: thermodynamically stable, highly reactive with respect to hydrolysis, adsorption and complex formation, and as a result largely unavailable as dissolved, free Fe^{3+}. The inorganic speciation of iron is dominated by a host of low solubility Fe(III) hydrolysis products ["Fe(III)'"] [1,37,55–57]. Over time these species undergo progressive dehydration and crystallization to more insoluble forms, eventually leading to the formation of stable, insoluble iron oxide minerals [1]. The Fe hydrolysis species precipitate as hydroxides at concentrations of ~ 0.7 nM [5]; as they age their solubility decreases further to levels in the range of 0.1–0.3 nM [56–58]. Fe(III)' will also adsorb to particle surfaces with high affinity, and hence are readily scavenged from the water column during particle settling [1].

Species of the more soluble and kinetically labile Fe(II) redox state are intermittently present in seawater as a result of Fe(III) reduction by a variety of processes in different ocean environments. Chemical and/or microbial reduction of Fe(III)' occurs on a large scale in anoxic basins and sediments (Sections 3.1.4 and 3.3.4) and on a microscopic scale within the fecal pellets of zooplankton. In the surface ocean reduction occurs via absorption of high visible-low UV light (photo-reduction) [51,59–65], and via biologically-mediated reactions at cell surfaces [12,66–68].

In contrast to their Fe(III) counterparts, inorganic Fe(II) species ["Fe(II)'"] are very soluble and Fe(II)–organic complexes only weakly bound [1]; the immediate result of the various reduction processes is intermittent increases in the availability of dissolved, labile Fe species. Once released in oxic waters, however, dissolved Fe(II)' are quickly reoxidized by dissolved oxygen or H_2O_2 [69] and reprecipitated as Fe(III) hydroxides or reincorporated into complexes with organic ligands. Nonetheless, the net result of this iron redox cycling is to increase the concentrations of dissolved Fe(II)' and Fe(III)' in seawater. These dissolved Fe species are the most important iron source to phytoplankton.

2.3. Interaction with Organic Compounds

The near-ubiquity of natural Fe-binding organic ligands and the critical importance of organic complexation in controlling the speciation and distribution of dissolved iron have been among the more startling realizations in chemical oceanography in the last decade. Measurements using competitive ligand equilibration/cathodic stripping voltammetry techniques have shown that >99% of dissolved Fe(III)' in seawater collected from a wide variety of ocean regions is bound to organic chelators [36,38,49,50,70]. In the North Pacific, two distinct ligand classes of differing binding strengths were identified, with a total concentration (~ 2 nM) far in excess of dissolved Fe concentrations [38]. These direct measurements are supported by experimental work showing that Fe(III)' solubility is greater in natural seawater than in either artificial seawater or NaCl solution of equal ionic strength, presumably because of the Fe-binding ligands present in natural seawater [57]. Biogeochemical models of Fe cycling also seem to require organic chelators (or a functional equivalent) to explain the observed distribution of dissolved iron in the ocean [71].

Complexation with organic ligands decreases the concentrations of reactive Fe(III)', increasing Fe solubility (with respect to Fe-hydroxide formation) by a significant, and perhaps biologically crucial, increment on the order of 0.2–0.5 nM [56,57], and protects dissolved Fe from particle scavenging. The benefit derived by organisms from this state of affairs is obvious, and there is considerable evidence to suggest that many of the Fe-binding ligands in seawater have been released by microorganisms, protozoa, and zooplankton [40,72].

3. IRON DISTRIBUTION AND CYCLING IN THE OCEAN

The distribution of dissolved and particulate iron in the ocean is highly hetero-geneous, a complex function of spatially and temporally variable external inputs, lateral and vertical redistribution by ocean currents, active biological uptake, recycling, and export, and a host of chemical and physical processes that can add or remove iron from the dissolved, biologically-available pool. Measured dissolved iron concentrations in the ocean range over six orders of magnitude, from as low as 0.03 nM in the surface waters of open ocean HNLC regions to 3 mM in hydrothermal vent fluids, with all other marine environments falling between these extremes.

3.1.   External Iron Sources

External sources of iron to the ocean include atmospheric dust deposition, river input, hydrothermal vents, and release from marine sediments via reductive dis-solution and resuspension. The magnitudes of these sources are highly uneven in space and time, and for chemical and physical reasons, only a small fraction of the iron they deliver is retained to become bioavailable in the surface waters of the open ocean.

3.1.1.   Atmospheric Dust Deposition

Atmospheric deposition of continentally-derived particles ("aerosols", "dust") is a major external input of iron to the oceans. In some open ocean areas, including the iron-limited subarctic Pacific, dust deposition is the dominant iron source to the surface layer. The atmospheric iron source is composed largely of alumino-silicate minerals derived from arid and semi-arid mid-latitude regions in the Northern Hemisphere [73–75]; the exception to this rule is downwind of heavily populated areas, where a significant fraction of aerosol iron may originate from anthropogenic sources [76–78].

Aerosol production rates vary in response to precipitation amount and wind strength [79–81]. Typically only relatively small particles (~1–100 μm diameter) become entrained in the atmosphere. The largest atmospheric particles are deposited quickly via gravitational settling near the source area, but smaller size classes (<10 μm) reach high altitudes and can be transported hundreds or thousands of kilometers before being deposited [82–84]. Deposition occurs through either the direct "dry" settling of particles or the entrainment of particles in rainfall, termed "wet" deposition.

Deposition fluxes are difficult to measure, and vary over space and time [82,85,86]. The relative importance of dry vs. wet deposition also varies seasonally and from place to place. For instance, wet deposition of aerosol iron dominates in the Atlantic intertropical convergence zone (ITCZ) due to high rainfall [87], while most of the aerosol flux to the northwestern Pacific in the spring is in the form of dry deposition [88]. Ocean basins directly downwind of desert source areas, notably the northwest Pacific, north Atlantic, and Arabian Sea, receive high aerosol fluxes [88–94]. In contrast, measured aerosol concentrations at stations in the remote South Pacific, South Indian, and Southern Oceans are 1–3 orders of magnitude lower [95–99]. Even within high-flux basins, large latitudinal and longitudinal gradients exist, with aerosol concentrations, and hence iron fluxes, declining with distance from dust sources and primary wind-transport pathways [82,87,88,93,100]. Aerosol observations using TOMS and METEOSAT satellites over the period 1979–1997 also reveal extreme interannual variability in the emission and transport of Saharan dust over the Atlantic, especially from the fringing Sahel region [101,102] (Fig. 2).

The fraction of iron in soils and rocks at the Earth's surface varies, but generally falls between 2.9% and 4.8%, with a widely-used average of 3.5% [82,86].

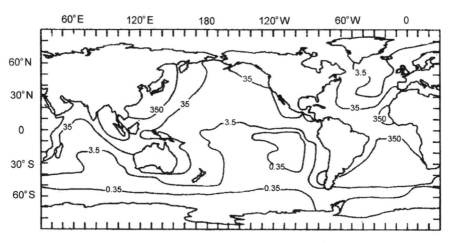

Figure 2 Atmospheric iron fluxes to the ocean, in $mg\ m^{-2}\ year^{-1}$. Adapted from Ref. [86], using an average Fe content for mineral dust of 3.5% [82].

Similar fractions have been measured in aerosol particles [103,104]. Applying this average composition to an estimated total aerosol flux of 472×10^{12} g year^{-1} [82], the total global atmospheric flux of iron to the oceans is on the order of $16-32 \times 10^{12}$ g year^{-1}.

Only a small fraction of the atmospheric flux of iron to the oceans ever becomes bioavailable, largely because of the low solubility of the particulate and colloidal Fe(III) phases that comprise the bulk of aerosol iron species. Nonetheless, bottle incubation experiments have shown that aerosol addition is an efficient stimulator of chlorophyll and biomass production in phytoplankton cultures [105,106], emphasizing the need to understand the factors and processes controlling the chemical speciation and solubility of aerosol iron before and after deposition.

Most aerosol particles in the atmosphere are exposed to low pH conditions (pH 1–5.5) during cycles of cloud formation and evaporation, which may enhance the release of dissolved Fe species from the particles [77,107–109] and facilitate high rates of photoreductive dissolution of particulate and colloidal Fe(III)' [59,110–112]. The net result is a significant, sometimes dominant component of Fe(II) species in atmospheric, solution-phase iron [77,108,109]. Complexation of Fe(III) with organic species in atmospheric water (e.g., oxalate, acetate, formate, humic compounds) may also increase the solubility and photoreactivity of aerosol iron [59,111,113–115]. During wet deposition, entrained aerosol iron is exposed to more neutral conditions, first in rainwater (pH 4–7) and then upon mixing with seawater (pH > 8), after which an estimated 0.3–6.8% of the aerosol iron remains in the dissolved phase. In theory, the high pH and ionic strength of seawater should promote a more complete conversion to particulate phases [1]; the observed solubility may be the result of organic complexation in the atmosphere or in seawater immediately after deposition. In contrast to wet deposited iron, the solubility of dry deposited iron in seawater is generally low ($\ll$1%) [77,103,108]. Iron derived from anthropogenically-influenced "urban" or coastal aerosols is more soluble (up to 12%) [78,108,112] than mineral aerosol iron due to a larger fraction of labile, exchangeable Fe, including Fe(II) [76,77,108,116].

In summary, it appears that although the atmospheric iron input to the global ocean is dominated by dry deposition ($\sim$70%), the input of soluble, bioavailable iron comes largely through wet deposition. Jickells and Spokes [82] combine these observations to estimate an overall solubility of atmospheric iron in seawater of 0.8–2.1% of the total Fe deposition flux, resulting in a soluble, bioavailable iron flux to the oceans of $13-67 \times 10^{10}$ g year^{-1}. This accounts for a large fraction of the iron input at many locations in the ocean [117–121].

3.1.2. River Input

The rivers and streams of the world discharge, on average, 37×10^{12} m^3 year^{-1} of water, carrying 4.6×10^{15} g year^{-1} of dissolved constituents and

$\sim$20 $\times$ 10^{15} g year^{-1} of suspended sediment. As with aerosol flux, river discharge is highly seasonal and geographically uneven. A large fraction of the global discharge is to the central Atlantic Ocean due to the action of very large rivers (Amazon, Orinoco, Congo, Mississippi); most of the global sediment load is borne by rivers draining the Himalayas [122,123].

Measured values of dissolved Fe in rivers vary widely, as illustrated by the Amazon and Danube, which have concentrations of 800–1000 and 21–156 nM, respectively [124,125]. Using an estimated riverine dissolved Fe concentration of 720 nM, de Baar and de Jong calculate a global flux of 26 $\times$ 10^9 mol dissolved Fe/year to the coastal zone [7]. However, experiments have shown that up to 90% of the "dissolved" fraction actually consists of small colloid particles ($<$0.4 μm) that tend to flocculate with organic matter to form larger particles when brought into contact with seawater, and are efficiently removed during estuarine mixing [126–128]. As a result, the dissolved Fe concentration of water that actually reaches the ocean is closer to 40 nM [126], yielding a global flux of only 1.5 $\times$ 10^9 mol Fe year^{-1} [7]. Likewise, most of the large riverine load of particulate iron (13 $\times$ 10^{12} mol year^{-1}) is deposited in deltas and estuaries and never reaches the open ocean [7].

3.1.3. Hydrothermal Sources

Fluids released from hydrothermal vents at mid-ocean ridges and in back-arc basins contain high concentrations of reduced iron acquired via reductive dissolution during circulation through basaltic rocks at high temperatures (350–400°C) and pressures (300–400 bar) [129–133]. Typical concentrations of dissolved Fe are 1–3 mmol L^{-1}, and at least at certain vents the Fe flux is dominated by Fe(III) oxyhydroxides. Though the gross hydrothermal input of iron to the deep ocean is large, an estimated 30–90 $\times$ 10^9 mol year^{-1} [7], most of the reduced Fe species in vent fluids are rapidly oxidized and precipitated upon mixing with cold ambient bottom water ($\sim$2°C) [134]. The net result is large ferromanganese deposits close to venting sites but a negligible flux of iron to deep ocean water [7].

3.1.4. Mobilization in Sediments

The flux of iron from marine sediments, though poorly constrained, is probably on the same order as the atmospheric flux, and is likely to be an important source term in ocean regions where dust input is low. A clear signal of this source is the strong increasing gradient of iron concentrations moving toward the continental margins [7,135]. Both deep marine clays and coastal and shelf muds are enriched in iron relative to crustal abundances, with average iron contents of $\sim$6% [7]. Roughly half of this iron exists in forms susceptible to reductive dissolution (oxide coatings, organics) and includes a component exported from overlying waters [136–139]. Diagenetic alteration of these iron-rich sediments releases high concentrations (10–100 μM) of Fe(II) into sediment pore waters

[136,140–143]. The vast majority of dissolved pore water iron diffusing through the sediment column will be immobilized, either through oxidation to insoluble Fe(III) species in oxic sediment layers or through sulfide mineral precipitation under euxinic conditions [144–146]. The small fraction that reaches the sediment–water interface is nonetheless enough to increase dissolved Fe concentrations at this boundary to 1–100 nM, well above the mean oceanic value. Much of this remaining dissolved Fe(II)' will then be reprecipitated as Fe(III)' or scavenged onto particles upon mixing with bottom water [138], with the remainder contributing to the dissolved iron inventory of deep waters [7]. The importance of the benthic flux as a source of iron to surface waters has been demonstrated for several parts of the ocean, especially along continental shelves [135,139, 147–149]. The spatial distribution of benthic iron sources is unknown, but is probably related to sedimentary oxygen profiles, which control the degree of Fe(II) production and escape from sediments [7].

3.2. Iron Cycling in the Ocean

3.2.1. Upwelling and Advection

Ocean currents and other mixing processes play a central role in redistributing iron horizontally and vertically throughout the ocean [150]. The effect of advective transport of iron from a source area to an iron-limited area can be clearly seen in the equatorial Pacific in the vicinity of the Galapagos Islands: Very low dissolved Fe concentrations ($\sim$0.05 nM) are measured east or up-current of the islands, while concentrations down-current of and within the island group itself are much higher (1–3 nM) [151,152]. In the coastal waters of California, which receive non-negligible river and dust inputs, pulses of shelf-derived Fe iron in spring upwelling events appear to influence the productivity and pattern of summer diatom blooms [147,153,154]. Even in iron-limited regions the advective action of mesoscale eddies can produce scattered patches of relatively Fe-enriched waters [155].

In remote regions of the open ocean that receive little atmospheric input, the vertical transport of iron-rich deep waters is particularly important [73,156]. The primary source of iron to the Fe-limited Southern Ocean is entrainment of Fe-enriched water during deep winter mixing, augmented by the summer upwelling of circumpolar deep water at the Polar Front and the southern front of the Antarctic Circumpolar Current [119,155,157].

Attempts to estimate global upwelling fluxes of iron to the euphotic zone have yielded sharply divergent results. The model of Fung et al. [120] calculated a total upwelling flux of 0.7×10^9 mol Fe year^{-1}, a small fraction of a global source term dominated by atmospheric dust input. Even in the dust-starved Southern Ocean, upwelling provided only 3.3% of the Fe input [120]. In contrast, Moore et al. estimated iron input to the surface layer from a variety of physical processes, including upwelling, entrainment, and turbulent mixing, and arrived at a much larger global flux of $\sim 14 \times 10^9$ mol Fe year^{-1}, although inclusion of physical loss terms significantly reduces this flux [121].

Watson's Fe budget for the HNLC regions attributes a relatively large fraction (19–99%) of the total Fe input to the surface ocean to upwelling processes, especially in the equatorial Pacific and Southern Ocean [156]. The iron cycle models of Archer and Johnson predict that 55–85% of oceanic carbon export production is supported by upwelled and entrained iron [154]. The differences among the estimates reflect the great uncertainty in these calculations, but all analyses suggest an important role for upwelled Fe, especially in the equatorial Pacific and Southern Ocean HNLC regions, where aerosol inputs are low relative to the rest of the ocean.

3.2.2. Biological Cycling, Export, and Regeneration

The high biological demand for iron relative to its abundance in the surface ocean results in a biogeochemical cycle that, at least in surface waters, is dominated by organisms. Fe uptake by prokaryotic and eukaryotic photosynthesizers, diazotrophs, and heterotrophic bacteria in the mixed layer is the largest single flux of the element in the ocean system, amounting to $21–26$ ($\times 10^9$) mol Fe year^{-1} [120,121]. Globally, the largest fraction of this Fe uptake appears to be performed by small phytoplankton ($\sim 73\%$), followed by diatoms ($\sim 22\%$) and diazotrophs ($\sim 5\%$). The large Fe uptake by diazotrophs relative to their contribution to primary production ($\sim 1\%$) reflects the high Fe requirements of N-fixation [121].

Iron taken up by microorganisms is subject to intense recycling in the surface layer, carried out predominately by a more or less perennial "small food web" of nanoplankton and microzooplankton that is based on the efficient reuse of nutrient resources [158–161,162]. The remineralization of iron is enhanced by passage through the digestive tracts of zooplankton, where low pH conditions promote iron dissolution [161,163]. Seasonally, or when favorable conditions of light and nutrient supply coincide, the background communities in many marine ecosystems are augmented by blooms of large eukaryotic phytoplankton, notably diatoms. These intermittent "large food webs" support the secondary production of large zooplankton and higher trophic levels. It is during and after large phytoplankton blooms that most export production of carbon, iron, and other nutrients to the deep ocean occurs, driven by the direct sinking of large deceased phytoplankton (e.g., diatom frustules) or the indirect pathway through zooplankton fecal pellets [7,164,165]. Though much of the nutrient detritus of a phytoplankton bloom will be recycled within the mixed layer, at least some fraction will pass into the deep ocean. Export production is often enhanced by the diurnal migration of zooplankton from surface feeding zones down to depth (500–1000 m), where most defecation occurs. Moore et al. have estimated that $\sim 78\%$ of the global iron uptake by phytoplankton is remineralized within the surface mixed layer, with the remainder lost to detrital sinking, mixed layer shoaling, and turbulent mixing [121].

3.2.3. Particle Scavenging

The Fe(III) species that make up the bulk of the free dissolved iron in the ocean are highly particle reactive and are readily scavenged from seawater by settling mineral particles and biogenic debris [7,71,166,167]. These particles may contain strong adsorptive agents, such as ferromanganese coatings, clay minerals, or organic functional groups with particular affinity for hydrolyzed trace metal species [7,167,168]. The rate of particle scavenging depends on the dissolved iron concentration and the abundance and size distribution of adsorptive particles.

Fine particles (<10 µm), with long residence times and minimal settling velocities, will scavenge Fe but remain more or less stationary in the water column, allowing for the development of a competitive equilibrium between Fe in adsorbed and dissolved states. Coarse particles are largely biogenic in origin and associated with phytoplankton blooms [7,169]. Relatively infrequent "rains" of large biogenic particles, with settling velocities of tens to hundreds of meters per day, have been observed to sweep out most of the small particles in the water column, along with the particle-bound iron [7,164,165]. Comparable scavenging events have been observed during and after heavy mineral dust deposition in ocean regions near dust sources [76,108,116,170,171]. Below the photic zone, particle scavenging is the major removal process for dissolved iron. Scavenging counterbalances Fe regeneration by the decomposition of organic matter, and sets a limit on dissolved iron concentrations and residence time in the deep ocean. It has been suggested that scavenging loss of Fe is responsible for the low deep water Fe : P ratios (relative to Fe : P in sinking organic matter) observed in both the Atlantic and the Pacific. Where these low Fe : P waters are ventilated at the surface, they may play a role in the development of iron limited, HNLC conditions [172].

3.3. Patterns in the Distribution of Iron in the Ocean

The global coverage of observations of iron concentrations in the open ocean is sparse and unevenly spread, and even in the better-sampled areas sampling is seldom adequate to capture the short-order spatial and temporal variability that characterizes much of the ocean. In coastal areas, especially along the Atlantic Rim and in the semi-enclosed seas of Europe, the range and variation in iron distribution has been better characterized. The North Atlantic and North Pacific basins have been sampled relatively frequently, in different seasons and under a variety of conditions, whereas vast tracks of ocean, especially in the Southern Hemisphere, are routinely characterized based on just a handful of measurements. The exception to this north–south imbalance is the Southern Ocean, where investigation of iron limitation in high nutrient, low chlorophyll waters has resulted in extensive sampling. In spite of these shortcomings, enough trustworthy data has been collected to identify the dominant patterns in the distribution of iron in the ocean. De Baar and de Jong, among others, provide an exhaustive compilation of iron observations from around the world [7]; for the purposes of this chapter, only general patterns will be discussed.

3.3.1. Dissolved Iron in the Open Ocean

The vertical distribution of dissolved iron in the vast majority of the ocean is similar to that of other biologically-assimilated nutrients, notably nitrate and phosphate, and characterized by the following key features: (a) depletion at the surface, driven by biological uptake; (b) a mid-depth maximum, correlated to the rate of export production from the photic zone and the result of microbial regeneration of exported material; (c) reasonably stable, invariant concentrations at depth [71]. Iron, however, differs from the macronutrients in notable, sometimes surprising ways, which raise basic questions about how its "nutrient-like" profile is maintained (Fig. 3).

Dissolved iron concentrations in the surface waters of the open ocean are the net result of the suite of dynamic source, removal, and cycling processes discussed in previous sections. The combined effects of biological uptake, particle scavenging, and the low solubility of $Fe(III)'$ restrict dissolved iron to nanomolar concentrations nearly everywhere; Fe recycling, mediated by zooplankton grazing and the microbial food web, along with Fe-binding organic ligands and the diurnal process of photoreduction, partially counterbalance the removal processes. Nonetheless, the residence time of dissolved iron in the surface layer is very short, on the order of 1–2 months [7]. Dissolved iron concentration gradients are evident moving toward or away from external sources. A classic example is the subarctic Pacific, where iron concentrations generally decrease from east to west as the strength of the Asian dust source declines [118]. The strong increasing gradient moving onshore toward islands, coastlines, and continental shelves is a related pattern, a particularly dramatic example being the ~ 12 nM increase in dissolved Fe measured in a Southern Ocean transect moving from open water toward the Kerguelen Island shelf [173]. Entire ocean regions with strong iron sources, such as the tropical and subtropical Atlantic, display elevated surface concentrations (0.4–1.4 nM) [156] relative to others with weak sources, such as the eastern equatorial Pacific, where dissolved iron is often <0.1 nM [151,152,174,175]. In summary, the majority of open ocean measurements of dissolved iron concentration fall between 0.03 and 1.0 nM [7].

Dissolved iron concentrations in the deep ocean, below the nutricline, are typically higher (0.3–1.2 nM) [7] and more stable than in the surface ocean, with an average concentration of ~ 0.7 nM [7,71]. This observation makes sense at first glance: the net effect of biological export from the surface and remineralization at depth would be a transfer of dissolved Fe to the deep water. Iron, however, is chemically much more similar to particle reactive metals such as manganese, aluminum, and lead than to nitrate or phosphate, and by this reasoning ought to be *depleted* at depth, not enriched, as a result of particle scavenging [7,71,166,167]. By the same token, the residence time of iron in the deep ocean is on the order of

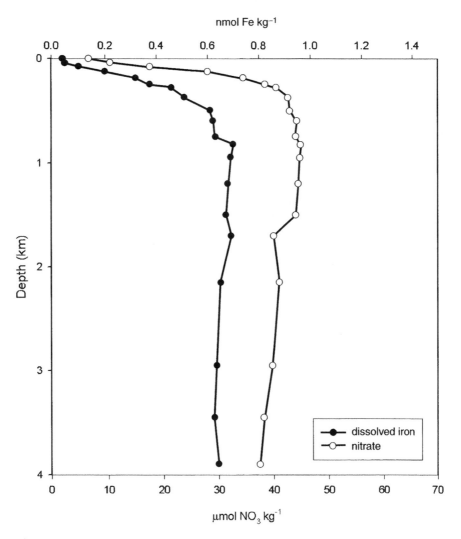

Figure 3 Vertical distribution of dissolved Fe and nitrate in the central Subarctic Pacific at Ocean Station Papa, 50°N, 145°W. Adapted from Ref. [118].

15–150 years [7,176,177], far shorter than the residence times of nutrient elements (10^3–10^5 years) [166]. One probe of this apparent paradox is to look for interocean fractionation of deep water dissolved iron concentrations. Because deep water moves from the Atlantic to the Pacific via thermohaline circulation on a timescale of 1000 years, nutrient elements tend to be enriched in the Pacific relative to the Atlantic [71]. Scavenged elements tend to be depleted in the Pacific because their water column removal rates greatly exceed the rate of thermohaline circulation [166,178].

Iron shows only minor interocean fractionation [7], fitting the profile of neither a nutrient nor a scavenged element. Some have suggested that this behavior is the result of the widespread organic complexation of iron (Section 2.3), which would maintain dissolved iron concentrations and discourage particle scavenging below a certain concentration [55,71,179].

3.3.2. Dissolved Iron in the Coastal Ocean, Shallow Seas, and Semi-enclosed Basins

Concentrations of dissolved iron in the coastal ocean, shallow seas overlying the continental shelves, and semi-enclosed basins are typically several orders of magnitude higher (1–100 nM) than in the open ocean, thanks in large part to the close proximity of major continentally-derived sources. Recent surveys of the western Mediterranean Sea have revealed significant, order of magnitude variations in both dissolved (0.13–4.8 nM) and particulate (0.8–14.5 nM) surface water iron concentrations between autumn and spring as a result of seasonal variations in aeolian deposition [180]. Iron concentrations in much of the coastal ocean can be expected to mirror the seasonal and episodic nature of source processes and biological activity.

Despite generally elevated Fe concentrations, a high degree of small-scale spatial heterogeneity has been documented in parts of the coastal ocean, notably in the central California upwelling zone (Section 4.2.3), where Fe-replete patches occur nearby other areas where Fe has been shown to limit phytoplankton growth [153,181]. Low iron concentrations, potentially limiting or near-limiting for coastal species, have at times been measured in other coastal regions [180,182,183], but widespread iron limitation in the coastal ocean has not been demonstrated.

3.3.3. Particulate Iron

The distribution of non-colloidal particulate iron (>0.4 μm) in the surface ocean is similar to that of dissolved iron, with strong increasing gradients moving toward the coast and continental shelf. Particulate concentrations are often several orders of magnitude greater than dissolved concentrations, ranging from extreme lows of ~0.1 nM in the remote gyres to 100–800 nM in coastal areas [7]. A majority of particulate iron (50–90%) is considered refractory, while the remainder (10–50%) is leachable with dilute acid, and may be more easily converted to bioavailable forms [7]. Processes such as photoreduction and zooplankton-mediated dissolution of biogenic particles may result in daily or episodic variations in particulate Fe abundance in the euphotic zone [59–65,161].

3.3.4. Suboxic/Anoxic Zones

Iron chemistry and distribution in a handful of suboxic and anoxic marine environments stand in marked contrast to conditions prevailing the predominantly oxic ocean. As oxygen levels decline toward zero, there is a gradual shift in the relative stability of Fe(II) and Fe(III), and a concomitant increase in dissolved iron concentrations as solid Fe(III)−hydroxides undergo reductive dissolution. At only mildly euxinic conditions ($\sim$10 μM sulfide), the ratio of Fe(II) : Fe(III) has undergone a profound shift relative to oxic conditions, from $\sim$$10^{-10}$ to 10^{7} [7]. Reduced Fe(II) species dominate, and dissolved iron concentrations increase because of the higher solubility of this redox state.

Under stronger euxinic conditions, concentrations of both iron and hydrogen sulfide increase until solid FeS phases precipitate, removing Fe from the water column. A prime example of this iron regime is found in the Black Sea [184,185]. Here oxic surface waters support low iron concentrations ($\sim$1 nM), increasing to $\sim$70 nM in the suboxic zone ($\sim$100 m) and to a maximum of $\sim$300 nM in the mildly anoxic conditions at 200 m. Fe concentrations at greater depths decline to $\sim$40 nM as a result of FeS precipitation. Similar profiles have been found in other anoxic basins, such as silled fjords [185,186] and the Baltic Sea during the summer [187].

4. IRON LIMITATION OF MARINE PRIMARY PRODUCTIVITY AND CONTROL ON ECOSYSTEM STRUCTURE

Previous sections of this chapter have discussed the biological requirement for iron, the chemical and physical limits on bioavailable iron supply, and the sources and internal cycling processes that determine the highly heterogeneous distribution of iron in the oceans. Although it is perhaps possible to make qualitative predictions about the primary productivity of various ocean regions based on iron distributions alone, primary productivity is a function of many factors, and it does not necessarily follow that the productivity of Fe-impoverished regions (such as the ocean gyres) is *iron limited*. To determine the extent to which iron limitation controls primary productivity, and hence carbon export to the deep ocean, it is necessary to consider the biogeochemical cycles of other major, potentially limiting plant nutrients (N, P, Si) and the processes that control their distributions (e.g., remineralization, circulation), as well as other limiting factors, such as light and zooplankton grazing (Section 4.2).

4.1. Evidence for the Role of Iron in Regulating Productivity in High Nutrient Low Chlorophyll Regions

Zero or near-zero nitrate concentrations and very low phosphate concentrations have been measured in summer in the subtropical gyres of all the major oceans and in the equatorial Atlantic and Indian Oceans, implying widespread N limitation of new production [156,188]. Other parts of the ocean, such as the North Atlantic above 50°N, the Arabian Sea, and various coastal upwelling regions, have elevated nutrient concentrations ($>$2 μM NO_3^-) but also support high chlorophyll concentrations ($>$0.5 mg m^{-3}). In contrast to both the highly productive and the nutrient depleted ocean regions mentioned earlier, some 20–40% of the surface ocean is characterized by the year-round presence of nitrate (>2 µM) and phosphate combined with low chlorophyll concentrations (<0.5 µg L^{-1}). These high nutrient, low chlorophyll ("HNLC") conditions characterize areas of the eastern equatorial Pacific, subarctic Pacific, and the Southern Ocean. Each of these otherwise very different ocean regions occurs in a zone of divergent upwelling, which provides enough nutrients to support much higher levels of new production than are currently observed in these regions [34,151,189–193].

HNLC regions are dominated by "small" ecosystems made up of pico- and nanophytoplankton (<5 µm) supported largely by recycled nutrients [34,151,191,194,195]. The availability of reduced N species (NH$_4^+$, urea) limits the uptake of nitrate by phytoplankton, contributing to the large standing stocks of NO$_3^-$ that are the first indicator of HNLC conditions. Grazing communities in HNLC regions are typically comprised of microzooplankton capable of rapid doubling-rates, along with a smaller mesozooplankton component [156,191]. Not surprisingly, carbon cycling in HNLC ecosystems is characterized by intense recycling in the euphotic zone and low *f*-ratios (0.07–0.2 in the equatorial Pacific) [196]. The relatively small export fluxes in these regions occur during episodic diatom blooms [191,197–202].

It has long been suggested that iron limitation plays a role in the HNLC regions [203,204]. The development of trace metal clean sampling techniques in the 1980s made it possible to investigate this hypothesis. New, accurate measurements indeed revealed that iron concentrations in the HNLC regions were extremely low, especially in the equatorial Pacific and Southern Ocean [34,151,152,174,175,205]. Other tell-tale signs included the distinct high productivity/low nitrate "wakes" seen downstream of land features (such as the Galapagos Islands) in the HNLC regions, suggesting that some factor associated with land was stimulating productivity [105,152,206]. The first experimental evidence of the importance of iron in controlling the productivity came from bottle incubation experiments conducted in the subarctic Pacific in the late 1980s [117,118]. These initial efforts have been confirmed with numerous additional bottle incubations and expanded to include six mesoscale iron enrichment experiments covering each of the three major HNLC regions (Fig. 4).

4.1.1. Bottle Incubation Experiments

Early bottle incubation experiments in all three of the HNLC regions demonstrated that nanomolar iron enrichment resulted in significant nutrient depletion, chlorophyll *a* increase, strong growth of larger phytoplankton, especially diatoms, and a somewhat delayed increase in net productivity, relative to control samples [105,117,118,207–211]. Later studies confirmed and refined these results [106,158,212–219], documenting early physiological responses by phytoplankton (increased ^{15}N uptake, nitrate reductase expression) to Fe addition [200,220], and showing that even picomolar additions of Fe could produce a measurable effect [174]. The incubation studies have demonstrated

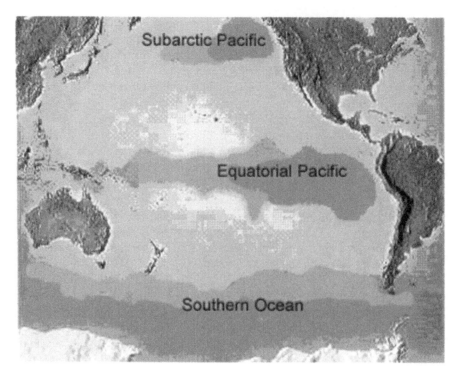

Figure 4   The major high nitrate, low chlorophyll (HNLC) regions of the ocean. The Southern Ocean HNLC region is circumpolar.

that the productivity and nutrient drawdown responses to Fe addition are accompanied by large shifts in the phytoplankton flora, from communities dominated by nano- and pico-sized phytoplankton, utilizing ammonium, to communities in which most of the biomass is composed of diatoms drawing on the large standing stocks of nitrate.

In vitro incubations differ from actual ocean conditions in important ways that limit the applicability of their results to marine ecosystems. The fixed light intensity used in these experiments is a poor mimic of the variable conditions in the ocean mixed layer, while the walls of the incubation bottles themselves may provide unnatural growth surfaces. Bottle experiments cannot hope to accurately represent that nature or scale of the community response [151]. The artificial restriction on the abundance of grazers in bottle incubations, for instance, has been offered as an alternative (and confounding) explanation for the phytoplankton blooms observed in these experiments [221–223]. In an effort to overcome the limitations of bottle experiments, researchers have conducted a series of mesoscale *in situ* Fe enrichment experiments, in which large areas of the surface ocean were seeded with iron and the ecosystem response measured over many days.

4.1.2. Mesoscale Enrichment Experiments

To date, seven *in situ* mesoscale iron enrichment experiments have been performed in the HNLC regions, each with its own acronym: IRONEX I and II in the equatorial Pacific south of the Galapagos Islands; SOIREE, EisenEx, and SOFeX in various sectors of the Southern Ocean south of the Polar Front; SEEDS and SERIES in the western and eastern gyres of the subarctic Pacific, respectively. The SOFeX experiment actually consisted of two separate enrichments, one in the high silica waters south of the Polar Front, the other in low silica waters to the north [224]. With important variations, the experiments have followed the same basic approach: enrichment of a large patch ($50-100$ km^2) of low iron, HNLC surface water with enough ferrous sulfate to raise ambient iron concentrations to $\sim 1-4$ nM, many times greater than pre-experiment levels; concurrent addition of the inert tracer sulfur hexaflouride (SF_6); monitoring of chemical, biological, and physical changes within the enriched patch for $2-4$ weeks in comparison to nearby control areas.

The ecosystem responses to iron enrichment in each of the experiments have been dramatic, and strikingly similar to one another. After seven such experiments, a "classical" response to iron addition can now be outlined:

(a)   After the initial enrichment, dissolved iron concentrations decline rapidly, returning to pre-experiment levels within a few days unless additional enrichments are performed [33–35,151,224,225].

(b)   Photosynthetic efficiency, phytoplankton growth rate, chlorophyll *a* concentration and primary productivity increase dramatically, reaching levels $2-6$ times greater than those of control areas [33–35,151,224–227].

(c)   After a lag time of several days, a large, sustained, diatom-dominated phytoplankton bloom develops, accounting for most of the observed increases in primary productivity and new production [33–35,224, 225,228,229].

(d)   Significant depletion of pCO_2, [33–35,224,225,229–231] and major nutrients (nitrate, phosphate, silicate) [33–35,224,225,228] occurs within surface waters of the patch.

Based on these results, the mesoscale enrichment experiments have established that iron plays a strong limiting role in each of the major HNLC regions of the ocean. However, the apparent uniformity of the findings reported above masks important differences in the outcomes of the experiments—differences that shed light on the structure and function of ecosystems in HNLC regions, and how they change in response to iron additions.

4.1.2.1. Biological response: Unlike all the other enrichment experiments, IRONEX I did not stimulate a sustained phytoplankton bloom. The biological response, consisting of increases in photosynthetic efficiency, primary productivity, and chlorophyll *a*, was largely confined to an extant "small" community dominated by *Synechococcus* bacteria, picoplankton, and autotrophic dinoflagellates; diatoms comprised only a small component of the community (~17% of biomass), and increased in abundance only in proportion to other groups. The bloom had essentially ceased after day 3 of the experiment [151,226]. Similar initial responses (days 1–6) in the extant phytoplankton community were observed in other enrichment experiments, notably IRONEX II, but were eventually superseded by large diatom blooms.

The early termination of the IRONEX I bloom was probably the result of several interrelated factors. The experiment's single iron addition was more or less exhausted by day 3, at which time the enriched patch was subducted below a low-salinity front to a depth of >30 m, where it remained for the duration of the study. Under these conditions, a diatom bloom could not develop. At the same time, stimulation of the extant phytoplankton community was met with a rapid response by microzooplankton grazers, which quickly and efficiently recycled the increased productivity [151,156]. Similar food web effects were observed in other enrichment experiments. In SOIREE, for instance, small grazers effectively limited bacterial abundance, in spite of increased productivity. However, zooplankton herbivory appears to have concentrated on cells of <20 μm diameter, with little impact on the larger diatoms [34]. In experiments that used multiple additions to sustain elevated iron concentrations for more than a few days, significant diatom growth occurred in spite of increased grazing activity [33,34,224,225].

Among the experiments that stimulated sustained blooms, initial differences within the biological communities strongly influenced the character of the eventual bloom. The biological response in IRONEX II was quickly dominated by fast-growing pennate diatoms [33], whereas in SOIREE the large (30–50 μm) chain-diatom *Fragilariopsis kerguelenis* accounted for ~75% of primary productivity [34]. During EisenEx and SOFeX North, diatoms of the genus *Pseudonitzschia* showed the largest increases in abundance [224,228]. Prior to each of the experiments, these species, and diatoms in general, had been minor components of the biota. In contrast, the subarctic Pacific region in which SEEDS was performed included a considerable contingent of micro-sized species (36% of chlorophyll *a*), chiefly the pennate diatom *Pseudonitzschia turgidula*, but the Fe-induced bloom was dominated by an entirely different species, the chain-forming centric diatom *Chaetoceros debilis*, which had been present but rare in the pre-addition assemblage. Despite cold water temperatures, *C. debilis* achieved a high net growth rate of 2.6 doublings/day at the peak of the bloom, almost twice the observed growth rate of the dominant pennate diatoms in IRONEX II and the net algal growth rate in SOIREE [33–35]. Although diatom blooms are now an expected response to sustained iron fertilization, the composition of such blooms, and any secondary ecological responses dependent on species composition, is not yet predictable.

4.1.2.2. Biogeochemical response: The rapid loss of dissolved iron following the first and only addition in IRONEX I, attributable to Fe(II) oxidation, physical mixing, particle scavenging, and biological uptake [151], prompted the use of multiple-additions in subsequent studies [33,34,224,228]. In SOIREE, this strategy eventually resulted in a stable, elevated Fe concentration ($\sim$1 nM) that lasted from the fourth addition on day 7 to the end of the experiment on day 13. These stable dissolved Fe concentrations may have been related to a substantial increase in the concentrations of Fe-binding ligands over the course of the experiment (3.5–8.5 nM) [34].

The failure of a diatom bloom to develop in IRONEX I was matched by a negligible drawdown of nitrate, phosphate, and silicate within the enriched patch [151]. In contrast, the major blooms induced by the other experiments were accompanied by substantial macronutrient uptake, ranging from 10% to 30% depletion during EisenEx [228], through $\sim$55% depletion of nitrate in SEEDS [35], to 80% depletion of silicate in IRONEX II [33]. These results highlight the importance of diatoms in controlling the biogeochemical outcomes of enrichment experiments. It is important to note, however, that Fe-induced macronutrient drawdown in these experiments was in most cases incomplete. Though depleted relative to controls, nitrate levels remained above limiting levels ($\sim$3 - 25 μM) at the end of all of the experiments [33–35,151,224,225,228]. Secondary silicate limitation may have played a role in suppressing the IRONEX II, SERIES, and SOFeX North diatom blooms, based on the near-limiting concentrations ($\sim$1 μM) measured toward the ends of these experiments [33,224,225], but did not appear to play a role in the other experiments. By the end of SOIREE, none of the macronutrients appear to have been depleted to growth-limiting levels, in spite of the fact that dissolved Fe concentrations in the patch remained ten times those in control areas. Meanwhile, diatoms were accumulating flavodoxin at increasing levels [34]. This evidence suggests that even at the enriched concentrations achieved during SOIREE and the other enrichment experiments, bioavailable iron levels remained limiting.

Notably absent from the list of general HNLC ecosystem responses to iron enrichments is an increase in the export of carbon from the surface layer. Each of the fully-developed Fe-enrichment responses has been characterized by CO_2 uptake and increased POC concentrations in the surface layer [33–35,224,225, 229,230]; however, in most experiments this did not translate into increased export flux relative to control areas. The exceptions, SOFeX, and SERIES, were the experiments monitored for the longest periods ($\sim$4 weeks), implying that the export flux response is delayed, and may have gone unobserved in the aftermaths of the shorter experiments. In SOFeX South, a small but significant enhancement in particle export was measured. A flux of $\sim$10 mmol m^{-2} day^{-1} inside the patch amounted to $\sim$2100 tons of POC exported to 100 m depth by the end of the monitoring period [224,232]. An autonomous drifter employed in the SOFeX North experiment also observed significant, two- to sixfold build-up and export of POC from the enriched patch, although vigorous mixing made this result difficult to interpret [233].

During the SERIES bloom decline, the surface layer ($<$25 m) lost 75–79% of the POC and PSi inventories that had accumulated during the bloom. These declines corresponded to an eightfold increase in the amount of material—mostly diatom cells, aggregates, and fecal pellets—reaching 50 m sediment traps. This export flux, totaling ~47 mmol m^{-2} POC, nonetheless accounted for only 18% of the POC loss from the surface layer, and an even smaller fraction (8%) was detected below the permanent thermocline at 125 m depth. The bulk of the particulate material produced by the bloom was lost to herbivory, dissolution (in the case of PSi), and above all, bacterial remineralization within and below the mixed layer [225] (Fig. 5).

4.1.3. Evidence of Iron Limitation from Other Regions

Although most iron limitation studies have thus far focused on open ocean HNLC regions, a growing body of work suggests that iron availability can play a controlling role in other marine systems. The strongest evidence for this role, surprisingly, comes from coastal and near-coastal systems, which are generally assumed to be Fe-rich thanks to their proximity to iron sources (rivers,

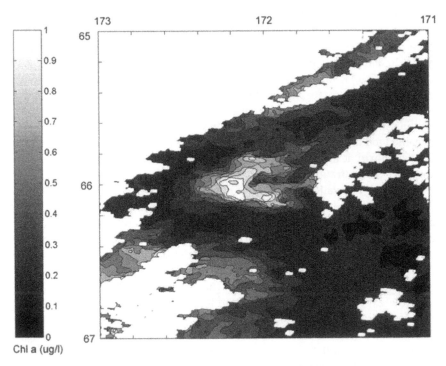

Figure 5 Sea-Viewing Wide Field-of-View Sensor (SeaWiFS) satellite image of an iron-induced phytoplankton bloom in the Southern Ocean. The bloom is apparent as a chlorophyll maximum near the center of the image (66°S, 172°E); large, contiguous white areas represent cloud cover. The image was captured on day 20 of the SOFeX South patch experiment. Coordinates are degrees South and East. Courtesy of the NASA SeaWiFS Project and the Goddard Earth Sciences Data and Information Services Center/Distributed Active Archive Center.

continental shelves, continental dust sources, etc.). Iron stress, however, is a function not only of ambient iron concentration but also the cellular iron requirements of phytoplankton species. In coastal regions, where many phytoplankton species are known to require relatively large amounts of iron to support optimal growth [5], iron stress can occur at concentrations that in open ocean regions would be considered replete.

Recent work in the California coastal upwelling region demonstrates that variation in iron distribution on fairly small geographic scales plays a critical role in determining community structure and the development of phytoplankton blooms. Hutchins et al. used field observations to define a spectrum of ecosystem types in the coastal zone (10–20 km offshore) based on Fe availability and phytoplankton assemblage, ranging from Fe-replete regimes in which diatom blooms dominate primary production to regimes under severe iron stress in which diatoms are absent, and the growth of even the smaller size classes is inhibited [181]. Other studies have related this heterogeneous Fe distribution to the presence or absence of a continental shelf iron source for upwelled water [148,234], while others explored seasonal variations in the delivery of iron to the surface in upwelled water, demonstrating that low iron conditions in the summer result in HNLC-like conditions in offshore areas [153] and that high dissolved Fe concentrations in upwelled water influence the extent of macronutrient drawdown in subsequent phytoplankton blooms [147].

Elsewhere, Hutchins et al. performed shipboard iron incubation experiments in the high-nutrient eastern boundary current regime off the west coast of South America, measuring low ambient iron concentrations ($\sim$0.1 nM) and demonstrating that a broad range of indicators, including photosynthetic efficiency, macronutrient depletion, and chlorophyll *a* increased in proportion to the amount of iron added [235]. Low, near-limiting concentrations of iron have been measured in the summer in the North Atlantic under stratified, post-bloom conditions [71], and off the Oregon coast [183], while Fe has been shown to affect fluorescence patterns and photochemical quantum efficiency among phytoplankton in the South Pacific gyre [236]. Combined with the evidence from the coastal upwelling systems, this has given rise to speculation that iron limitation, at least seasonally, is more widespread than previously assumed [156].

4.2. Interaction of Iron with Other Limiting Factors

Even in the HNLC regions, it is unlikely that iron alone limits phytoplankton production. A more accurate conceptual model of ocean productivity will probably involve *co-limitation* by multiple factors, several of which are discussed here.

Within the pico- and nanophytoplankton-dominated recycling ecosystems of the HNLC regions, microzooplankton populations have been observed to achieve very high growth rates, matching or exceeding those of their food sources [223]. As a result, these grazers can respond to and quickly decimate nanophytoplankton blooms, as appears to have been the case in IRONEX I [151]. While iron is undoubtedly a limiting factor for these "small" phytoplankton, as has been clearly demonstrated in enrichment experiments [33–35,151,226, 227], in many cases it is possible that microzooplankton grazing is the direct limit on the abundance of the dominant members of the extant community. By this line of reasoning, iron limitation, with its disproportionate suppression of the larger members of the phytoplankton flora, in effect creates an ecosystem in which grazing is the immediate limitation on productivity, if not new production. An addition of iron to this type of system would, among other things, stimulate the growth of diatoms, which are too large to be easily grazed by the fast-acting microzooplankton grazers; a bloom develops, and cannot be fully contained by the slower-responding mesozooplankton grazers [156].

The polar darkness that prevails for much of the year is the overwhelming feature of the Southern Ocean environment, the factor to which all other limits on phytoplankton growth are merely additive. Even in the Antarctic summer, when the sun is out for much of the 24 h period, the deep, turbulent mixed layer, frequent cloud cover, and low angle of the sun can combine to impose some degree of light limitation on Southern Ocean and subantarctic ecosystems [237–239]. To a lesser extent, light limitation is a limiting factor in the subarctic Pacific, especially in the winter [67]. Iron limitation compounds the growth limiting effect of low light conditions because of the importance of Fe in electron transfer and the synthesis of chlorophyll. Iron-stressed cells exhibit lower photosynthetic competence, and are less able to utilize limited light resources [156]. Conversely, it has been noted in several studies that iron use efficiency is proportional to illumination [30,217,240], implying that ambient iron concentrations that are adequate under high light conditions may be suboptimal in a low-light regime. In this way iron and light limitation can be expected to reinforce one another in situations where they coincide.

Low silicate concentrations have been measured in several HNLC regions, notably the eastern equatorial Pacific and subantarctic Southern Ocean north of the Polar Front [188,224,239,241,242], giving rise to the theory that silicate availability limits diatom growth, primary productivity, and carbon export flux in these regions [242,243]. The case for silicate rests on the special importance of diatoms, which construct their mineral shells from dissolved silicate, in controlling particle fluxes from otherwise efficient recycling ecosystems. As resistant, largely undigestable hard parts, siliceous frustules are likely be packaged with other recalcitrant materials into zooplankton fecal pellets or be directly exported, unrecycled, from the surface layer, resulting in a differential loss of silicate from the system [241]. Dugdale and Wilkerson [242] found that the magnitude of new production in the eastern equatorial Pacific upwelling zone is a function of silicate supply in upwelled waters rather than nitrate supply, and almost entirely supported by diatom growth. The authors suggest that silicate limits new production in the primary upwelling zone, a narrow band along the equator, while iron is the limiting factor in the equatorial currents to the north and south, including the region occupied in the IRONEX experiments [242]. Bottle enrichments with silica performed in conjunction with the SOFeX North Fe experiment in the subantarctic Southern Ocean stimulated massive increases in silica uptake, indicating that Si limitation in this low-silicate region (<5 μM) contributed to relatively low diatom abundance ($<50\%$) in the Fe-induced phytoplankton bloom [224]. The growth of other, non-siliceous groups was not limited, however, and it is unclear what effect bloom composition had on observed carbon export [224,233].

The intriguing prospect of Fe–Si co-limitation is supported by the observation that diatoms under iron stress often have Si:N ratios well above those of diatoms grown in Fe-rich conditions [43,156]. If this observation holds, it is conceivable that low iron conditions could induce silicate limitation, or conversely, high silicate availability could permit diatoms to grow at Fe concentrations that would otherwise be limiting [156]. Secondary limitation by silica appears to have played a part in the decline of iron-induced diatom blooms in the IRONEX II and SERIES enrichment experiments [33,225], but not, surprisingly, in SOFeX North, where a small but consistent silicate source to the Fe-enriched patch from physical mixing allowed the bloom to persist for >50 days [224].

The high iron requirement for nitrogen fixation and diazotrophic growth (Section 1.1) has led to the theory that iron limitation explains the low abundance of *Trichodesmium* spp. cyanobacteria (the dominate N-fixer in the ocean) in low nitrate, low chlorophyll ocean areas, such as the South Pacific gyre, which are far removed from atmospheric iron sources [244]. More generally it has been proposed that iron availability limits N-fixation on a global scale [244–246], with large impacts on the net flux of CO_2 into the oceans [244]. Recent evidence for this theory includes the 100-fold increase in *Trichodesmium* biomass measured after a summer dust storm in waters of the West Florida Shelf [247] and the low cellular iron quotas, quantum yields, and N-fixation rates measured in *Trichodesmium* cultures grown under low Fe-conditions [248]. Berman-Frank et al. further estimate that iron availability limits nitrogen fixation by *Trichodesmium* over 75% of the ocean [248].

The evidence presented in the preceding sections leads to the conclusion that iron limitation of productivity in the oceans is both more and less than it has been made out to be in previous accounts. Iron limitation is more widespread than once imagined, playing a critical role in not only the HNLC regions but also in many other ocean regions through interactions with other limiting factors. At the same time, it is clear that even within the HNLC regions, iron limitation is but one of several factors determining the character of an ecosystem, exerting a greater or lesser influence depending on conditions (light, availability of other nutrients, weather and currents, ecology) that may vary greatly over short distances and timescales. Moreover, iron enrichment experiments have demonstrated that iron limitation is inextricably linked to the composition and function of dynamic biological communities, adapted to low iron conditions, that both respond to and modulate the availability of iron.

4.3. Carbon Export and Iron Fertilization

In contrast to the primary productivity response, export production in HNLC regions appears to increase only modestly as a result of iron additions. Only in the more recent, longer-term enrichment studies (SERIES, SOFeX) did iron stimulation of primary productivity result in measurable increases in export production within the timeframe of the experiments. During a six day span toward the end of SERIES, POC export flux reached 47 mmol m^{-2} at 50 m and ~21 mmol m^{-2} at 125 m [225], translating into an export efficiency of 8% that is in many cases smaller than those following naturally-occurring blooms in HNLC regions [249,250]. If the measured POC flux at 125 m is extrapolated to the entire ~1000 m^2 SERIES enriched patch, the total C export over the 6 day period is only 252 tons C. During SOFeX South, a total addition of 1.3 tons of Fe resulted in ~2100 tons C exported to 100 m, an Fe-addition to C export ratio of only 1.3 × 10^{-4} [232].

This is bad news for geoengineering proposals (see Ref. [251]) to sequester anthropogenic carbon dioxide through iron fertilization of HNLC regions, which have relied on lab-based Fe:C ratios two orders of magnitude lower to bolster their claims of potential CO$_2$ uptake. Extrapolating from the SOFeX ratios, Buesseler and Boyd estimate that an ocean area of ~1 billion km^2, larger than the entire Southern Ocean, would have to be fertilized with over 1 million tons of Fe to transfer enough POC to 100 m to offset 30% of annual anthropogenic carbon emissions [231]. A further blow to proponents of iron fertilization came from the work of Law and Ling, who in the aftermath of the SOIREE bloom measured a substantial increase in N$_2$O production, enough to offset a significant portion of the radiative forcing (if any) resulting from Fe-stimulated carbon fixation [252]. Putting aside for a moment the host of serious and unpredictable ecological side effects of large-scale iron fertilization [251], the practical, economic, and scientific barriers alone would make such an enterprise inadvisable.

5. THE ROLE OF IRON IN REGULATING ATMOSPHERIC CO$_2$

5.1. The Iron Hypothesis

As originally articulated by Martin [253] 15 years ago, the iron hypothesis is a simple but powerful theory, in two parts: (a) Phytoplankton growth, and hence, carbon export, in the HNLC regions is presently limited by the insufficient availability of iron; (b) Increased delivery of iron via more vigorous transport and deposition of atmospheric dust, particularly to the Southern Ocean, enhanced biological productivity during past ice ages, and caused or contributed to the low CO$_2$ concentrations and cold climates inferred for these glacial periods. That iron availability, directly and through its control of ecosystem structure, limits productivity in many parts of the ocean has been convincingly established in numerous studies, and the more recent enrichment experiments have demonstrated at least a modest effect on export production (Section 4.3). The second tenet of Martin's hypothesis, however, is more problematic, and remains at the center of debate among those studying glacial–interglacial climate changes. This section will introduce aspects of the debate, and evaluate the iron hypothesis as a driver of global change.

5.2. Changes in Dust Input and Productivity in Glacial Periods

Records of atmospheric dust input to the ocean over glacial–interglacial cycles, from marine sediments, loess deposits, and cores from continental and polar ice sheets, are in basic agreement with one another, revealing far greater aerosol generation, transport, and deposition during glacial maxima than interglacials [254–257]. Based on an extensive global compilation of marine sediment records, Rea argued that these increases in dust deposition were restricted largely to the northern hemisphere, with little transport to the southern hemisphere from northern dust sources through the wet deposition belt of the ITCZ [256]. Nonetheless, Petit et al. have reconstructed substantial increases in dust deposition to Antarctica during the last four glacial periods based on the 420,000-year Vostock ice core records from East Antarctica [258]. These increases were large, 20–40 times greater than interglacial backgrounds, and were tentatively traced to a source region, the Patagonian plain of South America [258,259]. It seems reasonable to assume that corresponding increases in dust input occurred over large portions of the Southern Ocean. This conclusion is also supported by modeling efforts that reproduce the observed patterns and magnitudes of dust deposition to the oceans at the last glacial maximum [138,260–262].

Despite significant evidence for high dust deposition during glacials, the evidence in the marine sedimentary record for increased oceanic productivity and carbon export during glacial stages is mixed, and is complicated by differences among the various proxies used to reconstruct paleoproductivity. In the equatorial Pacific, for instance, different proxies, applied to sediment cores from various locations, yield conflicting results, suggestive of a high degree of mesoscale variability in space and time [263–266]. Similar complexity is apparent in the Southern Ocean, and here also there is mixed evidence indicating both region-wide increases and decreases in productivity during glacial stages [137,267–270]. Though some proxy records indicate that export production in the subantarctic zone was greater during glacials, consistent with a northward shift in the high productivity band [137,268,270], this enhanced export may not match interglacial export levels south of the polar front [268]. Making the tenuous assumption that the findings of the SERIES and SOFeX enrichment experiments can be applied to broader regions and timescales, it is possible that elevated primary productivity in HNLC regions during glacials was not matched by a corresponding significant increase in export production, or for that matter, long-term storage of atmospheric carbon.

5.3. Consequences for Atmospheric CO_2 and Global Climate

Could an increased supply of iron, driven by increased atmospheric dust deposition and perhaps facilitated by transport of "dust fertilized waters" to the Southern Ocean via thermohaline circulation [156,271], account for lower atmospheric CO_2 concentrations during glacial periods? Not surprisingly, the answer depends on who is being asked, and on the assumptions and methods used to address the question. Much discussion has focused on the apparent sequence in proxy records of the key processes thought to amplify orbital forcing and induce the transition between glacial and interglacial states. Interpreted from the Vostock ice core record, glacial terminations appear to occur in two stages, the first characterized by increases in temperature and atmospheric concentrations of CO_2 and CH_4, and decreases in dust input, the second by continued temperature rise, along with a rapid increase in methane and a decline in ice volume [258]. This implies that greenhouse gas concentrations led deglaciation as a feedback on orbitally-triggered climate changes, and that dust-driven increases in Fe input were a potential control on greenhouse gases via the biological pump. In a more detailed analysis of glacial termination II ($\sim$135 ky BP) in the Vostock record, Broecker and Henderson argued that Southern Ocean processes drove the observed, coupled increase in atmospheric CO_2 and Antarctic temperature over an 8000-year period, bounded at the outset by a drop in dust flux and at the end by a decline in global ice volume. However, because much of the decline in dust preceded the initiation of the change in CO_2 concentration, and because the CO_2 continued to increase for thousands of years after dust deposition had settled at interglacial levels, iron control of CO_2 concentration is, at first glance, considered unlikely [272].

The iron hypothesis can still explain the termination II ice core record with the help of two additional "corollaries", both of which are supported by available evidence. It has been observed that the dissolved iron concentration in surface waters is not linearly related to dust input [108,116,171]; as aerosol loading increases, the percent of aerosol iron that actually dissolves decreases, perhaps due to particle concentration effects, increased scavenging rates, or exhaustion of the available pool of unbound organic ligands. Whatever the cause, this saturation effect provides a plausible explanation for the lag-time between dust input declines and the onset of the CO_2 increase in the Vostock core [272]. The second "corollary" invokes the suspected role of iron availability in limiting nitrogen fixation as discussed in Section 4.2. High atmospheric iron inputs in glacial times to N-limited open ocean areas, such as the subtropical gyres, would presumably have resulted in higher rates of N-fixation and new production [273–275]. In theory, the several thousand year oceanic residence time of nitrate would have allowed this increased supply to be spread through the ocean via thermohaline circulation and upwelling, sustaining productivity even after dust inputs fell off, and preventing a rapid rise in atmospheric CO_2 [272]. This explanation does not account for the fact that an increased supply of fixed nitrate would probably lead to limitation by phosphate, which is nearly as scarce as nitrate over much of the ocean, and lacks a strong atmospheric source [82,275].

The iron hypothesis is just one of many plausible explanations for atmospheric carbon dioxide fluctuations over glacial–interglacial cycles. Competing theories invoke everything from increased surface stratification in the Southern Ocean [276] through changes in the flux of calcite to marine sediments [277] to a "silica hypothesis", in which increased atmospheric supply of silica during glacials stimulated diatom growth [278,279]. In all likelihood, atmospheric CO_2 concentrations over long timescales are the net result of many factors acting together or antagonistically.

Assuming for a moment that iron supply to the ocean does indeed play a role in regulating the carbon cycle on glacial–interglacial timescales, the question becomes how large a role that is. Models based on the current magnitude of HNLC conditions in the Southern Ocean have shown that utilization of 50% of the standing stock of nutrients from surface waters would result in a CO_2 drawdown to levels matching glacial minima [275,280]. This increase in macronutrient utilization is large, but consistent with some proxy evidence from the last glacial maximum [281] and comparable with nutrient drawdowns observed in present day Fe enrichment experiments. In the glacial Southern Ocean, biological nutrient uptake would also presumably be freed from secondary Si limitation by higher rates of atmospheric silica deposition [278]. Modeling by Watson et al., on the basis of observed uptake values for CO_2 and Si : C during SOIREE, suggested that the observed decline in iron input to the Southern Ocean at the last glacial–interglacial transition can account for the initial $\sim$40 ppm of CO_2 increase (roughly 50%), with other mechanisms responsible for the remaining increase to Holocene concentrations [282]. Other models, making different assumptions, yield different results; perhaps the only certainty at this point is that the role of iron in controlling the oceanic carbon cycle will be studied and debated for years to come.

6. SUMMARY AND CONCLUSIONS

Iron is an essential nutrient for all living organisms. Iron is required for the synthesis of chlorophyll and of several photosynthetic electron transport proteins and for the reduction of CO_2, SO_4^{2-}, and NO_3^- during the photosynthetic production of organic compounds. Iron concentrations in vast areas of the ocean are very low (<1 nM) due to the low solubility of iron in oxic seawater. Low iron concentrations have been shown to limit primary production rates, biomass accumulation, and ecosystem structure in a variety of open-ocean environments, including the equatorial Pacific, the subarctic Pacific and the Southern Ocean and even in some coastal areas.

Oceanic primary production, the transfer of carbon dioxide into organic carbon by photosynthetic plankton (phytoplankton), is one process by which atmospheric CO_2 can be transferred to the deep ocean and sequestered for long periods of time. Accordingly, iron limitation of primary producers likely plays a major role in the global carbon cycle. It has been suggested that variations in oceanic primary productivity, spurred by changes in the deposition of iron in atmospheric dust, control atmospheric CO_2 concentrations, and hence global climate, over glacial–interglacial timescales.

A contemporary application of this "iron hypothesis" promotes the large-scale iron fertilization of ocean regions as a means of enhancing the ability of the ocean to store anthropogenic CO_2 and mitigate 21st century climate change. Recent *in situ* iron enrichment experiments in the HNLC regions, however, cast doubt on the efficacy and advisability of iron fertilization schemes. The experiments have confirmed the role of iron in regulating primary productivity, but resulted in only small carbon export fluxes to the depths necessary for long-term sequestration. Above all, these experiments and other studies of iron biogeochemistry over the last two decades have begun to illustrate the great complexity of the ocean system. Attempts to engineer this system are likely to provoke a similarly complex, unpredictable response.

ACKNOWLEDGMENTS

Rochelle Labiosa, Zanna Chase, Pete Strutton, Joe Cackler, and Karrie Amsler provided valuable assistance during the preparation of this chapter.

ABBREVIATIONS

ATP	adenosine $5'$-triphosphate
EisenEx	Eisen ("iron") Experiment (Southern Ocean)
Fe(II)$'$	all inorganic Fe(II) species
Fe(III)$'$	all inorganic Fe(III) species, largely hydrolysis products
HNLC	high nutrient low chlorophyll
IRONEX	Iron Enrichment Experiments (Equatorial Pacific)
ITCZ	intertropical convergence zone
ky BP	thousand years before the present
NADPH	nicotinamide adenine dinucleotide (reduced)
POC	particulate organic carbon
PSi	particulate silica
SEEDS	Subarctic Pacific Iron Experiment for Ecosystem Dynamics Study
SERIES	Subarctic Ecosystem Response to Iron Enrichment Study
SOFeX	Southern Ocean Fe Experiment, including both North and South patches
SOIREE	Southern Ocean Iron Enrichment Experiment

REFERENCES

1. Sunda WG. In: Turner DR, Hunter KA, eds. The Biogeochemistry of Iron in Seawater. Chichester, West Sussex: John Wiley & Sons, 2001:41–84.
2. Redfield AC, Ketchum BH, Richards FA. In: Hill MN, ed. The Sea, Vol. 2. New York: Wiley, 1963:26–77.
3. Sunda WG. Mar Chem 1997; 50:169–172.
4. Sunda WG, Swift D, Huntsman SA. Nature 1991; 351:55–57.
5. Sunda WG, Huntsman SA. Mar Chem 1995; 50:189–206.
6. Lancelot C, Hannon E, Becquevort S, Veth C, de Baar HJW. Deep-Sea Res I 2000; 47:1621–1662.
7. de Baar HJW, de Jong JTM. In: Turner DR, Hunter KA, eds. The Biogeochemistry of Iron in Seawater. Chichester, West Sussex: John Wiley & Sons, 2001:123–253.
8. Ho TY, Quigg A, Finkel ZV, Milligan AJ, Wyman K, Falkowski PG, Morel FMM. J Phycol 2003; 39:1145–1159.
9. Raven JA. New Phytol 1988; 109:279–287.
10. Maldonado MT, Price NM. Mar Ecol Prog Ser 1996; 141:161–172.
11. Achilles KM, Church TM, Wilhelm SW, Luther GW, Hutchins DA. Limnol Oceanogr 2003; 48:2250–2255.
12. Kustka AB, Sanudo-Wilhelmy SA, Carpenter EJ, Capone D, Burns J, Sunda WG. Limnol Oceanogr 2003; 48:1869–1884.
13. Neilands JB. Annu Rev Nutr 1981; 1:27–46.
14. Granger J, Price NM. Limnol Oceanogr 1999; 44:541–555.
15. Soria-Dengg S, Reissbrodt R, Horstmann U. Mar Ecol Progr Ser 2001; 220:73–82.
16. Weaver RS, Kirchman DL, Hutchins DA. Aquat Micro Ecol 2003; 31:227–239.
17. Trick CG. Curr Microbiol 1989; 18:375–378.
18. Haywood MG, Holt PD, Butler A. Limnol Oceanogr 1993; 38:1091–1097.
19. Reid RT, Butler A. Limnol Oceanogr 1991; 36:1783–1793.
20. Reid RT, Live DH, Faulkner DJ, Butler A. Nature 1993; 366:455–458.
21. Wilhelm SW, Trick CG. Limnol Oceanogr 1994; 39:1979–1984.
22. Wilhelm SW. Aquat Micro Ecol 1995; 9:295–303.
23. Hudson RJM, Morel FMM. Limnol Oceanogr 1990; 35:1002–1020.
24. Soria-Dengg S, Horstmann U. Mar Ecol Prog Ser 1995; 127:269–277.
25. Anderson MA, Morel FMM. Limnol Oceanogr 1982; 27:789–813.
26. Eide D, Guerinot ML. ASM News 1997; 63:199–205.
27. Maranger R, Bird DF, Price NM. Nature 1998; 396:248–251.
28. Nodwell LM, Price NM. Limnol Oceanogr 2001; 46:765–777.
29. Hudson RJM, Morel FMM. Deep-Sea Res 1993; 40:129–151.
30. Sunda WG, Huntsman SA. Nature 1997; 390:389–392.
31. Price NM, Ahner BA, Morel FMM. Limnol Oceanogr 1994; 39:520–534.
32. Boyd PW, Muggli DL, Varela DE, Goldblatt RH, Chretien R, Orians KJ, Harrison PJ. Mar Ecol Prog Ser 1996; 136:179–193.
33. Coale KH, Johnson KS, Fitzwater SE, Gordon RM, Tanner S, Chavez FP, Ferioli L, Sakamoto C, Rogers P, Millero F, Steinberg P, Nightingale P, Cooper D, Cochlan WP, Landry MR, Constantinou J, Rollwagen G, Trasvina A, Kudela R. Nature 1996; 383:495–501.
34. Boyd PW, Watson AJ, Law CS, Abraham ER, Trull T, Murdoch R, Bakker DCE, Bowie AR, Buesseler KO, Chang H, Charette M, Croot P, Downing K, Frew R,

Gall M, Hadfield M, Hall J, Harvey M, Jameson G, LaRoche J, Liddicoat M, Ling R, Maldonado MT, McKay RM, Nodder S, Pickmere S, Pridmore R, Rintoul S, Safi K, Sutton P, Strzepek R, Tanneberger K, Turner S, Waite A, Zeldis J. Nature 2000; 407:695–702.

35. Tsuda A, Takeda S, Saito H, Nishioka J, Nojiri Y, Kudo I, Kiyosawa H, Shimomoto A, Imai K, Ono T, Shimomoto A, Tsumune D, Yoshimura T, Aono T, Hinuma A, Kinugasa M, Suzuki K, Sohrin Y, Noiri Y, Tani H, Deguchi Y, Tsurushima N, Ogawa H, Fukami K, Kuma K, Saino T. Science 2003; 300:958–961.

36. Wu J, Luther GW. Mar Chem 1995; 50:159–177.

37. Hudson RJM, Covault DT, Morel FMM. Mar Chem 1992; 38:209–235.

38. Rue EL, Bruland KW. Mar Chem 1995; 50:117–138.

39. Witter AE, Luther GW. Mar Chem 1998; 62:241–258.

40. Witter AE, Hutchins DA, Butler A, Luther GW. Mar Chem 2000; 69:1–17.

41. Vassiliev IR, Kolber Z, Wyman KD, Mauzerall D, Shukla VK, Falkowski PG. Plant Physiol 1995; 109:963–972.

42. Kirchman DL, Hoffman KA, Weaver R, Hutchins DA. Mar Ecol Progr Ser 2003; 250:291–296.

43. Takeda S. Nature 1998; 393:774–777.

44. LaRoche J, Geider RJ, Graziano LM, Murray H, Lewis KJ. J Phycol 1993; 29:767–777.

45. Geider RJ, LaRoche J. J Photosynth Res 1994; 39:275–301.

46. LaRoche J, Murray H, Orellana M, Newton J. J Phycol 1995; 31:520–530.

47. LaRoche J, Boyd PW, McKay RML, Geider RJ. Nature 1996; 382:802–805.

48. Bruland KW, Rue EL. In: Turner DR, Hunter KA, eds. The Biogeochemistry of Iron in Seawater. Chichester, West Sussex: John Wiley & Sons, 2001:255–289.

49. van den Berg CMG. Mar Chem 1995; 50:139–157.

50. Rue EL, Bruland KW. Limnol Oceanogr 1997; 42:901–910.

51. Morel FMM, Price NM. Science 2003; 300:944–947.

52. Kuma K, Katsumoto A, Nishioka J, Matsunaga K. Est Coast Shelf Sci 1998; 47:275–283.

53. Skoog A, Benner R. Limnol Oceanogr 1997; 42:1803–1813.

54. Wu J, Boyle EA, Sunda WG, Wen L-S. Nature 2001; 293:847.

55. Millero FJ, Yao W, Aicher J. Mar Chem 1995; 50:21–39.

56. Kuma K, Nishioka J, Matsunaga K. Limnol Oceanogr 1996; 41:396–407.

57. Liu X, Millero FJ. Mar Chem 2002; 77:43–54.

58. Liu X, Millero FJ. Geochim Cosmochim Acta 1999; 63:3487–3497.

59. Waite TD, Morel FMM. Environ Sci Technol 1984; 18:860–868.

60. Wells ML, Mayer LM. Deep-Sea Res 1991; 38:1379–1395.

61. Kuma K, Nakabayashi S, Suzuki Y, Matsunaga K. Mar Chem 1992; 38:133–143.

62. Waite TD, Szymczak R, Espey QI, Furnas MJ. Mar Chem 1995; 50:79–91.

63. Miller WL, King DW, Lin J, Kester DR. Mar Chem 1995; 50:63–77.

64. Voelker BM, Sedlak DL. Mar Chem 1995; 50:93–102.

65. Barbeau K, Rue EL, Trick CG, Bruland KT, Butler A. Limnol Oceanogr 2003; 48:1069–1078.

66. Jones GJ, Palenik BP, Morel FMM. J Phycol 1987; 23:237–244.

67. Maldonado MT, Price NM. Deep-Sea Res II 1999; 46:2447–2473.

68. Maldonado MT, Price NM. Limnol Oceanogr 2000; 45:814–826.

69. Millero FJ, Sotolongo S. Geochim Cosmochim Acta 1989; 53:1867–1873.

70. Gledhill M, van den Berg CMG. Mar Chem 1994; 47:41–54.
71. Johnson KS, Gordon RM, Coale KH. Mar Chem 1997; 57:137–161.
72. Macrellis HM, Trick CG, Rue EL, Smith G, Bruland KW. Mar Chem 2001; 76:175–187.
73. Duce RA, Tindale NW. Limnol Oceanogr 1991; 36:1715–1726.
74. Schutz LW, Prospero JM, Buat-Ménard P, Carvalho RAC, Cruzado A, Harriss P, Heidam NZ, Jaenicke R. In: Knapp AH, ed. The Long Range Atmospheric Transport of Natural and Contaminant Substances. Dordrecht: Kluwer, 1990:197–229.
75. Prospero JM, Ginoux P, Torres O, Nicholson SE, Gill TE. Rev Geophys 2002; 40:1002.
76. Chester R, Murphy KJT, Lin FJ, Berry AS, Bradshaw GA, Corcoran PA. Mar Chem 1993; 42:107–126.
77. Spokes LJ, Jickells TD, Lim B. Geochim Cosmochim Acta 1994; 58:3281–3287.
78. Hoffmann P, Dedik AN, Ensling J, Weinbruch S, Weber S, Sinner T, Gutlich P, Ortner HM. J Aerosol Sci 1996; 27:325–337.
79. Prospero JM, Nees RT. Nature 1986; 320:735–738.
80. Goudie AS, Middleton NJ. Climate Change 1992; 20:197–225.
81. Arimoto R. Earth-Sci Rev 2001; 54:29–42.
82. Jickells TD, Spokes LJ. In: Turner DR, Hunter KA, eds. The Biogeochemistry of Iron in Seawater. Chichester, West Sussex: John Wiley & Sons, 2001:85–121.
83. Prospero JM, Lamb PJ. Science 2003; 302:1024–1027.
84. Maring H, Savoie DL, Izaguirre MA, Custals L, Reid JS. J Geophys Res 2003; 108:8592.
85. Arimoto R, Duce RA. J Geophys Res 1986; 91:2787–2792.
86. Duce RA, Liss PS, Merrill JT, Atlas EL, Buat-Ménard P, Hicks BB, Miller JM, Prospero JM, Arimoto R, Church TM, Ellis W, Galloway JN, Hansen L, Jickells TD, Knapp AH, Reinhardt KH, Schneider B, Soudine A, Tokos JJ, Tsunogai S, Wollast R, Zhou M. Global Biogeochem Cycles 1991; 5:193–259.
87. Sarthou G, Baker AR, Blain S, Achterberg EP, Boye M, Bowie AR, Croot P, Laan P, de Baar HJW, Jickells TD, Worsfold PJ. Deep-Sea Res I 2003; 50:1339–1352.
88. Uematsu M, Wang ZF, Uno I. Geophys Res Lett 2003; 30:1342.
89. Uematsu M, Duce RA, Prospero JM, Chen L, Merrill JT, McDonald RL. J Geophys Res 1983; 88:5342–5352.
90. Prospero JM. In: Guerzoni S, Chester R, eds. The Impact of Desert Dust across the Mediterranean. Dordrecht: Kluwer, 1996:133–151.
91. Arimoto R, Duce RA, Ray BJ, Ellis WG, Cullen JD, Merrill JT. J Geophys Res 1995; 100:1199–1213.
92. Chiapello I, Bergametti G, Gomes L, Chatenet B, Dulac F, Pimenta J, Suares ES. Geophys Res Lett 1995; 22:3191–3194.
93. Ball WP, Dickerson RR, Doddridge BG, Stehr JW, Miller TL, Savoie DL, Carsey TP. J Geophys Res 2003; 108:8001.
94. Tindale NW, Pease PP. Deep-Sea Res 1999; 46:1577–1595.
95. Prospero JM, Uetmatsu M, Savoie DL. In: Riley JP, Chester R, Duce RA, eds. Chemical Oceanography, Vol 10. London: Academic Press, 1989:187–218.
96. Arimoto R, Ray BJ, Duce RA, Hewitt AD, Boldi R, Hudson A. J Geophys Res 1990; 95:22389–22405.
97. Ezat U, Dulac F. C R Acad Sci Ser II 1995; 320:9–14.
98. Wagenbach D, Gorlach U, Moser K, Munnich KO. Tellus 1988; 40B:426–436.

99. Dick AL. Geochim Cosmochim Acta 1991; 55:1827–1836.
100. Kim G, Church TM. Global Biogeochem Cycles 2002; 16:1046.
101. Chiapello I, Moulin C. Geophys Res Lett 2002; 29:1176.
102. Moulin C, Chiapello I. Geophys Res Lett 2004; 31:L02107.
103. Guieu C, Loye-Pilot MD, Ridame C, Thomas C. J Geophys Res 2002; 107:4258.
104. Zhang XY, Gong SL, Arimoto R, Shen ZX, Mei FM, Wang D, Cheng Y. J Atmos Chem 2003; 44:241–257.
105. Martin JH, Gordon RM, Fitzwater SE. Limnol Oceanogr 1991; 36:1793–1802.
106. Johnson KS, Coale KH, Elrod VA, Tindale NW. Mar Chem 1994; 46:319–334.
107. Zhu X, Prospero JM, Millero FJ, Savoie DL, Brass GW. Mar Chem 1992; 38:91–107.
108. Spokes LJ, Jickells TD. Aquat Geochem 1996; 1:355–374.
109. Zhu X, Prospero JM, Millero FJ. J Geophys Res 1997; 102:21297–21305.
110. Graedel TE, Weschler CJ, Mandich ML. Nature 1985; 317:240–242.
111. Sulzberger, Laubscher H. Mar Chem 1995; 50:103–115.
112. Zhu X, Prospero JM, Savoie DL, Millero FJ, Zika RG, Saltzmann ES. J Geophys Res 1993; 98:21297–21305.
113. Grgic I, Dovzan A, Berzic G, Hudnik V. J Atmos Chem 1998; 29:315–337.
114. Zuo Y. Geochim Cosmochim Acta 1995; 59:3123–3131.
115. Pehkonen SO, Siefert R, Erel Y, Webb S, Hoffman MR. Environ Sci Technol 1993; 27:2056–2062.
116. Guieu C, Thomas AJ. In: Guerzoni S, Chester R, eds. The Impact of Desert Dust Across the Mediterranean. Dordrecht: Kluwer, 1996:207–216.
117. Martin JH, Fitzwater SE. Nature 1988; 331:341–343.
118. Martin JH, Gordon RM, Fitzwater S, Broenkow WW. Deep-Sea Res I 1989; 36:649–680.
119. Löscher BM, de Jong JTM, de Baar HJW, Veth C, DeHairs F. Deep-Sea Res II 1997; 44:143–187.
120. Fung IY, Meyn SK, Tegen I, Doney SC, John JG, Bishop JK. Global Biogeochem Cycles 2000; 14:281–295.
121. Moore JK, Doney SC, Glover DM, Fung IY. Deep-Sea Res II 2002; 49:463–507.
122. Degens ET, Kempe S, Richey JE. In: Degens ET, Kempe S, Ritchey JE, eds. SCOPE 42: Biogeochemistry of Major World Rivers. Chichester: John Wiley and Sons, 1991:323–347.
123. Milliman JD, Syvitski JPM. J Geol 1992; 100:525–544.
124. Wu J, Luther GW. Geochim Cosmochim Acta 1996; 60:2729–2742.
125. Giueu C, Martin JM. Est Coast Shelf Sci 2002; 54:501–512.
126. Dai M, Martin JM. Earth Planet Sci Lett 1995; 131:127–141.
127. Wen LS, Santschi P, Gill G, Paternostro C. Mar Chem 1999; 63:185–212.
128. Guieu C, Huang WW, Martin JM, Yang YY. Mar Chem 1996; 53:255–267.
129. von Damm KL, Bray AM, Buttermore LG, Oosting SE. Earth Planet Sci Lett 1998; 160:521–536.
130. Seyfried WE, Mottl MJ. In: Karl DM, ed. The Microbiology of Deep-sea Hydrothermal Vents. Boca Raton, FL: CRC Press, 1995:1–34.
131. Elderfield H, Schultz A. Annu Rev Earth Planet Sci 1996; 24:191–224.
132. Schultz A, Elderfield H. In: Cann J, Elderfield H, Laughton T, eds. Mid Ocean Ridges. Cambridge: Cambridge University Press, 1999:171–209.
133. von Damm KL, Bischoff JL. J Geophys Res 1987; 92:11,334–11,346.

134. German CR, Holliday BP, Elderfield H. Earth Planet Sci Lett 1991; 107:101–114.
135. Boye M, Aldrich AP, van den Berg CMG, de Jong JTM, Weldhuis M, de Baar HJW. Mar Chem 2003; 80:129–143.
136. Canfield DE, Thamdrup B, Hansen JW. Geochim Cosmochim Acta 1993; 57:3867–3883.
137. Kumar N, Anderson RF, Mortlock RA, Froelich PN, Kubik P, Dittrich-Hannen B, Suter M. Nature 1995; 378:675–680.
138. Koschinsky A, Gaye-Haake B, Arndt C, Maue G, Spitzy A, Winkler A, Halbach P. Deep-Sea Res II 2001; 48:3629–3651.
139. Berelson W, McManus J, Coale K, Johnson K, Burdige D, Kilgore T, Colodner D, Chavez F, Kudela R, Boucher J. Continental Shelf Res 2003; 23:457–481.
140. Canfield DE. Geochim Cosmochim Acta 1989; 53:619–632.
141. Davison W, Grime GW, Morgan JAW, Clarkes K. Nature 1991; 352:323–325.
142. Burdige DJ, Earth-Sci Rev 1993; 35:249–284.
143. van Capellen Y, Wang Y. Am J Sci 1996; 296:197–243.
144. Wijsman JWM, Middelburg JJ, Heip CHR. Mar Geol 2001; 172:167–180.
145. Taillefert M, Hoover VC, Rozan TF, Theberge SM, Luther GW. Estuaries 2002; 25:1088–1096.
146. Rozan TF, Taillefert M, Trouwborst RE, Glazer BT, Ma SF, Herszage J, Valdes LM, Price KS, Luther GW. Limnol Oceanogr 2002; 47:1346–1354.
147. Fitzwater SE, Johnson KS, Elrod VA, Ryan JP, Coletti LJ, Tanner SJ, Gordon RM, Chavez FP. Continental Shelf Res 2003; 23:1523–1544.
148. Bruland KW, Rue EL, Smith GJ. Limnol Oceanogr 2001; 46:1661–1674.
149. Tankere SPC, Muller FLL, Burton JD, Statham PJ, Guieu C, Martin JM. Continental Shelf Res 2001; 21:1501–1532.
150. For the purposes of this chapter I will focus on the delivery of iron to the euphotic zone through advection and upwelling, but it is important to note that related processes can result in losses of iron from a region.
151. Martin JH, Coale KH, Johnson KS, Fitzwater SE, Gordon RM, Tanner SJ, Hunter CN, Elrod VA, Nowicki JL, Coley TL, Barber RT, Lindley S, Watson AJ, van Scoy K, Law CS, Liddicoat MI, Ling R, Stanton T, Stockel J, Collins C, Anderson A, Bidigare R, Ondrusek M, Latasa M, Millero FJ, Lee K, Yao W, Zhang JZ, Friedrich G, Sakamoto C, Chavez F, Buck K, Kolber Z, Greene R, Falkowski P, Chisholm SW, Hoge F, Swift R, Yungel J, Turner S, Nightingale P, Hatton A, Liss P, Tindale NW. Nature 1994; 371:123–129.
152. Gordon RM, Johnson KS, Coale KH. Deep-Sea Res II 1998; 45:995–1041.
153. Johnson KS, Chavez FP, Elrod VA, Fitzwater SE, Pennington JT, Buck KR, Walz PM. Geophysical Res Lett 2001; 28:1247–1250.
154. Archer DE, Johnson K. Global Biogeochem Cycles 2000; 14:269–279.
155. Measures CI, Vink S. Deep-Sea Res II 2001; 48:3913–3941.
156. Watson AJ, In: Turner DR, Hunter KA, eds. The Biogeochemistry of Iron in Seawater. Chichester, West Sussex: John Wiley & Sons, 2001:9–40.
157. Hiscock MR, Marra J, Smith WO, Goericke R, Measures C, Vink S, Olson RJ, Sosik HM, Barber RT. Deep-Sea Res II 2003; 50:533–558.
158. Hutchins DA, DiTullio GR, Bruland KW. Limnol Oceanogr 1993; 38:1242–1255.
159. Hutchins DA, Bruland KW. Mar Ecol Progr Ser 1994; 110:259–269.
160. Hutchins DA, Wang W, Fisher NS. Limnol Oceanogr 1995; 40:989–994.
161. Barbeau K, Moffett JW, Caron DA, Croot PL, Erdner DL. Nature 1996; 380:61–64.

162. Smetacek V, de Baar HJW, Bathmann UV, Lochte K, Rutgers van der Loeff MM. Deep-Sea Res II 1997; 44:1–21.
163. Barbeau K, Moffett JW. Env Sci Technol 1998; 32:2969–2975.
164. Deuser WG. Deep-Sea Res 1986; 33:225–246.
165. Fischer G, Ratmeyer V, Weber G. Deep-Sea Res II 2000; 47:1961–1997.
166. Whitfield M, Turner DR. In: Stumm W, ed. Aquatic Surface Chemistry. New York: Wiley, 1987:457–493.
167. Balistrieri L, Murray J, Brewer PG. Deep-Sea Res 1981; 28:101–121.
168. Honeyman BD, Balistrieri L, Murray JW. Deep-Sea Res 1988; 35:227–246.
169. Moran SB, Buesseler KO. J Mar Res 1993; 51:893–922.
170. Zhuang G, Duce RA. Deep-Sea Res I 1993; 40:1413–1429.
171. Guieu C, Chester R, Nimmo M, Martin JM, Guerzoni S, Nicolas E, Mateu J, Keyes S. Deep-Sea Res II 1997; 44:655–674.
172. Wu JF, Boyle E. Global Biogeochem Cycles 2002; 16:1086.
173. Bucciarelli E, Blain S, Tréguer P, Mar Chem 2001; 73:21–36.
174. Takeda S, Obata H. Mar Chem 1995; 50:219–227.
175. Coale KH, Fitzwater SE, Gordon RM, Johnson KS, Barber RT. Nature 1996; 379:621–624.
176. Bruland KW, Orians KJ, Cowen JP. Geochim Cosmochim Acta 1994; 58:3171–3182.
177. Landing WM, Bruland KW. Geochim Cosmochim Acta 1987; 51:29–43.
178. Orians KJ, Bruland KW. Nature 1985; 316:427–429.
179. Millero FJ. Earth Planet Sci Lett 1998; 154:323–329.
180. Sarthou G, Jeandel C. Mar Chem 2001; 74:115–129.
181. Hutchins DA, DiTullio GR, Zhang Y, Bruland KW. Limnol Oceanogr 1998; 43:1037–1054.
182. Gledhill M, van den Berg CMG, Nolting RF, Timmermans KR. Mar Chem 1998; 59:283–300.
183. Chase Z, van Green A, Kosro PM, Marra J, Wheeler PA. J Geophys Res 2002; 107:3174.
184. Lewis BL, Landing WM. Deep-Sea Res II 1991; 38:773–803.
185. Haraldsson C, Westerlund S. Mar Chem 1988; 23:417–424.
186. Oguz T, Murray JW, Callahan AE. Deep-Sea Res I 2001; 48:761–787.
187. Kremling K, Tokos JJS, Brügmann L, Hansen HP. Mar Poll Bull 1997; 34:112–122.
188. Levitus S, Burgett R, Boyer T. World Ocean Atlas 1994, Vol 3: Nutrients, NOAA Atlas NESDIS, US Dept of Commerce, Washington, DC, 1994.
189. Wong CS, Whitney FA, Iseki K, Page JS, Zeng J. Global Biogeochem Cycles 1994; 5:119–134.
190. McClain CR, Arrigo K, Tai KS, Turk D. J Geophys Res 1996; 101:3697–3713.
191. Chavez FP, Buck KR, Service SK, Newton J, Barber RT. Deep-Sea Res II 1996; 43:835.
192. Chavez FP, Barber R. Deep-Sea Res 1987; 34:1229–1243.
193. Fiedler PC, Philbrick V, Chavez FP. Limnol Oceanogr 1991; 36:1834–1850.
194. Booth BC, Lewin J, Postel JR. Progr Oceanogr 1993; 32:57–99.
195. Landry MR, Gifford DJ, Kirchman DL, Wheeler PA, Monger BC. Progr Oceanogr 1993; 32:239–258.
196. McCarthy JJ, Garside C, Nevins JL, Barber RT. Deep-Sea Res II 1996; 43:1065–1093.

197. Turner DR, Owens NJP. Deep-Sea Res II 1995; 42:907–932.
198. Smetacek V, Klaas C, Menden-Deuer S, Rynearson TA. Deep-Sea Res II 2002; 49:3835–3848.
199. Strass VH, Garabato ACN, Pollard RT, Fischer HI, Hense I, Allen JT, Read JF, Leach H, Smetacek V. Deep-Sea Res II 2002; 49:3735–3769.
200. de Baar HJW, Boyd PW. In: Hanson RB, Ducklow HW, Field JG, eds. The Dynamic Ocean Carbon Cycle: A Modern Synthesis of the Joint Global Ocean Flux Study. Cambridge: Cambridge University Press, 2000:61–140.
201. Longhurst A, Sathyendranath S, Platt S, Caverhill C. J Plankton Res 1995; 17:1245–1271.
202. Banse K, English DC. Mar Ecol Progr Ser 1997; 156:51–66.
203. Hart TJ. Discovery Rep 1934; 8:1–268.
204. Thomas WH. J Fish Res Board 1969; 26:1133–1145.
205. Fitzwater SE, Johnson KS, Gordon RM, Coale KH, Smith WO Jr. Deep-Sea Res II 2000; 47:3159–3179.
206. Chisholm SW, Morel FMM. Limnol Oceanogr 1991; 36:1507–1511.
207. Martin JH, Broenkow WW, Fitzwater S, Gordon RM. Limnol Oceanogr 1990; 35:775–777.
208. Martin JH, Gordon RM, Fitzwater SE. Nature 1990; 345:156–158.
209. de Baar HJW, Buma AGJ, Nolting GC, Cadee G, Jaques G, Treguer PJ. Mar Ecol Progr Ser 1990; 65:105–122.
210. Price NM, Andersen LF, Morel FMM. Deep-Sea Res I 1991; 38:1361–1378.
211. Coale KH. Limnol Oceanogr 1991; 36:1851–1864.
212. Ditullio GR, Hutchins DA, Bruland KW. Limnol Oceanogr 1993; 38:1361–1378.
213. de Baar HJW, de Jong JTM, Bakker DCF, Loscher BM, Veth C, Bathmann U, Smetacek V. Nature 1995; 373:412–415.
214. de Baar HJW, van Leeuwe MA, Scharek R, Goeyens L, Bakker KMJ, Fritsche P. Deep-Sea Res II 1997; 44:229–260.
215. Fitzwater SE, Coale KH, Gordon RM, Johnson KS, Ondrusek ME. Deep-Sea Res II 1996; 43:995–1015.
216. Boyd PW, Muggli DL, Varela DE, Goldblatt RH, Chretien R, Orians KJ, Harrison PJ. Mar Ecol Progr Ser 1996; 136:179–193.
217. van Leeuwe MA, Scharek R, de Baar HJW, de Jong JTM, Goeyens L. Deep-Sea Res II 1997; 44:189–207.
218. Sedwick PN, Blain S, Queguiner B, Griffiths FB, Fiala M, Bucciarelli E. Deep-Sea Res II 2002; 49:3327–3349.
219. Coale KH, Wang X, Tanner SJ, Johnson KS. Deep-Sea Res II 2003; 50:635–653.
220. Timmermans KR, van Leeuwe MA, de Jong JTM, McKay RML, Nolting RF, Witte HJ, van Ooyen J, Swagerman MJW, Kloosterhuis H, de Baar HJW. Mar Ecol Progr Ser 1998; 166:27–41.
221. Dugdale RC, Wilkerson FP. Global Biogeochem Cycles 1990; 4:13–19.
222. Banse K. Limnol Oceanogr 1990; 35:772–775.
223. Cullen JJ. Limnol Oceanogr 1991; 36:1578–1599.
224. Coale KH, Johnson KS, Chavez FP, Buesseler KO, Barber RT, Brzezinski MA, Cochlan WP, Millero FJ, Falkowski PG, Bauer JE, Wanninkhof RH, Kudela RM, Altabet MA, Hales BE, Takahashi T, Landry MR, Bidigare RR, Wang X, Chase Z, Strutton PG, Friederich GE, Gorbunov MY, Lance VP, Hilting AK, Hiscock MR, Demarest M, Hiscock WT, Sullivan KF, Tanner SJ, Gordon RM, Hunter CN,

Elrod VA, Fitzwater SE, Jones JL, Tozzi S, Kobilek M, Roberts AE, Herndon J, Brewster J, Ladizinsky N, Smith G, Cooper D, Timothy D, Brown SL, Selph KE, Sheridan CC, Twining BS, Johnson ZI. Science 2004; 304:408–414.

225. Boyd PW, Law CS, Wong CS, Nojiri Y, Tsuda A, Levasseur M, Takeda S, Rivkin R, Harrison PJ, Strzepek R, Gower J, McKay RM, Abraham E, Arychuk M, Barwell-Clarke J, Crawford W, Crawford D, Hale M, Harada K, Johnson K, Kiyosawa H, Kudo I, Marchetti A, Miller W, Needoba J, Nishioka J, Ogawa H, Page J, Robert M, Saito H, Sastri A, Sherry N, Soutar T, Sutherland N, Taira Y, Whitney F, Wong SE, Yoshimura T. Nature 2004; 428:549–553.

226. Kolber ZS, Barber RT, Bidigare RR, Chai F, Coale KH, Dam HG, Lewis MR, Lindley ST, McCarthy JJ, Roman MR, Stoecker DK, Verity PG, White JR. Nature 1994; 371:145–149.

227. Behrenfeld MJ, Bale AJ, Kolber ZS, Aiken J, Falkowski PG. Nature 1996; 383:508–510.

228. Gervais F, Riebesell U, Gorbunov MY. Limnol Oceanogr 2002; 47:1324–1335.

229. Cooper DJ, Watson AJ, Nightingale PD. Nature 1996; 383:511–513.

230. Watson AJ, Whitney FA, Iseki K, Page JS, Zeng J. Nature 1994; 371:143–145.

231. Buesseler KO, Boyd PW. Science 2003; 300:67–68.

232. Buesseler KO, Andrews JE, Pike SM, Charette MA. Science 2004; 304:414–417.

233. Bishop JKB, Wood TJ, Davis RE, Sherman JT. Science 2004; 304:417.

234. Firme GF, Rue EL, Weeks DA, Bruland KW, Hutchins DA. Global Biogeochem Cycles 2003; 17:1016.

235. Hutchins DA, Hare CE, Weaver RS, Zhang Y, Firme GF, DiTullio GR, Alm MB, Riseman SF, Maucher JM, Geesey ME, Trick CG, Smith GJ, Rue EL, Conn J, Bruland KW. Limnol Oceanogr 2002; 47:997–1011.

236. Behrenfeld MJ, Kolber ZS. Science 1999; 283:840–843.

237. Mitchell BG, Broody EA, Holm-Hansen O, McCain C, Bishop J. Limnol Oceanogr 1991; 36:1662–1677.

238. Nelson DM, Smith WO. Limnol Oceanogr 1991; 36:1650–1661.

239. Boyd PW, LaRoche J, Gall M, Frew R, McKay RML, J Geophys Res 1999; 104:13395–13408.

240. Geider RT, Greene RM, Kolber Z, McIntyre HL, Falkowsi PG. Deep-Sea Res I 40:1205–1224.

241. Dugdale RC, Wilkerson FP, Minas HJ. Deep-Sea Res I 1995; 42:697–719.

242. Dugdale RC, Wilkerson FP. Nature 1998; 391:270–273.

243. Jiang MS, Chai F, Dugdale RC, Wilkerson FP, Peng TH, Barber RT. Deep-Sea Res II 2003; 50:2971–2996.

244. Falkowski PG. Nature 1997; 387:272–275.

245. Reuter JGJ. Eos 1982; 63:445.

246. Letelier RM, Karl DM. Mar Ecol Progr Ser 1996; 133:263–273.

247. Lenes JM, Darrow BP, Cattrall C, Heill CA, Callahan M, Vargo GA, Byrne RH, Prospero JM, Bates DE, Fanning KA, Walsh JJ. Limnol Oceanogr 2001; 46:1261–1277.

248. Berman-Frank I, Cullen JT, Shaked Y, Sherrell RM, Falkowski PG. Limnol Oceanogr 2001; 46:1249–1260.

249. Buesseler KO. Global Biogeochem Cycles 1998; 12:297–310.

250. Buesseler KO, Ball L, Andrews J, Cochran JK, Hirschberg DJ, Bacon MP, Fleer A, Brzezinski M. Deep-Sea Res II 2001; 48:4275–4297.

251. Chisholm SW, Falkowski PG, Cullen JJ. Science 2001; 294:309–310.

252. Law CS, Ling RD. Deep-Sea Res II 2001; 48:2509–2527.
253. Martin JH, Paleoceanography 1990; 5:1–13.
254. Xhang XY, Lu HY, Arimoto R, Gong SL. Earth Planet Sci Lett 2002; 202:637–643.
255. Kohfeld KE, Harrison SP. Quatern Sci Rev 2003; 22:1859–1878.
256. Rea DK. Rev Geophys 1994; 32:159–195.
257. Reader MC, Fung I, McFarlane N. J Geophys Res 1999; 104:9381–9398.
258. Petit JR, Jouzel J, Raynaud D, Barkov NI, Barnola J-M, Basile I, Bender M, Chappellaz J, Davis M, Delaygue G, Delmotte M, Kotlyakov VM, Legrand M, Lipenkov VY, Lorius C, Pépin L, Ritz C, Saltzman E, Stievenard M. Nature 1999; 399:429–436.
259. Basile I, Grousset FE, Revel M, Petit JR, Biscaye PE, Barkov NI. Earth Planet Sci Lett 1997; 146:573–589.
260. Andersen KA, Armengaud A, Genthon C. Geophys Res Lett 1998; 25:2281–2284.
261. Mahowald N, Kohfeld KE, Hansson M, Balkanski Y, Harrison SP, Prentice IC, Schulz M, Rodhe H. J Geophys Res 1999; 104:15895–15916.
262. Werner M, Tegen I, Harrison SP, Kohfeld KE, Prentice IC, Balkanski Y, Rodhe H, Roelandt C. J Geophys Res 2002; 107:4744.
263. Paytan A, Kastner M, Chavez FP. Science 1996; 274:1355–1357.
264. Loubere P. Nature 2000; 406:497–500.
265. Higuerra JC, Berger WH. Geology 1994; 22:629–632.
266. Lyle M. Nature 1988; 335:529–532.
267. Mortlock RA, Charles CD, Froelich PN, Zibello MA, Saltzman J, Hays JD, Burckle LH. Nature 1991; 351:220–223.
268. Nürnberg CC, Bohrmann G, Schlüter M, Frank M. Paleoceanography 1997; 12:594–603.
269. Bareille, Labracherie M, Bertrand P, Labeyrie L, Lavaux G, Dignan M. J Mar Sys 1998; 17:527–539.
270. Chase Z, Anderson RF, Fleisher MQ. Paleoceanography 2001; 16:468–478.
271. Lefévre N, Watson AJ. Global Biogeochem Cycles 1999; 13:727–736.
272. Broecker WS, Henderson GM. Paleoceanography 1998; 13:352–364.
273. Gruber N, Sarmiento JL. Global Biogeochem Cycles 1997; 11:235–266.
274. Karl D, Letelier R, Tupas L, Dore J, Christian J, Hebel D. Nature 1997; 388:533–538.
275. Falkowski PG, Barber RT, Smetacek V. Science 1998; 281:200–206.
276. Francois R, Altabet MA, Yu EF, Sigman DM, Bacon MP, Frank M, Bohrmann G, Barielle G, Labeyrie LD. Nature 1997; 389:929–935.
277. Archer DE, Eshel G, Wiguth A, Broecker W, Pierrehumbert R, Tobis M, Jacob R. Global Biogeochem Cycles 2000; 14:1219–1230.
278. Harrison KG. Paleoceanography 2000; 15:292–298.
279. Dugdale RC, Wilkerson FP. Scientia Marina 2001; 65 (suppl 2):141–152.
280. Barnola JM, Raynaud D, Korotkevitch YS, Lorius C. Nature 1987; 329:408.
281. Sigman DM, Altabet MA, Francois R, McCorkle DC, Gaillard JF. Paleoceanography 1999; 14:118–134.
282. Watson AJ, Bakker DCE, Ridgwell AJ, Boyd PW, Law CS. Nature 2000; 407:730–733.
283. Haygood MG, Holt PD, Butler A. Limnol Oceanogr 1993; 38:1091–1097.
284. Albrecht-Gary AM, Crumbliss AL. In: Sigel A, Sigel H, eds. Metal Ions in Biological Systems, Vol 35: Iron Transport and Storage in Microorganisms, Plants, and Animals. New York: Marcel Dekker, 1998:239–327.

8

The Biogeochemistry of Cadmium

François M. M. Morel and Elizabeth G. Malcolm

Department of Geosciences, Princeton University, 153 Guyot Hall,
Princeton, New Jersey 08544, USA

1.	Introduction	196
2.	Concentrations: Sources and Sinks	197
	2.1. Cadmium in the Atmosphere	197
	2.2. Cadmium in Rivers and Soils	198
	2.3. Cadmium in the Oceans	199
3.	Chemical Speciation	201
4.	Biological Effects	204
	4.1. Toxicity	204
	4.2. Detoxification	205
	4.2.1. Cadmium Binding	205
	4.2.2. Export	207
	4.3. Cadmium Use in Phytoplankton	207
5.	The Biogeochemistry of Cadmium as an Algal Nutrient in the Sea	209
	5.1. Mass Balance in Surface Seawater	209
	5.2. Cellular Quotas	211
	5.3. Uptake and Growth Rates	211
	5.4. Remineralization	212
6.	Cadmium as a Paleotracer	215
7.	Envoi	216
	Acknowledgments	217
	Abbreviations	217
	References	217

1. INTRODUCTION

Cadmium is one of the all-star pollutants, along with a few other elements such as mercury, arsenic, and lead. This distinction results from the relative abundance of Cd in the environment and of its great toxicity to all living things, including man. Much as the "Minamata Bay disease" in the 1950s focused the attention of environmental scientists and toxicologists on mercury, the earlier outbreak of the "itai–itai disease" in the Jinzu river basin in Japan focused their attention on Cd, the direct cause of this disease. As a result, much attention has been directed to the sources and transport of Cd in the environment and to its noxious effects on various organisms.

For many years, however, oceanographers have had a different interest in Cd. They have discovered that the distribution of Cd in the oceans is similar to that of the major algal nutrients nitrogen and phosphorus. In view of the remarkable correlation between Cd and P concentrations observed throughout the oceans (Fig. 1), the concentration of Cd—as recorded, for example, in the $CaCO_3$ tests of fossil foraminifera—can be used as a proxy for that of P. Recently, it has become apparent that the nutrient-like behavior of Cd in the

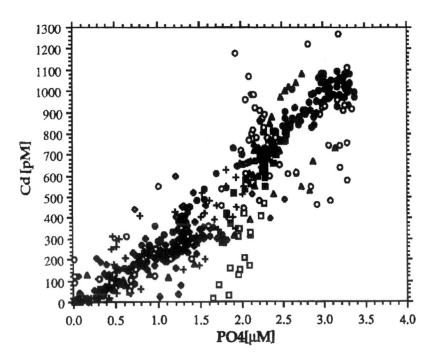

Figure 1 Concentrations of cadmium vs. phosphate measured in ocean water samples worldwide. Circles = Pacific, diamonds = Atlantic, squares = Antarctic, triangles = Indian Ocean. Reproduced by permission from Ref. [66].

oceans results from the fact that Cd itself is a nutrient for marine phytoplankton, playing a role similar to other micronutrients such as Zn and Co. And so we have come full circle and are now interested in the transport and transformations of Cd in the environment and in its biochemical effects because of its beneficial role in the growth of marine microorganisms and in oceanic primary production. As a result, this chapter on the biogeochemistry of cadmium focuses heavily on the ocean and the organisms in it, particularly the phytoplankton, which use the metal as a nutrient and control its geochemistry.

2. CONCENTRATIONS: SOURCES AND SINKS

Most Cd in the environment is found as a companion of Zn with which it shares its external electronic configuration and many chemical properties. The principal industrial source of Cd is as a byproduct of Zn mining, the major Zn ores, sphalerite and wurtzite (ZnS) containing typically a fraction of a percent (0.2–0.7%) of CdS. Early industrial use of Cd was mostly for electroplating and in bearing alloys [1]. About two thirds of the modern use of the metal is for nickel–cadmium batteries [2], the rest going to pigments, metal plating, and the plastic industry.

Because of concern about its environmental toxicity, the use of Cd has decreased in recent years. A few studies of lake sediments and snowpacks have documented the rise in Cd pollution as a result of the industrial revolution and its sharp decrease since the 1970s (Fig. 2) [3–5]. According to the data of Fig. 2, the present inputs of Cd to remote northern regions through aeolian transport of aerosols are near their pre-industrial value. As the industrial use of Cd declines, much of the remaining Cd pollution results from its presence as a contaminant in Zn-containing materials.

2.1. Cadmium in the Atmosphere

Anthropogenic sources of Cd to the atmosphere include the metal processing industry (primarily smelting), waste incineration and fossil fuel combustion [6] (Fig. 3). A 1983 budget estimated that human activities accounted for ~85% of Cd inputs to the atmosphere [7]. The natural sources to the atmosphere include volcanoes and wind-blown soil. Both natural and anthropogenic sources release Cd to the atmosphere as a component of aerosol particles.

The atmospheric residence time of Cd averages ~1 week, in which time the aerosols can be transported locally or for hundreds of kilometers before removal. This occurs by dry deposition of aerosols [8] and wet deposition after incorporation into cloud droplets or falling precipitation [9–11]. High elevation forests have also been found to receive additional Cd deposition due to direct impaction of cloud droplets onto the forest canopy [12].

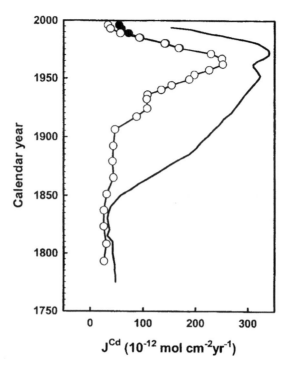

Figure 2 Cadmium accumulation rates. Open circles are estimated from a sediment core from Lake Tantare, Quebec, Canada. Filled circles are predicted final accumulation rate in surface sediments after post-depositional transfer of Cd is halted when sediments are buried below 8 cm. Line from measurements of a Greenland snowpack. Reproduced by permission from Ref. [3].

2.2. Cadmium in Rivers and Soils

The mean pre-industrial concentration of dissolved Cd in rivers has been estimated to be 45 pM [13]. As a result of anthropogenic contamination, the mean for dissolved Cd measured in all rivers worldwide in the early 1980s was four times higher (200 pM) [14]. Many river systems have much higher Cd concentrations as a result of pollution from industry and historic mining. For example, the mean dissolved Cd concentrations in the Lot and Garonne rivers (which feed the Gironde estuary in France; see below) were 1.3 and 0.8 nM, respectively (measured from 1992–1998; [15]).

The average Cd content of the lithosphere is $\sim$0.1 mg kg^{-1}, while the global mean soil concentration has been estimated at 0.5 mg kg^{-1}. Han et al. [16] have estimated that the cumulative amount of Cd extracted from mining from 1860 to 2000 distributed throughout the world's soils equals an addition of 0.3 mg kg^{-1} Cd to pre-industrial soil concentrations. Local sources of Cd

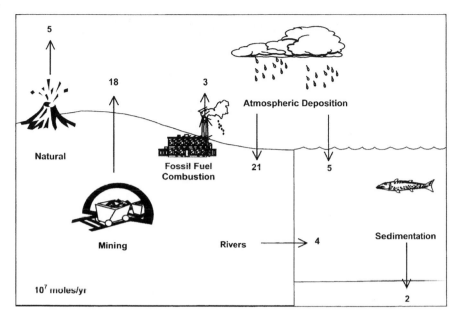

Figure 3 Flux estimates of the global cadmium cycle. All fluxes are $\times 10^7$ mol year^{-1}. Emissions to the atmosphere are from Ref. [2], mining/smelting flux also from Ref. [16]. Atmospheric deposition and net river input (gross—loss to estuaries) to the ocean are from Ref. [18]. Loss of oceanic cadmium to marine sediments is scaled up from a cadmium accumulation rate of 0.006 mmol cm^{-2} year^{-1} for the Pacific from Ref. [18] and is highly uncertain. (A mass balance is not expected since the atmospheric and riverine inputs include anthropogenic increases that are recent compared to the residence time of Cd in the oceans.) Atmospheric deposition to land was calculated based on the steady-state assumption that emissions to the atmosphere equal losses to the ocean and land.

contamination to soils include commercial fertilizer, sewage sludge application to agricultural land and atmospheric deposition of anthropogenic emissions [17].

2.3. Cadmium in the Oceans

The riverine and atmospheric inputs to the oceans appear to be about equal but these fluxes are poorly known. With a total input of 10^8 mol year^{-1} and an oceanic inventory of 10^{12} mol, the average oceanic residence time of Cd is $\sim 10,000$ years [18]. The bulk of the cadmium in seawater is in the dissolved phase and only a tiny fraction in suspended solids. The particulate concentration is in the range $0.04 - 4$ pM, corresponding roughly to a concentration in the solid phase of $0.1 - 1$ μmol Cd : mol C (Fig. 4) [19].

The vertical distribution of dissolved cadmium concentration in the oceans follows a profile typical of phytoplankton nutrients: from a few picomolar at the

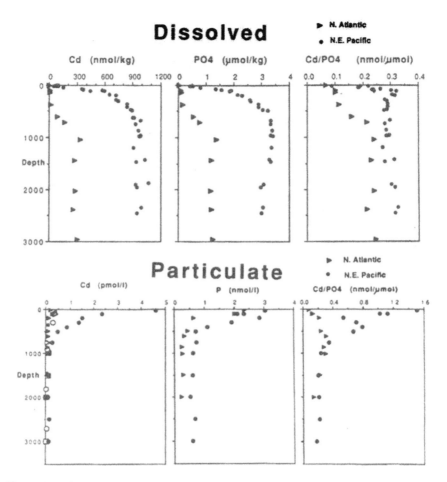

Figure 4   Dissolved and particulate concentrations of cadmium and phosphate from Pacific (filled circles) and Atlantic (all other symbols) stations. Reproduced by permission from Ref. [19].

surface, it increases to a broad maximum of a few nanomolar at the bottom of the main thermocline (1000 m in Fig. 4) and remains more or less constant at greater depths. The particulate cadmium profile, in contrast, shows a maximum at the surface (Fig. 4) [19]. Such profiles result from Cd uptake by photosynthetic organisms at the surface, and remineralization at depth through decomposition of sinking biomass in the water column. The excellent correlation that exists between dissolved cadmium and phosphate concentrations in the oceans (Fig. 1) must reflect a similar rate of water column remineralization in addition to a near complete scavenging by microorganisms at the surface.

Like that of other nutrients, the cycle of cadmium in the oceans is thus dominated by: (i) a downward flux of particulate cadmium caused by sinking of biogenic particles; and (ii) an upward flux of dissolved cadmium resulting from diffusive and advective (upwelling) transport from the high concentration in the deep to the low concentration at the surface [18].

Two major patterns of horizontal distribution can be discerned for Cd in the oceans: (i) a much smaller concentration in the deep Atlantic compared to the deep Pacific and Indian Oceans, as is characteristic of all algal nutrients; (ii) a decrease in surface concentrations from the coasts of continents to the open oceans, which is observed for many elements. Cadmium concentrations in the deep Atlantic ($\sim$0.3 nM) are roughly three times lower than in the deep Pacific or Indian Oceans (0.9 nM; Fig. 4) about the same factor as observed for phosphate [20–22]. This higher accumulation in the deep waters of the Pacific than in those of the Atlantic results from the simultaneous downward transport of the elements that are incorporated in settling biomass and the horizontal transport by major oceanic currents, which run toward the Atlantic at the surface and toward the Pacific in deep waters [23].

In surface coastal waters, the concentration of Cd, like that of most trace elements, is typically elevated, reaching values of 200 pM. Such elevated surface concentrations near continents can result from coastal upwelling (as along the Eastern Pacific Coast) bringing cadmium and other trace elements from deep water to the surface, along with major algal nutrients. It can also result from continental/riverine inputs. In much of the Western North Atlantic, the surface concentration of Cd appears to result from the mixing of high Cd water from the North American Shelf with low Cd water from the Sargasso Sea [22] (Fig. 5).

3. CHEMICAL SPECIATION

Cadmium is present in water as the hydrated divalent cation, Cd^{2+}, which has a completely filled d shell (d^{10}). With only the +II redox state stable in water, the aquatic chemistry of Cd is dominated by its ability to coordinate strongly to a wide range of ligands. In oxic waters, chloride is the most effective inorganic ligand for binding Cd^{2+} [24]. The result is a gradient from the free hydrated ion to chloro complexes as the principal unchelated species as one moves from freshwaters to seawater (Fig. 6). Because Cd^{2+} is a relatively large ion (ionic radius $= 0.97$ Å), the water molecules in its first hydration sphere are not held very tightly and can exchange rapidly with other ligands (the first order rate constant for water loss, k_{-w}, is $3 \times 10^8 \text{ s}^{-1}$). Thus, Cd^{2+} exhibits fast reaction kinetics in water.

Under anoxic conditions, the speciation of cadmium is dominated by coordination with reduced sulfur compounds including sulfide and polysulfides. In sediments, the solubility of Cd must be controlled by its adsorption on oxides or the formation of minerals: $CdCO_3$ under oxic conditions and CdS under anoxic conditions [24]. The high affinity of Cd^{2+} for the surface of

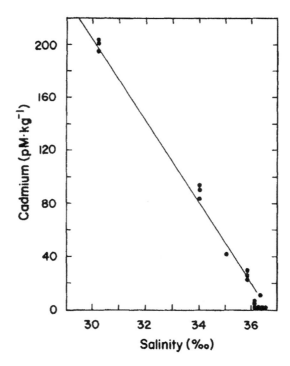

Figure 5   Transect of surface Cd concentrations (pM) as a function of salinity from the North American Shelf to the Sargasso Sea. The linear relation between Cd concentration and salinity is consistent with a mixing of low salinity water and high Cd concentration with high salinity water with a low Cd concentration. Reproduced by permission from Ref. [67].

hydrous ferric oxide (HFO) results in 50% of the Cd adsorbed on 1 mM HFO at pH 6.7 [25]. The solubility products of $CdCO_3$ and CdS, are low: $10^{-13.7}$ and 10^{-27}, respectively.

Chelation of Cd by strong organic chelators in natural waters has been demonstrated by electrochemical means. Paradoxically the best data come from studies in the Central North Pacific where the total Cd concentration is extremely low [26]. Anodic stripping voltammetry analysis of surface samples showed consistently that ~70% of the Cd is bound to some uncharacterized chelator with an apparent binding constant of 10^{12} M^{-1} (Fig. 7). This Cd-binding ligand is only found in the top 200 m of the water column at concentrations between 20 and 100 pM. Because the organic complexation of Cd is much less extensive than that of Zn [27], the unchelated forms of both metals are present at similar concentrations (Cd′ = 1 pM ~Zn′) although the total Cd concentration in Pacific surface waters is only about one hundredth of that of Zn (Cd_T = 1–4 pM ≪ Zn_T = 50–300 pM).

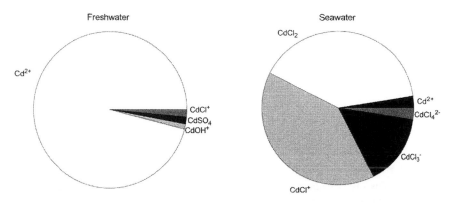

Figure 6 Predicted major chemical species of cadmium in typical freshwater and seawater.

The combination of the relatively high affinity of Cd^{2+} for particles and of the formation of strong chloro complexes at high chloride concentrations results in a particularly interesting behavior of Cd in estuaries. Unlike that of most other trace metals, the dissolved Cd concentration in estuaries usually reaches a marked maximum at mid salinities [28,29]. This is illustrated in Fig. 8 for the Gironde

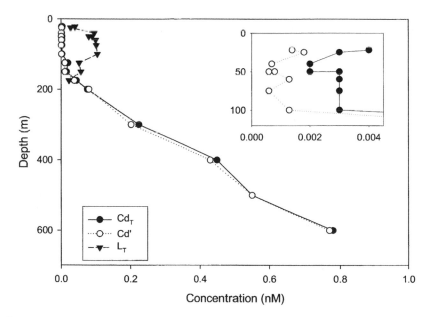

Figure 7 Vertical distribution of total dissolved cadmium (Cd_T), unchelated cadmium (Cd') and cadmium-complexing ligands (L_T) in the central North Pacific. Inset shows details of 0–150 m depths. Redrawn from Ref. [26].

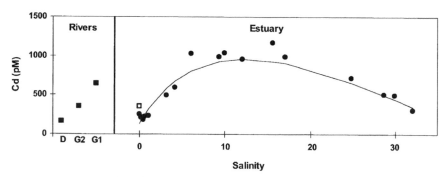

Figure 8 Dissolved cadmium in the Gironde Estuary System. D = Dordogne River, G1 = Garonne River, G2 = Garonne River at high flow. Open square is the water-discharge weighted mean of the Dordogne and Garonne Rivers. Circles represent estuarine samples. Line shows the fit of a mass balance and chemical equilibrium model to measurement data. Redrawn from Ref. [29].

estuary near Bordeaux, France. The increase in dissolved Cd with salinity in the headwaters of the estuary is explained by the desorption of Cd from the particulate phase caused by increasing chloride complexation. At higher salinities, when most of the reactive particulate Cd has been desorbed, dilution of the riverine Cd by Cd-poor seawater leads to a gradual decrease in dissolved Cd concentration. As shown on Fig. 8, the variations in dissolved Cd with salinity can be modeled on the basis of mass balance and thermodynamic equilibrium between Cd^{2+}, Cl^-, and particles [29].

4. BIOLOGICAL EFFECTS

4.1. Toxicity

There is an abundant literature on the biochemistry of Cd as a toxic element in a variety of organisms from bacteria to humans. Like all other reactive trace metals, Cd can be toxic simply because of unspecific reactions with protein ligands. For example, reaction of Cd^{2+} with cysteine thiol groups, for which it has a great affinity, can denature enzymes and make them inactive. More specific toxic effects of Cd result from blockage of certain physiological functions when Cd^{2+} substitutes for other metals, Ca^{2+} or Zn^{2+}, in particular. The ionic radius of Cd^{2+} and Ca^{2+} are very similar and cadmium can interfere with Ca metabolism or replace Ca in structural functions [30,31].

In humans the principal route of exposure to Cd is through tobacco smoke with the net result that smokers have a Cd concentration in blood 4–5 times that of non-smokers [30]. Food is the main secondary source of Cd intake as a result of Cd accumulation in agricultural soils from fertilizers and atmospheric deposition. The Environmental Protection Agency reference dose for Cd in drinking water is $0.5 \ \mu g \ kg^{-1} \ day^{-1}$ [32]. Some populations have received substantial exposure from industrial sources, as exemplified below in the case of the itai–itai disease in Japan.

The two principal causes of Cd toxicity in humans are skeletal demineralization and renal dysfunction. There is also some evidence that Cd may be a carcinogen (through inhalation exposure) and that it may cause reproductive or developmental health effects [32].

Even at a low level of exposure, the replacement of Ca by Cd in bones results in poor mineralization and leads to fragility and risk of bone fracture. Because Cd^{2+} is a good chemical analog for Zn^{2+}, Cd can also substitute for Zn in the active center of many Zn-enzymes and block their activity. It seems, for example, that the substitution of Cd for Zn in alkaline phosphatase—an important Zn enzyme in bone formation—may contribute to bone demineralization [31].

The mechanisms for kidney damage, which is the primary damage caused by Cd exposure, are less clear. Cd that is initially taken up in the intestines, is distributed into the blood and causes increase in production of metallothioneins (MT, see below) in the liver, which sequesters the Cd, and protects against acute exposure. With prolonged exposure, the Cd–MT complex is released from the liver, filtered through the glomerular membrane in the kidneys, and taken up in the renal tubules where it accumulates. Here the Cd–MT is broken down, releasing the Cd and resulting in toxic effects, apparently through interference with Ca and Zn regulation [33].

The itai–itai disease, the most infamous case of Cd toxicity to humans, was first reported in the mid 1940s in the Jinzu River basin in Japan. It was a combination of bone and kidney disease. The disease was particularly painful ("itai–itai" translates into "ouch! ouch!") because of the fragility of the bones which could not sustain any pressure. Prolonged exposure to Cd resulted from ingestion of contaminated rice, as rice fields were being irrigated by river water contaminated with the waste from a zinc mine [34,35].

4.2. Detoxification

A cell can ill-afford high internal concentrations of free ionic cadmium (or any other reactive metal that would bind indiscriminately to proteins). The internal concentration of Cd^{2+} in most cells is thus tightly controlled by a variety of binding and/or export mechanisms.

4.2.1. Cadmium Binding

All eukaryotes and some prokaryotes such as cyanobacteria synthesize metal binding proteins—known collectively as metallothioneins—in response to high concentrations of several metals including Cd. These are cysteine-rich, low molecular mass polypetides whose thiol functional groups have high affinity for soft cations such as Cd^{2+} [36]. All plants, including phytoplankton, synthesize class

III MTs known as phytochelatins (PCs). Originally named "cadistins" for their clear biochemical connection to cadmium, PCs have the general formula $(\gamma\text{-glu-cys})_n\text{-gly}$, with n varying from 2 to perhaps 11. The "gamma" peptide bond links the carboxyl from the side chain of glutamate to the amine of cysteine (Fig. 9). The terminal glycine is sometimes replaced by another amino acid or simply absent.

PCs are synthesized from glutathione by an enzyme, PC synthase. Binding of Cd (or another metal) to the enzyme and/or to glutathione is necessary to promote the reaction of a glutathione (γ-glu-cys-gly) with another glutathione (to form the dimer, n = 2) or with another PC (to form higher oligomers). In phytoplankton the dimer is normally dominant [37–39] (Fig. 9). Intracellular binding of the metal by PC eventually results in deactivation of the PC synthase. Several metals have been shown to induce PC synthesis but cadmium is generally found to be the most effective inducer.

In phytoplankton, which utilize Cd as a nutrient, PCs may serve as a Cd buffer as well as a detoxification mechanism. As seen in Fig. 9, PCs are synthesized even under extremely low metal concentrations. In culture experiments where the concentration of Zn is decreased to very low values and Cd replaces Zn for essential biochemical functions (see below), the cellular PC concentration

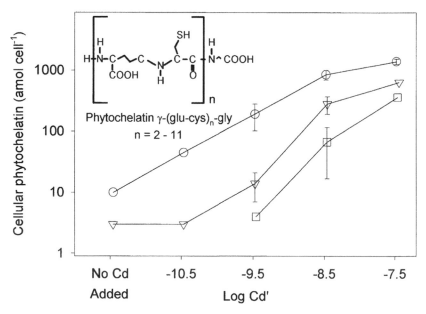

Figure 9 Concentrations of PC in *T. weissflogii* laboratory cultures (in attomoles per cell) as a function of the log of the unchelated Cd^{2+} ion concentration (in moles per liter). Symbols correspond to PC chain length, circle: n = 2, triangle: n = 3, square: n = 4. Reproduced by permission from Ref. [48].

actually increases [37]. This *increase* in PC in response to a *decrease* in Zn concentration in the medium would be paradoxical if PCs only served to detoxify metals. But this increase in PC concentration makes sense if PCs serve to store and buffer the increased cellular Cd taken up to replace the missing Zn.

4.2.2. Export

Export systems for several trace metals, including Cd, have been extensively studied in heterotrophic bacteria. These export systems confer metal-resistance to the organism (a convenient phenotype for the selection of mutants) and they are usually plasmid-encoded. In gram-negative bacteria, high Cd concentrations activate *czc* genes, which encode a three protein complex that exports Cd^{2+} (along with Zn^{2+}, Co^{2+}, and Ni^{2+}) through the inner membrane, the periplasm and the outer membrane and functions as an H^+ antiporter. In gram-positive bacteria, Cd resistance is also effected by a Cd export system. In this case, the *cad* genes encode an ATP-dependent transmembrane transporter of Cd^{2+} [40].

Yeast, which, like microalgae, produce PCs in response to high Cd concentrations transport the Cd–PC complex from the cytoplasm to the vacuole via a transport molecule in the vacuolar membrane [41]. In marine diatoms, radiotracer studies have shown that Cd is exported at high external Cd concentration ($Cd' = 5$ nM [42,43]). It seems that this export system may be similar to that of yeast and export the Cd–PC complex outside of the cell.

4.3. Cadmium Use in Phytoplankton

There are only a few studies of the biological utilization of Cd and they have been restricted to marine phytoplankton. It may be that Cd is only a useful element in marine phytoplankton; or it may be that Cd has more widespread biological uses, which have been masked so far by the similar, perhaps redundant, and more obvious role of Zn in the model organisms studied in the laboratory. In any case the discovery of the biological role of Cd in phytoplankton owes much to the obvious clue provided by its vertical concentration profiles in the oceans [19–22,44] (Fig. 4).

The first evidence that cadmium had a beneficial biological function came from growth data in laboratory cultures of the diatom *Thalassiosira weissflogii* [43,45]. As shown in Fig. 10, cultures of this coastal diatom grow slowly when the unchelated Zn concentration in the medium is reduced to about $Zn' = 3$ pM. pM. These same cultures grow much faster when Cd is added to the medium at unchelated concentrations ≥ 5 pM [46]. This effect, which is particularly obvious at low Co concentrations, has now been observed in other families of marine phytoplankton. For example, Cd enhances the growth rate of the cosmopolitan coccolithophore *Emiliana huxleyi* when the unchelated Zn and Co concentrations in the medium are below 1 pM (Fig. 10) [42]. From similar laboratory studies, it appears that Zn, Cd, and Co can substitute for each other in many marine eukaryotic phytoplankton [47–51].

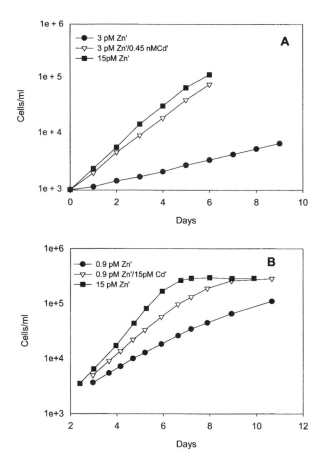

Figure 10 Growth rate of (A) diatom *T. weissflogii* and (B) coccolithophore *E. huxleyi* laboratory cultures under different unchelated zinc (Zn′) and unchelated cadmium (Cd′) concentrations [(A) T. W. Lane, F. M. M. Morel, unpublished results; (B) Y. Xu, F. M. M. Morel, unpublished results].

Laboratory studies of various species of marine microalgae grown in the presence of low concentrations of trace metals, including Cd, show an average cellular Cd quota of 2 μmol Cd (mol C)$^{-1}$, similar to the cellular quotas of other essential micronutrients such as Cu and Co [52]. In sharp contrast, marine cyanobacteria are highly sensitive to Cd toxicity, in particular at low Zn concentrations [53]. Thus, conditions that lead to a beneficial effect of Cd on the growth of eukaryotic phytoplankton (e.g., Zn′ <10 pM and Cd′ >5 pM) result in growth inhibition of *synechococcus*, one of the dominant species of photosynthetic bacteria in the sea.

The underlying biochemical mechanisms for beneficial effects of Cd have not all been elucidated. For example, we do not know how Cd accelerates the growth of *E. huxleyi* as illustrated in Fig. 10(B). It seems most likely, of course, that Cd replaces Zn (and/or Co) as a cofactor in some Zn enzyme(s). There is some suggestion that such replacement of Zn by Cd may occur for alkaline phosphatase or superoxide dismutase. In the case of diatoms, however, it is clear that a major role of Cd is as a metal center in carbonic anhydrase (CA). This role was first hinted at by the observation that slow growth rates obtained at low Zn' could be accelerated by increasing the CO_2 in the medium [46,54]. The enzyme CA, which catalyzes the interconversion of HCO_3^- and CO_2, and is known to be involved in inorganic carbon acquisition in microalgae was thus suspected. CA is one of the more abundant Zn enzymes known in biological systems.

The ambient concentration of CO_2 in seawater ($\sim$10 μM) is low compared to the half saturation constant of RubisCO (ribulose 1,5-bisphosphate carboxylase/oxygenase) the enzyme that fixes the inorganic carbon into PG3 (3-phosphoglycerate) in the first step of the Calvin cycle (the dark reaction of photosynthesis). Microalgae have thus evolved various carbon concentrating mechanisms (CCM) to augment the CO_2 concentration at the site of fixation [55]. These mechanisms all involve interconversion between CO_2 and HCO_3^- at some point. But at neutral pH the hydration/dehydration reaction of CO_2/HCO_3^- has a half life of $\sim$30 s, much too slow for a cellular process (for example, diffusion from one end of the cell to the other takes on the order of 10 ms), and requires catalysis. The enzyme CA is an extraordinary effective catalyst: some CA catalyze the CO_2/HCO_3^- reaction at a rate that nearly reaches the limit imposed by the diffusion of molecules [56].

The cadmium carbonic anhydrase (CdCA) of *T. weissflogii* has been isolated and sequenced [M. A. Saito, T. W. Lane, H. Park, and F. M. M. Morel, in preparation]. It exhibits no homology with any of the other known CAs but a highly homologous protein is coded in the genome of the diatom *Thalassiosira pseudonana* [57]. Near-edge X-ray spectroscopy has shown that sulfur is involved in the coordination of Cd but the enzyme has not yet been fully characterized. As expected, the expression of CdCA in *T. weissflogii* is modulated by both pCO_2 and Zn' in the medium [46,51]: the concentration of the enzyme increases at low pCO_2 and low Zn' (Fig. 11).

Since the role of Zn in the active center of CA depends on the acid–base properties of the $\equiv$ZnOH complexes, it is not surprising that Cd, whose outer electron configuration (d^{10}) is identical to that of Zn, should be able to catalyze the same reaction. Indeed, some level of activity has usually been measured in various CAs where Cd has been substituted for Zn [56]. So far, the diatom CdCA is the first and only known native Cd enzyme, however.

5. THE BIOGEOCHEMISTRY OF CADMIUM AS AN ALGAL NUTRIENT IN THE SEA

5.1. Mass Balance in Surface Seawater

On the basis of the concentrations of Cd in the dissolved and particulate phases and known physical and biological rates, we can establish a first order mass

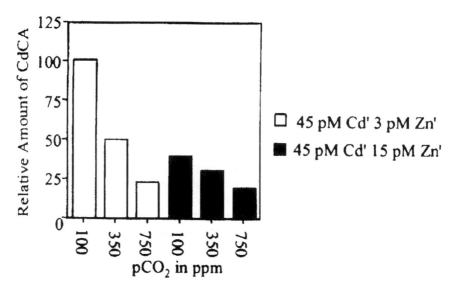

Figure 11   Relative amount of CdCA produced by *T. weissflogii* laboratory cultures grown under differing concentrations of carbon dioxide and unchelated zinc (Zn'). Reproduced by permission from Ref. [46].

balance for Cd in surface seawater at a given locale. Fig. 12 presents the results of such an exercise for the North Pacific Gyre. The downward Cd flux can be quantified on the basis of the concentration of cadmium in surface particles and estimates of the downward flux of biogenic particles, obtained from sediment traps or various methods for measuring export primary production—i.e., the fraction of the biomass formed photosynthetically by unit time at the surface that is exported below the mixed layer.

   An average cellular Cd quota of 2 μmol Cd (mol C)$^{-1}$ and a phytoplankton biomass of 0.4 μmol C L^{-1} correspond to a particulate Cd concentration of 0.8 pM in the euphotic zone. Along with photosynthetically fixed carbon, this particulate Cd turns over at a rate of $\sim$0.5 day^{-1} : 3/4 is remineralized and 1/4 (i.e., 0.1 pM day^{-1}) exported to deep waters with the sinking biomass. The exchange between the particulate and dissolved pools leads to a residence time of a few days for the dissolved Cd pool at the surface. The downward flux of Cd in the particulate pool must be balanced by an upward diffusive/ advective flux. A typical vertical diffusivity of 2 m^2 day^{-1} would require a vertical gradient in dissolved Cd of 5 pM m^{-1} (5×10^{-9} mol m^{-3} $\times$ 1 m^{-1} $\times$ 2 m^2 day^{-1} = 10 nmol m^{-2} day^{-1}) to match the settling flux. This gradient is roughly what one can estimate from vertical concentration profiles as shown in Fig. 4. Since the concentration of dissolved Cd is $\sim$2 pM, the average residence time of the Cd in the surface water is $\sim$20 days.

   If the cycling of Cd in seawater is mainly the result of its use as a micronutrient by phytoplankton, we should be able to relate the concentrations and

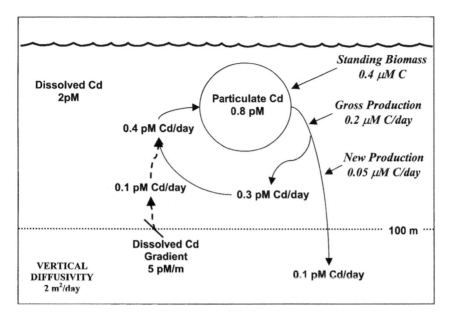

Figure 12 First order mass balance for cadmium in surface seawater of the North Pacific Gyre.

fluxes shown on Fig. 12 to the cellular processes that have been studied in the laboratory. Since most of the laboratory data to date pertain to diatoms, we consider a surface water where diatoms are the dominant eukaryotic phytoplankton. The processes involved in the cellular metabolism of Cd include its uptake from the medium, its intracellular storage, and its use as a center in metalloenzymes, particularly CA as depicted in Fig. 13. In addition Cd is remineralized along with other nutrients upon decomposition of the phytoplankton biomass.

5.2. Cellular Quotas

The particulate Cd concentrations measured in field samples of surface seawater (Fig. 4) are in the range 0.1–1 mmol Cd : mol P, equivalent to 1–10 μmol Cd : mol C. They are a good match with the cellular Cd quotas of phytoplankton measured in laboratory cultures under conditions of low (i.e., "natural") metal concentrations which are also in the range 0.1–1 mmol Cd : mol P and average 0.2 mmol Cd : mol P. The biological utilization of cadmium can thus account quantitatively for the particulate concentration and, hence, for the draw-down of cadmium from surface waters observed in the field.

5.3. Uptake and Growth Rates

In marine phytoplankton, cadmium uptake appears effected by a specific transport system as well as by leakage through the transport systems of other

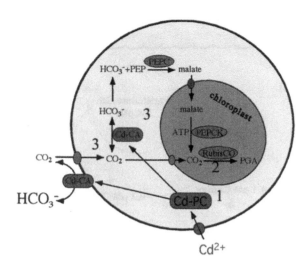

Figure 13 Hypothetical model of cadmium metabolism in a marine diatom showing the buffering of Cd by PC binding and its use in periplasmic and cytoplasmic carbonic anhydrases as part of C_4 metabolism.

metals, which may be important for organisms living in coastal waters. The half-saturation constant for Cd uptake by coastal diatoms in culture has been estimated at $K_{Cd'} \sim 3$ nM [43]. This half saturation constant is very high compared to ambient concentrations of unchelated Cd, underscoring the fact that trace metal transport systems in marine phytoplankton normally function far from saturation [58]. The net result is that the uptake rate is proportional to the Cd' concentration in the medium.

In the laboratory, cells growing at a growth rate of 2 day^{-1} in the presence of Cd' = 10 pM must take up Cd at a rate of 4 µmol Cd (mol C)$^{-1}$ d^{-1} to maintain a cellular quota of 2 µmol Cd (mol C)$^{-1}$. Under the conditions of Fig. 12 (Cd' = 1 pM), the corresponding uptake rate is calculated to be 0.16 pM Cd day^{-1} (=4 µmol Cd (mol C)$^{-1}$ day^{-1} × 1/10 × 0.4 µM C), nearly half of the uptake rate chosen to provide a mass balance in Fig. 12. This is really not a bad first order reconciliation of the data, particularly since we know that open ocean species are better adapted to low trace metal concentrations than the neritic species for which the lab data were obtained. The mass balance model of Fig. 12 implies that the phytoplankton in the surface waters of the open ocean can grow at a rate of 0.5 day^{-1} at somewhat lower concentrations of unchelated metals, including cadmium.

5.4. Remineralization

The remarkable correlation in the vertical concentration profiles of Cd and P in the oceans imply that the two elements are remineralized at proportionally similar rates [20]. One consequence of these similar profiles is that the Cd and P must also be transported at proportionally similar rates to the ocean surface by diffusive/advective processes. Thus the similar rates of vertical cycling of Cd and P resulting from biological uptake and remineralization should lead, on average, to a constant proportion between the two elements in both the water column and particulate organic matter. Indeed, the Cd/P ratio of 2–3 $\times$ 10^{-4} mol mol^{-1} observed in almost all deep water samples, which represents an integration of the particulate flux from the surface, corresponds very closely to the Cd quotas obtained in cultures and to the concentrations measured in natural phytoplankton assemblages.

The rather remarkable correspondence between the concentrations and rates obtained in laboratory cultures and in the field certainly support the notion that the biogeochemical cycling of Cd in the ocean is dominated by its use as a nutrient for phytoplankton. But because cadmium has not, in the past, been considered a biologically useful element, there are very few field measurements or experiments that provide direct information on its role as an algal nutrient. There are nonetheless two types of field data we can examine to see if the geochemical behavior of Cd can be explained by its biological function: (i) Is the biological uptake of Cd in a given locale modulated by pCO$_2$ and Zn in a manner that is consistent with its use in CA? (ii) Is the ocean-wide distribution of Cd also consistent with such a biochemical function?

The best evidence of the biological use of Cd in CA in the sea is the dependence of the particulate Cd concentration on pCO$_2$ and Zn'. An inverse correlation between Cd uptake by phytoplankton and pCO$_2$ can been seen in waters off Monterey in the Pacific Ocean [59]. There, within 50–100 miles from the coast, patches of recently upwelled water with high pCO$_2$ (700 ppm) are found in close proximity with patches of older water where the nutrients have been utilized and the pCO$_2$ lowered (down to 200 ppm). These waters exhibit a remarkable inverse correlation between pCO$_2$ and particulate Cd concentration [Fig. 14(A)]. The causality of the relationship between pCO$_2$ and particulate Cd concentration has been established in incubation experiments with field samples: decreasing pCO$_2$ in incubation bottles resulted in an increase in particulate Cd. These surface waters also exhibit a negative correlation between particulate Zn and Cd concentrations [Fig. 14(B)]. It thus appears that the ambient organisms in the surface waters off Monterey regulate Cd uptake as a function of pCO$_2$ and available zinc. These results precisely parallel the laboratory regulation experiments, and are thus entirely consistent with the biological use of Cd as a metal center in CA (Fig. 11).

We do not know the biochemical role of Cd in all species of phytoplankton whose growth is enhanced by Cd addition. But we know that this enhancement of growth is always observed at low Zn (and usually Co) concentrations and that Cd must thus somehow replace Zn in some biochemical function. We should thus expect that the uptake of Cd by phytoplankton should be enhanced at low ambient Zn concentrations. This expectation is borne out in the correlation

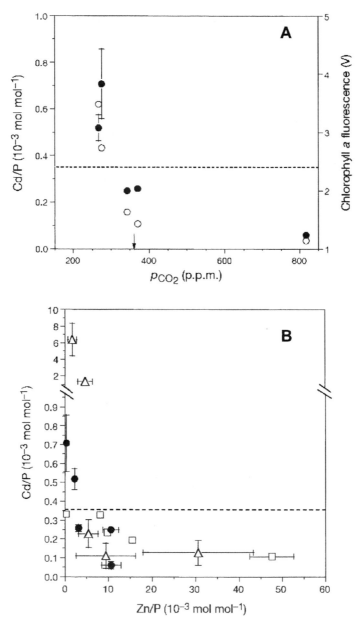

Figure 14 Phytoplankton measurements from the Monterey Bay and Santa Barbara Channel off the California coast. (A) Phytoplankton cadmium/phosphorus ratios (closed circles) and chlorophyll *a* fluorescence (open circles) vs. carbon dioxide concentration. (B) Cadmium/phosphorus vs. zinc/phosphorus for phytoplankton assemblages >0.45 μm (circles), 5–53 μm (squares) and >53 μm (triangles). Reproduced by permission from Ref. [59].

between trace metal and major nutrient concentrations in the surface waters of the North Pacific [50,60]. As seen in Fig. 15, the dissolved Zn concentrations in these waters reach very low values as a result of phytoplankton uptake before phosphate is exhausted. When the Zn concentrations are very low, the concentration of Cd (and Co) becomes highly correlated to that of P, both elements becoming exhausted at about the same point.

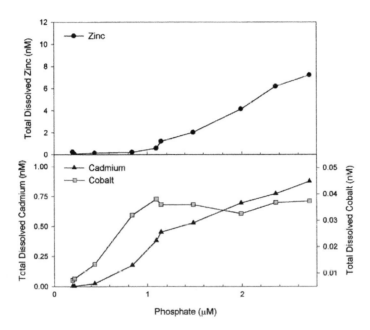

Figure 15 The sequential draw down of Zn (top), Cd and Co (bottom) in the North Pacific, suggestive of biochemical substitution in the phytoplankton community. Modified from Refs. [50,68], data from Ref. [60].

6. CADMIUM AS A PALEOTRACER

In concluding this chapter, we come back to the unusual importance of Cd in oceanography that results from its use as a paleotracer. The original discovery of the remarkable correlation between the concentrations of Cd and P [20] has been amply confirmed over the past 30 years (witness the extensive data displayed on Fig. 1). Boyle and his coworkers [61–64] have shown that the accumulation of Cd in the calcite shells of foraminifera is proportional to the Cd concentration in the surrounding seawater. Thus the Cd/Ca ratio measured in the shells of fossil foraminifera (after extensive cleaning procedures) provides a measure of the Cd concentration in seawater at the time and at the depth at which the animal was alive. Because foraminifera are ubiquitous in the oceans and fossils can be found in sediments that are millions of years old, the distribution of Cd concentrations in seawater can be reconstructed over geologic times. On the basis of the Cd–P correlation, one can then infer the distribution of nutrient concentrations in the ancient oceans. Measurements of Cd/Ca ratios in the aragonite skeletons of corals provide more detailed records of nutrient concentrations for the modern era. The Cd concentration in corals also reflects the ambient concentration of Cd in seawater, but the coefficient of proportionality is apparently more variable among species than in foraminifera.

On the basis of such record of past nutrient concentrations one can infer past changes in the major circulation patterns of the oceans and link them to global climate events such as ice ages. As explained earlier, the differences in the Cd and P concentrations in the deep Pacific and Atlantic oceans (Fig. 4) reflect the general pattern of the surface and deep currents. One can thus infer, on various scales, the relative proportions of water masses of various origins on the basis of the nutrient distribution in seawater. In this manner, for example, Boyle and Keigwin [64] concluded that the formation of North Atlantic Deep Water was much reduced during the peak of the last ice age (the last glacial maximum) because the nutrient concentration in the deep Atlantic was then much higher than now. A study of the Cd/Ca ratio in corals in the Cariaco Basin off Venezuela provides a more recent example of such paleosleuthing, on a more regional scale and for relatively modern events [65]. A sharp decrease in the Cd/Ca ratio in corals after 1950 compared to the first part of the 20th century indicate a dramatic reduction in upwelling in the Cariaco Basin and thus a change in regional climate. The limited historical data of sea surface temperature that are available support this assertion but are subject to systematic instrumental errors.

7. ENVOI

Thirty years ago, cadmium was only interesting to a few environmental scientists and toxicologists. For a while, it did capture the headlines as the cause of a horrible and painful disease; but it was not a particularly pleasant object of study and one might just have wished to forget its existence. But, as we have seen, the extensive studies of the biogeochemistry of Cd in the oceans over the past couple of decades have proved extraordinarily interesting and fruitful. They have provided us with new insights into the functioning of the oceans, both in the present and in the past. For example, some of the first clues about the probable causes of glacial cycles have come from the study of cadmium in sedimentary material. These biogeochemical studies have also lead to a revolution in our understanding of the biochemical role of cadmium, no longer solely a noxious element but also one that is essential (or at least useful) for life. The cadmium carbonic anhydrase of diatoms is surely the first enzyme ever discovered as a result of the large scale geochemical behavior of an element. So cadmium is in many ways a model trace element for biogeochemistry: its global cycle provides clues to its biochemical role and, in turn, its role as a metal center in enzymes explains its geochemical cycling. The unexpectedly interesting biogeochemistry of Cd might thus serve as a model for other elements.

ACKNOWLEDGMENTS

This work was supported in part by the Center for Environmental Bioinorganic Chemistry (NSF#CHE 0221978). We thank Matthew Reuer for helpful comments on the manuscript, Yan Xu and Todd Lane for providing unpublished data and Robert Sherrell for allowing us to use a figure from his Ph.D. thesis.

ABBREVIATIONS

ATP	adenosine 5'-triphosphate
CA	carbonic anhydrase
CCM	carbon concentrating mechanism
Cd'	unchelated Cd^{2+}
HFO	hydrous ferric oxide
MT	metallothionein
PC	phytochelatin
PEP	phosphoenolpyruvate
PEPC	phosphoenolpyruvate carboxylase
PEPCK	phosphoenolpyruvate carboxykinase
PG3	3-phosphoglycerate
PGA	phosphoglyceric acid
RubisCO	ribulose 1,5-bisphosphate carboxylase/oxygenase
Zn'	unchelated Zn^{2+}

REFERENCES

1. Nriagu JO. Cadmium in the Environment. New York: John Wiley & Sons, 1980:35–70.
2. Rydh CJ, Svard B. Sci Total Environ 2003; 302:167–184.
3. Alfaro-De La Torre MC, Tessier A. Geochim Cosmochim Acta 2002; 66:3549–3562.
4. Boutron CF, Cadelone J-P, Hong S. Sci Total Environ 1995; 160/161:233–241.
5. Von Gunten HR, Sturm M, Moser RN. Environ Sci Technol 1997; 31:2193–2197.
6. Pirrone N, Costa P, Pacyna JM. Water Sci Technol 1999; 39:1–7.
7. Nriagu JO. Nature 1989; 338:47–49.
8. Malcolm EG, Keeler GJ. Environ Sci Technol 2002; 36:2815–2821.
9. Malcolm EG, Keeler GJ, Lawson ST, Scherbatskoy TD. J Environ Monit 2003; 5:584–590.
10. Rea AW, Lindberg SE, Keeler GJ. Atmos Environ 2001; 35:3453–3462.
11. Atasi KZ, Fujita G, Le Platte G, Hufnagel C, Keeler G, Graney J, Chen T. Water Sci Technol 2001; 43:223–229.

12. Lawson ST, Scherbatskoy TD, Malcolm EG, Keeler G. J Environ Monit 2003; 5:578–583.
13. Meybeck M. Etablissement des flux polluants. Agence de l'Eau Adour-Garonne, 1992.
14. Martin J-M, Whitfield M. Trace Metals in Sea Water. 9, NATO Conference Series IV, Marine Sciences. New York City: Plenum Press, 1983:265–296.
15. Audry S, Blanc G, Schafer J. Sci Total Environ 2004; 319:197–213.
16. Han FX, Banin A, Su Y, Monts DL, Plodinec MJ, Kingery WL, Triplett GE. Naturwissenschaften 2002; 89:497–504.
17. Senesi GS, Baldassarre G, Senesi N, Radina B. Chemosphere 1999; 39:343–377.
18. Chester R. Marine Geochemistry. London: Academic Division of Unwin Hyman Ltd, 1990:1–698.
19. Sherrell RM. The Trace Metal Geochemistry of Suspended Oceanic Particulate Matter. PhD thesis, Massachusetts Institute of Technology/Woods Hole Oceanographic Institution Joint Program, Cambridge, 1989.
20. Boyle E, Sclater F, Edmond JM. Nature 1976; 263:42–44.
21. Boyle E, Huested SS, Jones SP. J Geophys Res 1981; 86:8048–8066.
22. Bruland KW. Earth Planet Sci Lett 1980; 47:176–198.
23. Broecker WS, Peng T-H. Tracers in the Sea. Palisades, New York: Eldigio Press, 1982:1–690.
24. Morel FMM, Hering JG. Principles and Applications of Aquatic Chemistry. 2nd ed. New York: John Wiley & Sons, 1993:1–588.
25. Benjamin MM, Leckie JO. J Colloid Interf Sci 1981; 79:209–221.
26. Bruland KW. Limnol Oceanogr 1992; 37:1008–1017.
27. Bruland KW. Limnol Oceanogr 1989; 34:269–285.
28. Shiller AM, Boyle EA. Geochim Cosmochim Acta 1991; 55:3241–3251.
29. Kraepiel AML, Chiffoleau J-F, Martin J-M, Morel FMM. Geochim Cosmochim Acta 1997; 61:1421–1436.
30. Jarup L. Nephrol Dial Transplant 2002; 17:35–39.
31. Brzoska MM, Moniuszko-Jakoniuk J. Food Chem Toxic 2001; 39:967–980.
32. U. S. EPA. Integrated Risk Information System (IRIS) on Cadmium. National Center for Environmental Assessment Office of Research and Development, 1999.
33. Nordberg G, Group C. J Trace Elem Exp Med 2003; 16:307–319.
34. Kasuya M. Water Sci Technol 2000; 42:147–155.
35. Nogawa K. In: Nriagu JO, ed. Cadmium in the Environment, Part II: Health Effects. New York: John Wiley & Sons, Inc., 1981:1–37.
36. Murasugi A, Wada C, Hayashi Y. J Biochem 1981; 90:1561–1564.
37. Ahner BA, Lee JG, Price NM, Morel FMM. Deep-Sea Res I 1998; 45:1779–1796.
38. Ahner B, Morel FMM. Prog Phycol Res 1999; 13:1–31.
39. Ahner BA, Kong S, Morel FMM. Limnol Oceanogr 1995; 40:649–657.
40. Silver S. J Ind Microbiol Biotechnol 1998; 20:1–12.
41. Oritz DF, Rusticitti T, McCue KF, Ow DW. J Biol Chem 1995; 270:201–205.
42. Lee JG, Ahner BA, Morel FMM. Environ Sci Technol 1996; 30:1814–1821.
43. Lee JG, Roberts SB, Morel FMM. Limnol Oceanogr 1995; 40:1056–1063.
44. Sherrell RM, Boyle EA. Earth Planet Sci Lett 1992; 111:155–174.
45. Price NM, Morel FMM. Nature 1990; 344:658–660.
46. Lane TW, Morel FMM. Proc Natl Acad Sci 2000; 97:4627–4631.
47. Yee D, Morel FMM. Limnol Oceanogr 1996; 41:573–577.

48. Morel FMM, Milligan AJ, Saito MA. Treatise on Geochemistry. Oxford: Pergamon, 2003.
49. Saito MA, Moffett JW, Chisholm SW, Waterbury JB. Limnol Oceanogr 2002; 6:1627–1636.
50. Sunda WG, Huntsman SA. Limnol Oceanogr 1995b; 40:1404–1417.
51. Morel FMM, Cox EH, Kraepiel AML, Lane TW, Milligan AJ, Schaperdoth I, Reinfelder JR, Tortell PD. Funct Plant Bio 2002; 29:301–308.
52. Ho TY, Quigg A, Finkel ZV, Milligan AJ, Wyman K, Falkowski PG, Morel FMM. 2003; 39:1145–1159.
53. Saito MA, Sigman DM, Morel FMM. Inorg Chim Acta 2003; 356:308–318.
54. Morel FMM, Reinfelder JR, Roberts SB, Chamberlain CP, Lee JG, Yee D. Nature 1994; 369:740–742.
55. Falkowski PG, Raven JA. Aquatic Photosynthesis. Capital City Press, 1997:1–375.
56. Coleman JE. Biology and Chemistry of the Carbonic Anhydrases. Vol. 429. The New York Academy of Sciences, Ann New York Acad Sci, 1984:26–48.
57. DOE Joint Genome Institute. http://genome.jgi-psf.org/diatom, 2004.
58. Hudson RJM, Morel FMM. Deep-Sea Res 1993; 40:129–150.
59. Cullen JT, Lane TW, Morel FMM, Sherrell RM. Nature 1999; 402:165–167.
60. Martin JH, Gordon RM, Fitzwater S, Broenkow WW. Deep-Sea Res 1989; 36:649–680.
61. Boyle E. Paleoceanography 1988; 3:471–489.
62. Delaney ML. Foraminiferal trace elements: uptake, diagenesis, 100 m.y. paleochemical history. Ph.D. thesis, Massachusetts Institute of Technology/Woods Hole Oceanographic Institution Joint Program, Woods Hole, 1983.
63. Hester K, Boyle E. Nature 1982; 298:260–261.
64. Boyle EA, Keigwin L. Nature 1987; 330:35–40.
65. Reuer MK, Boyle E, Cole JE. Earth Planet Sci Lett 2003; 210:437–452.
66. de Baar HJW, Saager PM, Nolting RF, van der Meer J. Mar Chem 1994; 46:261–281.
67. Bruland KW, Franks RP. Trace Metals in Sea Water. 9, NATO Conference Series IV, Marine Sciences. New York: Plenum Press, 1983:395–414.
68. Sunda WG, Huntsman SA. Limnol Oceanogr 2000; 45:1501–1516.

9

The Biogeochemistry and Fate of Mercury in the Environment

Nelson J. O'Driscoll[1], Andrew Rencz[2], and David R. S. Lean[1]

[1]*Biology Department, Faculty of Science, University of Ottawa,*
P.O. Box 450, Stn. A., Ottawa, Ontario K1N 6N5, Canada
[2]*Geological Survey of Canada, 601 Booth Street,*
Ottawa, Ontario, K1A 0E8, Canada

1.	Introduction	222
2.	Mercury Speciation in the Environment	222
3.	Processes Affecting Atmospheric Transport and Fate	223
	3.1. Atmospheric Emissions and Mercury Deposition	223
	3.2. Atmospheric Reactions	224
4.	Processes Affecting Aquatic Transport and Fate	226
	4.1. Mercury Oxidation and Reduction	226
	4.2. Mercury Methylation and Demethylation	228
	4.3. Binding and Sedimentation	228
5.	Effects of a Changing Landscape on Mercury Fate	229
	5.1. Wetlands	229
	5.2. Deforestation	230
6.	Big Dam West Lake Mercury Mass Balance	230
	6.1. Introduction	230
	6.2. Experimental Findings	231
7.	General Conclusions	233
	Acknowledgments	234

Abbreviations   234
References   235

1. INTRODUCTION

Mercury is a toxic environmental contaminant and is a primary issue of concern for several government agencies (e.g., United States Environmental Protection Agency, Environment Canada, and Health Canada). Mercury transport and fate are global issues since, due to its dispersion and transformation into bioavailable forms, it can bioaccumulate not only in contaminated sites but also in remote freshwater lakes [1–3]. For example, Kejimkujik Park (Nova Scotia, Canada) has no known local anthropogenic inputs of mercury and yet contains fish and loons that have some of the highest blood mercury concentrations in North America [4,5].

This chapter will review information relevant to the biogeochemical cycle of mercury, specifically (i) the speciation chemistry of mercury, (ii) the fate of mercury in air, water and sediment, and (iii) the effects of landscape changes on mercury fate. The chapter will end with a summary of a recent multi-disciplinary mercury mass balance conducted in Kejimkujik Park (Nova Scotia, Canada).

2. MERCURY SPECIATION IN THE ENVIRONMENT

Several recent government reports stress that the key to creating effective inter-national policy for regulating regional and global mercury fate is a more detailed knowledge of the mercury cycle [6,7]. There are several forms of mercury in the environment and a variety of processes that control its form and fate (Fig. 1). The three major forms (or species) of mercury in the environment are elemental mercury (Hg^0), inorganic mercury (Hg^{2+}), and methyl mercury (CH_3Hg^+).

Elemental mercury is volatile and sparingly soluble in water [8]. Due to the high vapor pressure of Hg^0 it is the dominant form of mercury in the atmosphere and has a long residence time (1–2 years) [9]. In the ionic form, mercury can exist in two oxidation states: Hg_2^{2+} (the mercurous ion), and Hg^{2+} (mercuric ion). The mercuric ion is highly soluble and reactive, such that it binds to particles in the atmosphere or various ligands in lake water (particularly with reduced sulfur functional groups) [10]. The mercurous form has a relatively short half-life due to disproportionation in which it reacts with itself to form elemental mercury and mercuric ion [11].

Organic mercury compounds (methyl mercury, dimethyl mercury, ethyl mercury, and phenyl mercury) are defined as mercury bound to carbon based structures. Methyl mercury (MeHg) is the predominant form of mercury that

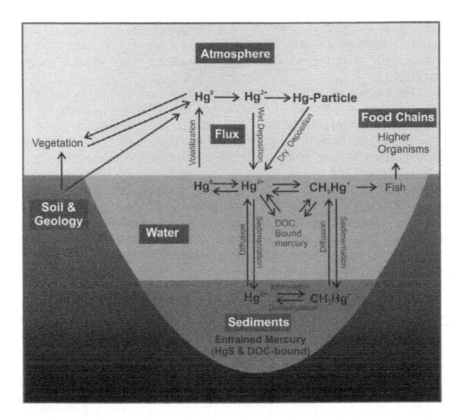

Figure 1 Conceptual diagram of mercury cycling in a freshwater lake ecosystem. Adapted from Ref. [86] with permission of SETAC Press.

bioaccumulates in the food chain. Such that, low methyl mercury levels in lake water (<0.1 ng L^{-1}) can result in high concentrations in fish [1,12]. The specific state (solid, liquid, or gas) and form of mercury compounds present in the atmosphere or lake water will vary depending on the biological, chemical, and physical conditions present.

3. PROCESSES AFFECTING ATMOSPHERIC TRANSPORT AND FATE

3.1. Atmospheric Emissions and Mercury Deposition

While atmospheric concentrations of mercury are low ($\sim 1-2$ ng m^{-3}), the total mass of mercury stored in the atmosphere is quite large. Fitzgerald [13] estimated the mass of mercury in the atmosphere to be 5000 metric tons. Current research using core samples from sediment, peat bog and glacial ice as a record of metal deposition indicate that mean amounts of total mercury in the atmosphere have been increasing over the past 100 years with some research indicating a recent decline over the past 10 years [14–17]. Fitzgerald [14] has suggested that mercury concentrations have been increasing at a rate of 0.6% over the past century, while other researchers have found the increase to be closer to 1% per year [18].

The sources contributing to this observed increase are a subject of much debate. Pacyna and Pacyna [19] determined that 1900 tons of mercury are emitted to the atmosphere annually from anthropogenic sources. Porcella et al. [20] estimated the value to be closer to 4000 tons year^{-1}. Mason et al. [21] estimated mercury release from natural sources (volcanoes, volatilization from soils, vegetation, and oceans) to account for approximately 2000 tons year^{-1}. However, mercury releases from natural sources are particularly difficult to quantify, as they are often diffuse (such as unaltered geology and vegetation) [22,23]. Part of the reason for conflicting estimates of natural and anthropogenic emissions is the difficulty in separating natural sources from re-mobilized past and current human releases. More research is needed to conclusively differentiate between natural releases and re-emissions of anthropogenic mercury loadings from these pools.

Previous research has suggested that the oxidation of insoluble Hg^0 to its more reactive and water-soluble form (Hg^{2+}) provides the mechanism for mercury removal from the atmosphere in wet and dry deposition. However, recent research by Van Loon et al. [24] showed that the solubility of elemental mercury in precipitation is 10,000 times greater when complexed with sulfite. The volume-weighted concentration of mercury in wet deposition throughout the United States has been found to range between 4 and 30 ng L^{-1} with annual mercury mass deposited ranging between 2 and 22 $\mu g \, m^{-2}$ [25]. In comparison to elemental mercury, particulate mercury (PM) and reactive gaseous mercury (RGM, ionic gaseous mercury) are generally present in very low concentrations in the atmosphere, however, these forms may be critical to deposition rates.

3.2. Atmospheric Reactions

The formation of reactive mercury species in the atmosphere has been implicated as a critical process controlling mercury deposition from the Arctic and Antarctic atmosphere. The measurement of mercury depletion events (MDEs) in the atmosphere of the high Arctic, Antarctic, and lower latitudes (European Arctic-Svalbard) during polar sunrise [26–28] has prompted a significant level of research investigating the role of RGM in mercury deposition. Recent work suggests that photo-induced conversion of Hg^0 to RGM (by halogen radicals such as bromine derived from sea salt) and deposition to snow may account for the rapid removal of mercury from the atmosphere observed in the Polar Regions [29].

In temperate regions, mercury speciation processes are important as they facilitate the movement of mercury to Polar Regions in a process referred to as the "Grasshopper Effect" (Fig. 2). Elemental mercury transport has erroneously

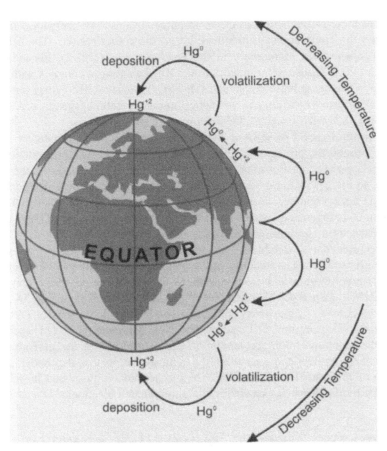

Figure 2   Movement of mercury to the Polar Regions by the "grasshopper effect".

been equated to other semi-volatile compounds such as PCBs that participate in a global distillation phenomenon that transfers chemical emissions from equatorial, subtropical and temperate regions toward the Poles [30,31]. However, in contrast to PCBs, the global distillation of mercury is dependent upon the oxidation state of mercury. Atmospheric oxidation and reduction reactions are critical to determining deposition, and redox reactions in the aqueous phase predominantly control rates of volatilization [32] (Section 4.1). Following mercury deposition in an oxidized form, it can be reduced in a lake to the more volatile elemental form and be released back to the atmosphere. The idea that redox chemistry primarily controls the temperature dependent deposition of mercury is supported by the work of Van Loon et al. [24], which details the chemistry of atmospheric mercury reactions with sulfite. The authors noted strong temperature dependence for the reduction of mercury by sulfite. Their work suggests that mercury will be deposited in wet deposition more efficiently in colder regions due to decreasing rates of mercury reduction in the atmosphere. In this way, mercury can "hop" through the environment in the direction of the colder regions. The net result is a concentration of mercury in the Polar Regions (Fig. 2). Recently, Van Loon et al. [33] also determined that the Hg^0 can form a complex with sulfite which is a more soluble compound ($Hg \cdot SO_2$). As such, Hg^{2+} may not be the dominant form of mercury in wet deposition and small changes in atmospheric chemistry may greatly affect the rates of mercury deposition.

Several recent reviews on the atmospheric chemistry of mercury [11,34] have outlined the principal gas phase, aqueous phase (in wet precipitation) and particulate phase reactions that are important to mercury speciation in the atmosphere. In the gas phase, oxidation reactions can primarily result in the formation of HgO from a number of oxidized species (e.g., O_3, H_2O_2, NO_3, etc.). The formation of halogenated species (e.g., $HgCl_2$ or $HgBr_2$ forms of RGM) has been suggested as a mechanism for MDEs in the Polar Regions [29,35]. There is very limited data available on gas phase reduction reactions involving HgO and HgX_2 species. The majority of the literature focuses on reduction reactions in wet precipitation and clouds. Reduction in the aqueous phase may occur by reaction with H_2O_2 (above pH 5.5), or SO_3^{2-}, and oxidation by O_3, H_2O_2 (below pH 5.5), organoperoxy compounds, and hypochlorite species. Particulate phase reactions are not well studied, however, Schroeder et al. [11] speculated that the reactions might be similar to those observed in the gaseous phase. Once mercury is deposited from the atmosphere to a water body, another series of reactions control its fate. These reactions are discussed in Section 4 with particular emphasis on solar radiation-mediated reactions.

4. PROCESSES AFFECTING AQUATIC TRANSPORT AND FATE

4.1. Mercury Oxidation and Reduction

As discussed in Section 3.2, redox reactions are very important to the fate of mercury in the atmosphere, and the same is true for the aqueous phase. The creation of dissolved gaseous mercury (DGM) in freshwater lakes and its loss to the atmosphere by volatilization is an important process in the mercury cycle. DGM is believed to consist primarily of Hg^0 formed from Hg^{2+} through the process of reduction [36]. DGM is the form in which mercury volatilizes from water to air, and volatilization is one of the primary means of mercury removal from an ecosystem. Several researchers have demonstrated that volatilization is a significant part of the mercury cycle. Rolfhus and Fitzgerald [37] found that mercury volatilization from Long Island Sound, New York, was equivalent to 35% of total annual mercury inputs to the system, and studies on the Great Lakes (Canada–US) show that mercury volatilization is equivalent to as much as 50% of the total mercury inputs [38,39]. Other studies [21,40] have found that mercury volatilization and wet deposition were close to being in balance. Volatilization has also been shown to be an important factor in the global distribution of mercury, and Mason et al. [21] indicate that mercury volatilization from the ocean surface may account for ~30% of the total global mercury emissions to the atmosphere.

The processes responsible for the formation of DGM in lakes (reduction) and the conversion of DGM to inorganic mercury (oxidation) are facilitated by solar radiation. Photo-reduction of mercury has been observed by many researchers in both saltwater [41–44] and freshwaters [32,40,45–47], in temperate lakes and rivers [32,40,46–49], Artic lakes [50], and southern wetlands [51], yet the mechanisms by which it occurs are not well understood. Both abiotically- and biotically-mediated mechanisms for photo-reduction have been suggested in the literature. Nriagu [52] outlined various abiotic mechanisms including homogeneous photolysis, reduction by inorganic particulates and organic molecules, as well as transient reductants. More recently Zhang and Lindberg [45] have suggested that iron(III) mediated photo-reduction is a significant mechanism in DGM formation. An alternative to reduction by redox intermediates is direct reduction of mercury by humic substances. Allard and Arsenie [53] determined that reduction by DOC is possible except at very low pH or with high chloride concentrations.

Several researchers have also suggested that bacteria mediate mercury reduction [54,55]. Siciliano et al. [56] recently examined the role of microbial reduction and oxidation processes in regulating DGM diel (over a 24 h period) patterns in freshwater lakes. The authors demonstrate that photochemically produced hydrogen peroxide regulates microbial oxidation processes and may account for the diel patterns observed in DGM data. Overall, the mechanisms responsible for mercury reduction and the relative contributions of biotic and abiotic processes are still unclear but solar radiation appears to be a common instigator of photo-reduction.

Photo-oxidation is the reversal of photo-reduction, that is, the transformation of DGM into inorganic mercury. As with photo-reduction, both abiotic [45,57] and biotic [56] mechanisms for photo-oxidation have been proposed. Lalonde et al. [57], who discovered that DGM can be photo-oxidized, claimed that chloride ions stabilize Hg(I) in solution and decrease the Hg(I)/Hg(0) potential such that electron transfer to semiquinones may take place. They determined that photo-oxidation of Hg(0) follows pseudo-first-order kinetics with a rate constant of $0.25\ \mathrm{h}^{-1}$ for freshwater and $0.6\ \mathrm{h}^{-1}$ for saline waters.

While the relative importance and precise mechanisms of these competing processes are currently unknown, it is likely that the balance of photo-reduction and photo-oxidation controls DGM dynamics in freshwaters. This is supported by the work of O'Driscoll et al. [32], which shows that DGM dynamics can be accurately modeled as a reversible first order reaction between oxidation states. O'Driscoll et al. [47] also determined that DOC concentration and levels of photo-reducible mercury are central to determining the rate of DGM production. While the results in Ref. [32] indicate that current predictive models do not accurately predict mercury volatilization, new research and refined methodologies (such as continuous analysis [36]) may lead to quick advances in future research.

4.2. Mercury Methylation and Demethylation

While a few researchers have observed photoinduced and non-photoinduced abiotic mercury methylation [58–60]; biotic methylation [mediated by sulfate reducing bacteria (SRB)] is believed to be the dominant process in most freshwater lakes [61]. In most lakes methylation occurs primarily in anoxic areas such as the surficial sediments or in surrounding wetlands. Some research indicates that the roots of aquatic macrophytes are also important sites of mercury methylation [62]. Rates of biotic methylation are a function of environmental variables affecting Hg^{2+} availability as well as the population sizes of methylating microbes. The balance of methylation and the reverse processes of demethylation determine the level of MeHg in an ecosystem. Biotic oxidative demethylation in sediment has been attributed primarily to SRBs and methanogens with degradation rates of $0.02–0.5 \, ng^{-1} \, d^{-1}$ reported for the Florida Everglades [63].

In addition to biotic demethylation, Sellers et al. [64] observed that photodemethylation is a significant process in ecosystems, with rates of $25 \, pg \, L^{-1} \, d^{-1}$ in surface water. Since then, several mechanisms have been proposed which implicate hydroxyl radicals as the principal demethylating agents [65,66]. While methylation and demethylation control the concentration of MeHg in a lake, the bioavailability of MeHg is dependent upon several factors including binding by ligands and removal to sediments.

4.3. Binding and Sedimentation

In addition to volatilization, mercury can be removed from a lake by the process of sedimentation. The primary processes affecting mercury sedimentation include binding to reduced sulfides, dissolved organic matter (DOM), particulates, and microorganisms. In reducing environments, inorganic mercury can react with sulfides to form an HgS precipitate [67]. The role of DOM in the fate of mercury is complicated. Several researchers have found that substantial amounts of total and methyl mercury are complexed by DOM [68,69]. Mercury bound to negatively charged functional groups on the organic complexes can flocculate at low pH and accumulate in sediments [70]. While many researchers have investigated the role of sulfides and DOM in mercury cycling, little work has been performed on the importance of biological surfaces as mercury binding agents. A recent paper by Daughney et al. [71] showed that mercury binding to bacteria is an important part of freshwater mercury dynamics. Mercury in sediments is also not permanently removed from the ecosystem since it can be converted to MeHg or Hg^0 by various anoxic processes (Section 4.2.), which is then available for flux to the water column or bioaccumulation by benthic organisms [12].

5. EFFECTS OF A CHANGING LANDSCAPE ON MERCURY FATE

5.1. Wetlands

A large amount of published research has shown that the amount of wetlands surrounding a lake is likely to affect the amount of MeHg inputs from the lake catchment [72–77]. In the Experimental Lakes Area (ELA) reservoir project in Ontario, it has been found that wetland areas of catchments provide 26–79 times more MeHg per unit area to downstream water than areas that contain no wetlands [78]. Similar findings in Sweden, Wisconsin, and the Adirondack lakes have shown the importance of wetlands as a MeHg source [79,80]. Rudd [78] estimates that the MeHg production from wetlands surrounding the catchments studied must equal at least 4.4 mg ha^{-1} year^{-1} to support the amount of MeHg exported. St. Louis et al. [81] determined in an experimental reservoir that MeHg is created very quickly in peat after an initial flooding event (increases of 40-fold in the first year). However, over the longer term (>3 years), MeHg production decreases and accumulation in biota may be significantly affected by in-lake methylation rates in small reservoirs. Runoff, soil erosion, flooding and drying cycles, and microbial activity are all factors that may change the speciation of mercury in wetlands.

The presence of wetlands in a lake's catchment basin has a profound effect on mercury and MeHg concentrations as it is linked to several other chemical changes within that lake. Lake color and dissolved organic carbon (DOC) concentration is a function of the % wetlands in the drainage basin [18,82,83]. It is likely that the correlation between percentage wetlands and mercury input is due to DOC-mediated transport of mercury species. Several studies have found correlations between DOC and mercury transport. Krabbenhoft et al. [80] observed that the Allequash Creek wetland in Northern Wisconsin released substantial amounts of DOC to a stream in fall. It was also observed that there was a twofold increase in Hg concentration in the stream; however, no effects were observed on MeHg concentrations. Pettersson et al. [72] found that the amount of MeHg associated with humic substances taken from a mire in Northern Sweden increased with increasing concentrations of humic substances. Lee and Hultberg [84] studied runoff from a catchment in the Lake Gardsjon watershed, where it was determined that MeHg was associated with humic substances and that they are transported together from the catchment by runoff to lake water. O'Driscoll and Evans [69] observed that substantial amounts of methyl mercury can bind to DOC which serves as a transporting agent from wetlands. Increasing mercury and MeHg in lake inflows due to DOC mediated transport will lead to an increase in mercury bioaccumulation in fish and wildlife. Carter et al. [85] found that size adjusted mean mercury concentrations in yellow perch from Kejimkujik lakes were positively correlated ($r = 0.54$, $p = 0.007$) with the total organic carbon concentration of lake waters. This is consistent with recent observations of a strong correlation between MeHg and DOC in the lake waters of Kejimkujik Park [86].

5.2. Deforestation

Several researchers have examined the effects of forestry on water quality and mercury fate. Carignan et al. [87] found that DOC concentrations were up to three times higher and K^+, Cl, and Ca^{2+} concentrations were up to six times higher in lakes with recently logged drainage basins than in reference lakes. Garcia and Carignan [88,89] found that methyl mercury concentrations were higher in the zooplankton of lakes with logged drainage basins than in non-logged lakes and that mercury concentrations in northern pike were significantly higher in lakes with logged drainage basins (3.4 μg g^{-1} wet wt.) than in reference lakes (1.9 μg g^{-1}). However, the mechanism that results in increased bio-accumulation with logging was at that time unclear.

Several researchers have suggested that logging in areas of the Amazon and Central Quebec may increase mercury inputs to water bodies by increasing soil erosion and DOC loading [90,91], however no direct mechanisms have been shown. O'Driscoll et al. [47] found that the rate of DGM photo-production was significantly higher for the non-logged lakes than for the logged lakes. This was the first proven mechanism by which forestry may increase mercury bioaccumulation. We postulate that this relationship is due to different concentrations of photo-reducible mercury arising from variations in DOC structure and dissolved ions between logged and non-logged lake sites. The data indicates that logging may reduce a lake's ability to produce DGM and thus may ultimately reduce mercury evasion. A reduction in mercury evasion may result in an increase in the mercury pool of lakes with extensive logging in their drainage basins.

6. BIG DAM WEST LAKE MERCURY MASS BALANCE

6.1. Introduction

To determine the relative importance of the different mercury flux pathways, a multidisciplinary team was assembled to determine all known mercury flux processes within Big Dam West Lake (BDW) in Kejimkujik National Park (Nova Scotia, Canada) [92]. The team was comprised of geologists, chemists, biologists, GIS experts, microbiologists, atmospheric scientists, and ecologists. BDW Lake has been the site of a considerable amount of previous mercury research and because Kejimkujik Park is a long-term acid rain monitoring site. As such it is one of Canada's main meteorological stations where mercury deposition and air concentrations are measured biweekly [25]. There is also a clear indication that mercury contamination is a serious problem as loons in Kejimkujik Park have been identified as having some of the highest blood mercury concentrations in North America [4]. While a large amount of mercury research has been performed in Kejimkujik, the underlying cause of the mercury bioaccumulation problem was not clear prior to our investigation. This was likely due to the large number of processes that can affect the speciation and transport of mercury through an ecosystem (as outlined in previous sections).

While several researchers have attempted mass balances for mercury in rivers and lakes [74,75,93,94], none have attempted to collect quantitative mercury measurements with a whole-ecosystem approach. The purpose of the study was to examine a complete set of mercury fluxes in the BDW Lake basin. A detailed explanation of the mass balance model, mass flux equations and the calculation of error is available in other publications [86,92].

6.2. Experimental Findings

Our conceptual model divided the BDW lake basin into three land cover types (based on remote sensing data): (i) terrestrial, (ii) wetland, and (iii) lake (Fig. 3). Average annual mass movements for total mercury were calculated using the collected data [86]. Wet precipitation was the only source of mercury inputs considered to the terrestrial system accounting for 184 g ($\sigma = 20.2$ g) of mercury deposited over land. The total outputs from the terrestrial system accounted for 372 g ($\sigma = 36.7$ g), of that, 35% (132 g, $\sigma = 0.0$ g) was incorporated into vegetation and, 13% (49 g, $\sigma = 0.0$ g) was volatilized from the soil surface. Although we were unable to measure mercury runoff directly, 191 g would be necessary in order to balance the inputs and outputs of the wetland component.

Of the mean 203 g ($\sigma = 36.7$ g) of mercury input to wetlands, 94% (191 g, $\sigma = 36.7$ g) was due to terrestrial runoff and 6% (12 g, $\sigma = 0.0$ g) was due to wet deposition. The total outputs from the wetland accounted for 203 g ($\sigma = 40.6$ g), 72% (147 g, $\sigma = 36.7$ g) was removed by outflow to the lake, 16% (32 g, $\sigma = 3.6$ g) was deposited to sediment and 12% (24 g, $\sigma = 0.0$ g) was volatilized from the wetland surface.

Of the mean 160 g ($\sigma = 38.4$ g) of mercury inputs directly to the lake, 92% (147 g, $\sigma = 36.7$ g) was due to inflow from wetlands, 7% (11 g, $\sigma = 0.0$ g) was due to wet precipitation and, 1% (2 g, $\sigma = 1.7$ g) was due to groundwater inflow. The total outputs from the lake accounted for 206 g ($\sigma = 54.1$ g); 75% (153 g, $\sigma = 50.5$ g) was removed by outflow, 14% (30 g, $\sigma = 3.6$ g) was deposited to sediments, and 11% (23 g, $\sigma = 0.0$ g) was volatilized from the lake surface. It should be noted that total outputs were larger than inputs by 46 g, however this is not significant in light of the deviation observed in these values.

In comparison to our findings, Henry et al. [94] performed a mass balance on Onondaga Lake, NY, and found that annual total mercury inputs accounted for 14.116 kg. Of the total inputs, 96.3% (13.6 kg) was due to terrestrial inflows, 3.1% (0.44 kg) was atmospheric deposition, 0.4% (0.056 kg) was sediment flux, and 0.1% (0.02 kg) was groundwater. Of the 13.916 kg of outputs, sedimentation accounted for 79.8% (11.1 kg), outflow 20.1% (2.8 kg), and volatilization 0.1% (0.016 kg). These results compare well with what was observed in this study with inflows contributing the majority of the total mercury, followed by

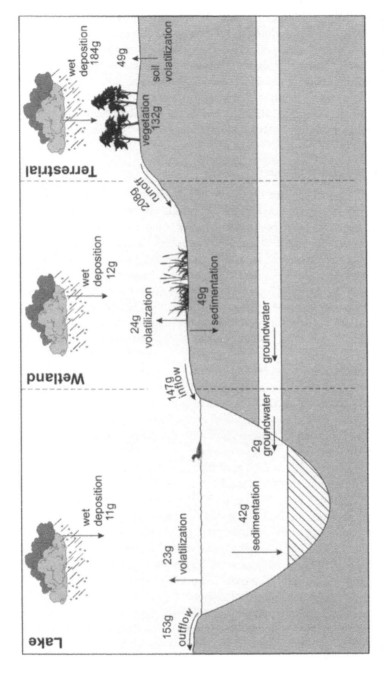

Figure 3 Mean annual mass movements of mercury calculated for Big Dam West Lake (Kejimkujik Park, Nova Scotia, Canada). Adapted from Ref. [86] with permission of SETAC Press.

precipitation and groundwater. However, the outputs of the two lakes are quite different. Outflow is the primary sink in BDW Lake, whereas sedimentation is the dominant sink in Onondaga Lake. The lakes are in fact opposite in terms of outflow vs. sedimentation. The level of volatilization is substantially higher in BDW as opposed to Onondaga Lake (11% compared to 0.1%, respectively).

The area within the BDW drainage basin is largely dominated by the terrestrial ecosystem. Of the 31.18 km^2 in the BDW watershed, 88.7% is terrestrial, 5.9% is wetland, and 5.5% is lake surface. The 207 g of mercury deposited in precipitation follows a similar distribution. The terrestrial catchment dominates the mass movement of total mercury in the BDW watershed. This mercury is then available for methylation as it moves through the wetland portions of the catchment. O'Driscoll et al. [92] reported the highest % of MeHg (30–40% of the total Hg) occurring in the wetland areas of this site.

There are several loss processes for mercury from this ecosystem. Terrestrial vegetation was found to play a significant role in mercury movement as evidenced by the large amount of mercury uptake found in this study (132 g). The importance of terrestrial vegetation is not fully explored since three important mercury fluxes were not measured: (i) uptake from roots, (ii) volatilization from leaves, and (iii) litter fall.

In addition, volatilization of mercury was found to be an important loss process in the BDW watershed. The magnitude of volatilization appears to be approximately double the direct wet deposition over lake and wetlands, and 27% of the direct wet deposition to the terrestrial catchment. Over the entire basin area the mass of mercury volatilized is 46% of the mass deposited by wet deposition.

7. GENERAL CONCLUSIONS

Research characterizing mercury speciation is an essential first step in the development of models to predict its transport and fate on regional and global scales. This review illustrates that several key global issues are likely to have major impacts on the speciation of mercury. These include: (i) climate change, (ii) ozone depletion, and (ii) changes in topography.

The effects of climate change on the speciation and fate of mercury in Polar ecosystems is particularly important. Not only is mercury increasing in the atmosphere but atmospheric deposition will be favored in colder climates due to changes in atmospheric redox chemistry. This means that mercury released in equatorial areas will undergo a global distillation via a process similar to the "grasshopper effect" observed with semi-volatile organic pollutants.

Solar radiation drives a number of chemical transformations of mercury. These include: (i) atmospheric speciation and deposition, (ii) oxidation–reduction in both freshwater and seawater, and (iii) methyl mercury degradation. Both biotic and abiotic redox reactions are influenced. While microbes have been thought to dominate methyl mercury production, abiotic formation cannot be ruled out. As such, future changes in ultraviolet radiation may have dramatic effects on the speciation and fate of mercury.

Human activity has altered the mercury cycle in many remote areas by increasing emissions to the atmosphere and changing the topography through flooding and forestry practices. We know from previous research that there is a correlation between percentage wetlands and methyl mercury loading to a lake. We have recently shown that forestry practices may limit the ability of a lake to volatilize mercury, thereby favoring mercury accumulation in lakes with logged catchments.

The work performed in Kejimkujik Park, Nova Scotia demonstrates that substantial mercury bioaccumulation can occur in remote areas where no abnormal sources of mercury exist. The mass balance preformed at BDW Lake in Kejimkujik Park, showed that movement of atmospheric mercury from the terrestrial catchment to wetland areas is the primary source of methyl mercury to the lake. Mercury volatilization was found to be an important process in this basin with annual volatilization equaling 46% of the mercury deposited by precipitation over the entire lake basin. This paper demonstrates that mercury speciation must be known to reliably predict the effect of anthropogenic influences on a regional and global scale.

ACKNOWLEDGMENTS

Funding provided in the form of scholarships to Dr. Nelson O'Driscoll [National Science and Engineering Council (NSERC) and Ontario Graduate Scholarships in Science and Technology (OGSST)], research grant funding to Dr. David Lean (NSERC), Toxic Substances Research Initiative (TSRI, Health Canada) and Metals in the Environment (MITE, Natural Resources Canada) funding to Dr. Andy Rencz.

ABBREVIATIONS

BDW	Big Dam West Lake
CEC	Commission for Environmental Cooperation
DGM	dissolved gaseous mercury
DOC	dissolved organic carbon
DOM	dissolved organic matter
ELA	Experimental Lakes Area
GIS	geographic information system
MDE	methyl mercury
MeHg	methyl mercury
NADP	National Atmospheric Deposition Program
PCB	polychlorinated biphenyl
PM	particulate mercury
RGM	reactive gaseous mercury

SRB   sulfate reducing bacteria
USEPA   United States Environmental Protection Agency
UV   ultraviolet

REFERENCES

1. Sorensen EMB. In: Sorensen EMB, ed. Metal Poisoning in Fish. Boca Raton, Florida: CRC Press, 1991:285–330.
2. Lathrop RC, Rasmussen PW, Knauer DR. Water Air Soil Pollut 1991; 56:295–307.
3. Cabana G, Tremblay A, Kalff J, Rasmussen JB. J Fish Aquat Sci 1994; 51:381–389.
4. Evers DC, Kaplan JD, Meyer MW, Reaman PS, Braselton WE, Major A, Burgess N, Scheuhammer AM. Environ Toxicol Chem 1998; 17:173–183.
5. Burgess NM, Evers DC, Kaplan JD. In: Burgess N, Beauchamp S, Brun G, Clair T, Roberts C, Rutherford L, Tordon R, Vaidya O, eds. Mercury in Atlantic Canada: A Progress Report. Atlantic Region, Sackville, NB: Environment Canada, 1998:96.
6. Unites States Environmental Protection Agency. Mercury Study Report to Congress. Office of Air Quality Planning and Standards, Office of Research and Development. Washington DC: US Government Printing Office, 1997. EPA-452/R-97-003.
7. International Joint Commission & Commission for Environmental Cooperation. Addressing Atmospheric Mercury Science and Policy, Research Triangle Park, North Carolina, USA, 2003.
8. Clever HL, Johnson SA, Derrick ME. J Phys Chem Ref Data 1985; 14:631–680.
9. Lindqvist O, Rodhe H. Tellus 1985; 27B:136–159.
10. Carpi A, Lindberg SE. Environ Sci Technol 1997; 31:2085–2091.
11. Schroeder WH, Yarwood G, Niki H. Water Air Soil Pollut 1991; 56:653–666.
12. Morel FMM, Kraepiel AML, Amyot M. Ann Rev Ecol Sys 1998; 29:543–566.
13. Fitzgerald WF. Global Biogeochemical Cycling of Mercury. Presented at the DOE/FDA/EPA Workshop on Methylmercury and Human Health, Bethesda, MD, 1994.
14. Fitzgerald WF. Water Air Soil Pollut 1995; 80:245–254.
15. Martinez-Cortizas A, Pontevedra-Pombal X, Garcia-Rodeja E, Novoa-Munoz JC, Shotyk W. Science 1999; 284:939–942.
16. Fitzgerald WF, Engstrom DR, Mason RP, Nater EA. Environ Sci Technol 1998; 32:1–7.
17. Schuster PF, Krabbenhoft DP, Naftz DL, Cecil LD, Olson ML, Dewild JF, Susong DD, Green JR, Abbott ML. Environ Sci Technol 2002; 36:2303–2301.
18. Watras CJ, Huckabee JW, eds. Mercury Pollution: Integration and Synthesis. Boca Raton, FL: CRC Press. Lewis Publishers, 1994.
19. Pacyna JM, Pacyna PE. Global Emissions of Mercury to the Atmosphere: Emissions from Anthropogenic Sources. Norwegian Institute for Air Research (NILU) Report, Kjeller, Norway, 1996.
20. Porcella DB, Ramel C, Jernelov A. Water Air Soil Pollut 1997; 97:205–207.
21. Mason RP, Fitzgerald WF, Morel FMM. Geochim Chosmochim Acta 1994; 58:3191–3198.
22. Zehner RE, Gustin MS. Environ Sci Technol 2002; 36:4039–4045.
23. St. Louis VL, Rudd JWM, Kelly CA, Hall BD, Rolfhus KR, Scott KJ, Lindberg SE, Dong W. Environ Sci Technol 2001; 35:3089–3098.
24. Van Loon L, Mader E, Scott SL. J Phys Chem A 2000; 104:1621–1626.

25. NADP (National Atmospheric Deposition Program). Annual Summary, 2002. http://nadp.sws.uiuc.edu/lib/data/2002as.pdf
26. Schroeder WH, Anlauf KG, Barrie LA, Lu JY, Steffen A, Schneeberger DR, Berg T. Nature 1998; 394:331–332.
27. Ebinghaus R, Kock HH, Temme C, Einax JW, Lowe AG, Richter A, Burrows JP, Schroeder WH. Environ Sci Technol 2002; 36:1238–1244.
28. Berg T, Sekkesaeter S, Steinnes E, Valdal AK, Wibetoe G. Sci Tot Environ 2003; 304:43–51.
29. Lindberg SE, Brooks S, Lin CJ, Scott KJ, Landis MS, Stevens RK, Goodsite M, Richter A. Environ Sci Technol 2002; 36:1245–1256.
30. Wania F, Mackay D. Environ Sci Technol 1996; 30:390A–396A.
31. Gouin T, Mackay D, Jones KC, Karner T, Meijer SN. Environ Pollut 2004; 128:139–148.
32. O'Driscoll NJ, Beauchamp S, Siciliano SD, Lean DRS, Rencz AN. Environ Sci Technol 2003; 37:2226–2235.
33. Van Loon LL, Mader EA, Scott SL. J Phys Chem A 2001; 105:3190–3195.
34. Lin C, Pehkonen SO. Atmos Chem 1999; 33:2067–2079.
35. Ariya PA, Khalizov A, Gidas A. J Phys Chem A 2002; 206:7310–7320.
36. O'Driscoll NJ, Siciliano SD, Lean DRS. Sci Tot Environ 2003; 304:285–294.
37. Rolfhus KR, Fitzgerald WF. Geochim Cosmochim Acta 2001; 65:407–418.
38. Mason RP, Sullivan KA. Environ Sci Technol 1997; 31:942–947.
39. Watras CJ, Morrison KA, Host JS. Limnol Oceanogr 1995; 40:556–565.
40. Amyot M, Mierle G, Lean DRS, McQueen DJ. Environ Sci Technol 1994; 28:2366–2371.
41. Amyot M, Gill GA, Morel FMM. Environ Sci Technol 1997; 31:3606–3611.
42. Baeyens W, Leermakers M. Mar Chem 1998; 60:257–266.
43. Costa M, Liss PS. Mar Chem 1999; 68:87–95.
44. Lanzillotta E, Ferrara R. Chemosphere 2001; 45:935–940.
45. Zhang H, Lindberg SE. Environ Sci Technol 2001; 35:928–935.
46. Amyot M, Mierle G, Lean D, McQueen D. Geochem Cosmochem Acta 1997; 61:975–987.
47. O'Driscoll NJ, Lean DRS, Loseto L, Carignan R, Siciliano SD. Environ Sci Technol (Web Release Date: 27-Mar-2004, DOI: 10.1021/es034702a)
48. Vandal GM, Mason RP, Fitzgerald WF. Water Air Soil Pollut 1991; 56:791–803.
49. Amyot M, Lean DRS, Poissant L, Doyon M. Can J Fish Aquat Sci 2000; 57(Suppl. 1):155–163.
50. Amyot M, Lean D, Mierle G. Environ Toxicol Chem 1997; 16:2054–2063.
51. Krabbenhoft DP, Hurley JP, Olsen ML, Cleckner LB. Biogeochem 1998; 40:311–325.
52. Nriagu JO. Sci Tot Environ 1994; 154:1–8.
53. Allard B, Arsenie I. Water Air Soil Pollut 1991; 56:457–464.
54. Barkay T, Turner RR, Van den Brook A, Liebert C. Microbial Ecol 1991; 21:151–161.
55. Vandal GM, Fitzgerald WF, Rolfhus KR, Lambourg CH. In: Mercury as a Global Pollutant: Toward Integration and Synthesis. Chelsea, MI: Lewis Publishers, 1994:529–538.
56. Siciliano SD, O'Driscoll NJ, Lean DRS. Environ Sci Technol 2002; 36:3064–3068.
57. Lalonde JD, Amyot M, Kraepiel AM, Morel FMM. Environ Sci Technol 2001; 35:1367–1372.

58. Gardfeldt K, Munthe J, Stromberg D, Indqvist O. Sci Tot Environ 2003; 304:127–136.
59. Falter R. Chemosphere 1999; 39:1051–1073.
60. Lean DRS, Siciliano SD. J Phys IV France 2003; 107:743–747.
61. Gilmore CC, Henry EA. Environ Pollut 1991; 71:131–169.
62. Mauro JBN, Guimaraes JRD, Melamed R. Appl Organometal Chem 1999; 13:631–636.
63. Marvin-Dipasquale MC, Oremland RS. Environ Sci Technol 1998; 32:2556–2563.
64. Sellers P, Kelly CA, Rudd JWM, MacHutchon AR. Nature 1996; 380:694–697.
65. Chen J, Pehkonen SO, Lin C. Water Res 2003; 37:2496–2504.
66. Gardfeldt K, Sommar J, Stromberg D, Feng X. Atmos Environ 2001; 35:3039–3047.
67. Radosevich M, Klein DA. Bull Environ Contam Toxicol 1993; 51:226–233.
68. Ravichandran M. Chemosphere 2004; 55:319–331.
69. O'Driscoll NJ, Evans RD. Environ Sci Technol 2000; 34(18):4039–4043.
70. Jackson TA, Parks JW, Jones PD, Woychuck RN, Sutton JA, Hollinger JD. Hydrobiologia 1982; 92:473–487.
71. Daughney CJ, Siciliano SD, Rencz AN, Lean D, Fortin D. Environ Sci Technol 2002; 36:1546–1553.
72. Pettersson C, Bishop K, Lee Y, Allard B. Water Air Soil Pollut 1995; 80:971–979.
73. Hurley JP, Benoit JM, Babiarz CL, Shafer MM, Andren AW, Sullivan JR, Hammond R, Webb DA. Environ Sci Technol 1995; 29:1867–1875.
74. Lee YH, Bishop KH, Munthe J, Iverfeldt A, Verta M, Parkman H, Hultberg H. Biogeochemistry 1998; 40:125–135.
75. Driscoll CT, Holsapple J, Schofield CL, Munson R. Biogeochemistry 1998; 40:137–146.
76. St Louis VL, Rudd JWM, Kelly CA, Beaty KG, Bloom NS, Flett RJ. Can J Fish Aquat Sci 1994; 5:1065–1076.
77. St Louis VL, Rudd JWM, Kelly CA, Beaty KG, Flett RJ, Roulet NT. Environ Sci Technol 1996; 30:2719–2729.
78. Rudd JWM. Water Air Soil Pollut 1995; 80:697–713.
79. Driscoll CT, Scholfield CL, Munson R, Holsapple J. Environ Sci Technol 1994; 28:136–143.
80. Krabbenhoft DP, Benoit JM, Andren AW, Babiarz CL, Hurley JP. Water Air Soil Pollut 1995; 80:425–433.
81. St Louis VL, Rudd JWM, Kelly CA, Bodaly RA, Paterson MJ, Beaty KG, Hesslein RH, Heyes A, Majewski AR. Environ Sci Technol 2004; 38:1348–1358.
82. D'Arcy P, Carignan R. Can J Fish Aquat Sci 1997; 54:2215–2227.
83. Dillon PJ, Molot LA. Water Res 1997; 33:2591–2600.
84. Lee YH, Hultberg H. Environ Technol Chem 1990; 9:833–841.
85. Carter J, Drysdale C, Burgess N, Beauchamp S, Bruin G, D'Entremont, A, eds. Mercury concentrations in yellow perch (Perca flavescens) from 24 lakes at Kejimkujik National Park, July, 2001, Parks Canada- Technical Reports in Ecosystem Science, Report Number 031, 2001.
86. O'Driscoll NJ, Rencz AN, eds. A multidisciplinary study of mercury cycling in a wetland dominated ecosystem: Kejimkujik Park. Nova Scotia: SETAC Publishers. In press.
87. Carignan R, D'Arcy P, Lamontagne S. Can J Fish Aquat Sci 2000; 57:105–117.
88. Garcia E, Carignan R. Can J Fish Aquat Sci 1999; 56:339–345.

89. Garcia E, Carignan R. Can J Fish Aquat Sci 2000; 57:129–135.
90. Carmouze JP, Lucotte M, Boudou A, eds. Mercury in the Amazon; Importance of Human and Environment, Health Hazards (Synthesis and Recommendations), Institut de Recherche pour le Development, Paris, 2001.
91. Lamontagne S, Carignan R, D'Arcy P, Prairie YT, Pare D. Can J Fish Aquat Sci 2000; 57(Supp. 2):118–128.
92. O'Driscoll N, Clair T, Rencz A, eds. Cycling of Mercury in Kejimkujik National Park: Toxic Substance Research Initiative #124 Summary December 2001.Occasional Report # 18, Environment Canada, Sackville, NB, Catalogue No. CW69-12/18-2001E., 2001.
93. Quemerais B, Cossa D, Rondea UB, Pham TT, Gagnon P, Fortin B. Environ Sci Technol 1999; 33:840–849.
94. Henry EA, Dodge-Murphy LJ, Bigham GN, Klein SM, Gilmour CC. Water Air Soil Pollut 1995; 80:509–518.

10

Biogeochemistry and Cycling of Lead

William Shotyk and Gaël Le Roux

*Institute of Environmental Geochemistry, University of Heidelberg,
Im Neuenheimer Feld 236, D-69120 Heidelberg, Germany*

1.	Introduction	240
2.	Chemistry of Lead and Behavior in the Environment	242
	2.1. Summary of Basic Chemical Properties	242
	2.2. Abundance and Occurrence	243
	2.3. Measuring Lead Concentrations	244
3.	Lead Isotopes and Their Measurement	245
	3.1. Stable Isotopes	245
	3.2. Measurements of Stable Lead Isotopes	246
	3.3. Intermediate Decay Products of U–Th Decay Series	247
4.	Ancient and Modern Uses of Lead	247
	4.1. Ancient and Medieval Uses	247
	4.2. Modern Uses	248
5.	Emissions of Lead to the Environment	250
	5.1. Lead in Natural vs. Anthropogenic Atmospheric Particles	251
	5.2. Atmospheric Lead from Alkyllead Fuel Additives	251
6.	Inputs and Fate of Anthropogenic Lead in the Biosphere	252
	6.1. Lead Concentrations in Soils	252
	6.2. Cumulative Impact of Anthropogenic, Atmospheric Lead	253
	6.3. The Fate of Anthropogenic Lead in Soils	254
	6.4. Lead Concentrations in Solution	255
7.	Temporal Trends in Atmospheric Lead Deposition	255
	7.1. Lead in Sediments	255

7.2. Lead in Bryophytes   256
7.3. Lead in Tree Rings and Bark Pockets   256
7.4. Peat Bog Archives   256
7.5. Relative Importance of Gasoline Lead vs. Other Sources
       of Industrial Lead   257
7.6. The Cumulative Input of Anthropogenic Lead   257
7.7. Lead in Polar Snow and Ice   259
7.8. Lead in Atmospheric Aerosols Today   260
8.   Environmental Lead Exposure and Human Health   263
8.1. Blood Lead Levels (BLLs) and Their Significance   263
8.2. Mechanism of Lead Poisoning   264
8.3. Predominant Sources of Lead Exposure   264
8.4. Other Sources of Lead Exposure   265
9.   Summary and Conclusions   267
Acknowledgments   267
Abbreviations   268
References   268

1. INTRODUCTION

The environmental geochemistry of Pb has probably stimulated more scientific interest than all other metallic elements combined. According to Jaworski [1], "the local, regional, and global biogeochemical cycles of lead have been affected by man to a greater degree than those of any other toxic element". Why? The main reasons are simple—lead is an extremely useful metal and is relatively simple to work with. Because it melts at a relatively low temperature (327°C), however, it is easily emitted to the atmosphere during smelting and refining. And because it has been used since Antiquity, environmental contamination by Pb is probably as old as civilization itself.

Lead is malleable, ductile, dense, and resistant to corrosion. Thus, it has found a tremendous range of commercial and industrial applications in manufacturing and building since ancient times. In fact, the value of lead to mankind has been considered so great that it has even been termed a "precious metal" [2], an appellation which it much deserves. In addition, occurrences of lead ores are comparatively easy to find. Galena (PbS), the main Pb sulfide mineral, has a diagnostic color, lustre, and cleavage. Also, Pb is readily extracted from sulfide ores by oxidation and the technology for processing them was well-known already in ancient times [3]. The low melting point of Pb makes it possible to process even with very primitive technology. The discovery of cupellation ~5000 years ago, when the ancients learned how to separate silver from lead ores, was a great stimulus for lead production worldwide [3]. Because of the widespread occurrence of its ores and the great value of both Pb and Ag, enormous quantities of lead have been produced ever since. As a consequence, anthropogenic emissions of Pb to the atmosphere, to soils, sediments, and waters, have been very extensive. In fact, there is probably no place on the surface of the Earth that is devoid of anthropogenic Pb [1].

Although Pb is one of the most useful of all the metals, it is also one of the most toxic [4]. Because it serves no biological role and occurs naturally at the surface of the Earth only in trace concentrations, there continues to be great interest in the environmental fate of anthropogenic Pb, and possible effects on human and ecosystem health [5]. Probably the single most compelling paper ever written about the geochemistry of Pb in the environment is the seminal work by Patterson [6] which still reads well today, 40 years after it was published. As western philosophy is sometimes considered footnotes to Plato, modern studies of the environmental geochemistry of Pb may be viewed as footnotes to Patterson.

With respect to scientific study, one great advantage of Pb compared to other metals of environmental interest is the number of stable isotopes (^{204}Pb, ^{206}Pb, ^{207}Pb, ^{208}Pb) which can be used to help "fingerprint" the predominant sources of both natural and anthropogenic Pb, to identify predominant pathways, and to study the fate of this metal in the environment [7]. This opportunity is made possible by the wealth of information about the isotopic composition of Pb in rocks, minerals, and lead ores compiled over the past decades by economic geologists, geochronologists and isotope geologists [8–11]; many of the results most relevant to environmental studies have been summarized by Sangster et al. [12]. In addition, ^{210}Pb, an unstable daughter product of the decay of ^{238}U which is supplied naturally to the air via the decay of ^{222}Rn, is a valuable tracer of atmospheric scavenging, soil migration, adsorption, biological uptake and other processes affecting the behavior and fate of Pb in the environment [13].

Given the voluminous literature about the geochemistry of Pb in the environment (5000 papers in 1978 alone, according to Nriagu [14]), we have concentrated our efforts on new developments in the following areas:

1. *The isotopic composition of Pb* is a powerful tool, which allows natural sources of anthropogenic Pb to be distinguished from anthropogenic ones. If quantitative measurements of Pb concentrations have allowed us to study the Pb problem in black and white, precise measurements of Pb isotope ratios allow us to see in color. Tremendous advances have been made regarding analytical developments for measuring Pb and Pb isotope ratios at ultra-trace concentrations, and some of these results are summarized here.

2. *Temporal trends in atmospheric Pb contamination*, including the fluxes and predominant sources of Pb using sediments, biomonitors, peat bogs, and polar ice. The historical trends in the predominant sources of atmospheric Pb are valuable indicators of recent changes

in environmental policies such as the phasing out of leaded gasoline. In addition, long-term records of atmospheric Pb are essential to determine the natural background rates, for comparison with modern values, but also to help understand the geological processes which controlled these fluxes.

3. *Fate of anthropogenic Pb in the terrestrial biosphere.* By determining the isotopic composition of Pb in different soil horizons and measuring Pb in soil solutions, it is possible to begin to understand the ultimate fate of industrial Pb, which has contaminated the surface layers of soils worldwide.

Our summary has been stimulated by two recent findings. First, even though most attention during the past decades has been given to anthropogenic emissions of Pb from the use of gasoline Pb additives, this is just a part of the environmental Pb story, and some of the greatest episodes of environmental Pb contamination pre-date by a wide margin the introduction and use of gasoline Pb additives. Second, recent studies have identified deleterious health effects in children at blood Pb concentrations much lower than previously believed to be important. Taken together, these new findings suggest that concerns regarding the sources, transport, and fate of Pb in the environment are more important than ever. In fact, while great progress has been made in reducing atmospheric Pb emissions, the greatest part of this success has been due to the gradual replacement of gasoline Pb additives. Further reductions in atmospheric Pb emissions, and additional progress in limiting human exposure to Pb, is going to be more challenging as well as more expensive.

Readers interested in previous studies of Pb in the environment are referred to the books by Nriagu [15], Boggess and Wixson [16], the Committee on Lead in the Human Environment [17], and Harrison and Laxen [18]. The chemistry and geochemistry of lead in the aquatic environment has been outlined in Refs. [19–21], and lead in soils in Refs. [22,23].

2.   CHEMISTRY OF LEAD AND BEHAVIOR IN THE ENVIRONMENT

2.1.   Summary of Basic Chemical Properties

Lead ($Z = 82$, atomic weight 207.19) is dull gray, soft, and weak but dense (11.35 g cm^{-3}). With the electronic configuration (Xe) $4f^{14}5d^{10}6s^26p^2$, Pb exhibits three formal oxidation states: (IV), (II), and (0). At Earth surface conditions, Pb(II) is by far the most common oxidation state (Fig. 1).

Lead readily loses two electrons, yielding Pb^{2+}, which hydrolyzes above pH 7. While Pb^{2+} is the dominant inorganic species throughout the pH and Eh range of most natural waters (Fig. 1), sulfate, carbonate, chloride, and phosphate complexes of Pb can also be important, depending on the concentrations of these species and the pH of the solution [24]. Lead carbonate is an effective control on the solubility of Pb in oxic soils and sediments above pH values of ~6 (Fig. 1). In

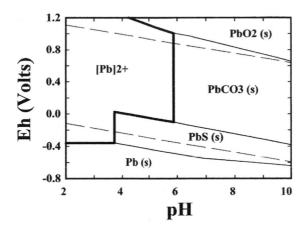

Figure 1 pH–Eh diagram for inorganic species of lead.

zones of active bacterial sulfate reduction such as anoxic sediments, the solubility of Pb is regulated by the formation of Pb sulfide. In neutral to alkaline soils, and in anoxic sediments, therefore, the mobility of Pb should be very limited. The behavior and fate of Pb in acidic soils, however, is more complicated: while solubility controls may be lacking, Pb^{2+} forms stable complexes with natural organic ligands, and is strongly adsorbed onto the surfaces of Mn and Fe hydroxides [25]. The solubilities of Pb phosphates are extremely low [25], and this can be a further control on Pb migration, especially in agricultural soils with added phosphorus.

2.2. Abundance and Occurrence

Typical Pb concentrations in rocks, soils, sediments, and fossil fuels ($\mu g/g$) are: basalt, 3; granite, 24; sandstone, 10; limestone, 6; shale, 23; soil, 35; coal; 10; sediment, 19. Manganese nodules may contain several hundred $\mu g/g$ Pb. In the Earth's crust, the average Pb concentration is estimated to be 15 $\mu g/g$, with ~ 17 $\mu g/g$ in the upper continental crust, and 13 $\mu g/g$ in the lower continental crust [26]. Lead is one of the most abundant of the heavy trace elements, for three reasons [27]. First, the cosmic abundance of Pb is relatively high. Second, in addition to common Pb (^{204}Pb), which is non-radiogenic, Pb is derived from the radioactive decay of ^{238}U, ^{235}U, and ^{232}Th. In fact, roughly one-half of the ^{206}Pb, one-third of the ^{207}Pb, and one-fifth of the ^{208}Pb, is derived from the respective U and Th isotopes since the Earth differentiated into a separate crust and mantle ~ 4.5 billion years ago [28]. Third, Pb is enriched in crustal rocks, partly because the ionic radius of Pb^{2+} (132 pm) is so similar to that of K^+ (133 pm), allowing divalent Pb to substitute for K in potassium feldspars and micas. Lead can also substitute readily for Sr^{2+} and Ba^{2+}, which have similar ionic radii to divalent Pb, but Pb has even been found to substitute in other silicates for smaller cations such as Ca^{2+} and Na^+. While Goldschmidt [29] classified

Pb as a chalcophile element because of its occurrence in galena (PbS), the lithophile character of Pb should not be overlooked. Micas and potassium feldspar, on average, contain $25-30$ $\mu g/g$. The abundance of Pb in silicates has been summarized by Wedepohl [30].

About 240 minerals of Pb are known [30], but galena (PbS), cerussite ($PbCO_3$), and anglesite ($PbSO_4$), are the most important economically. Probably 90% of the world's primary Pb is recovered from mining and refining galena. Boulangerite ($5PbS \cdot 2Sb_2S_3$), Bournonite ($PbCuSbS_3$), and Jamesonite ($Pb_4FeSb_6S_{14}$) also are important Pb sulfides.

2.3. Measuring Lead Concentrations

Quantitative determination of Pb in environmental and biological samples is made possible using a variety of well-established techniques such as wave dispersive X-ray fluorescence spectrometry (XRF) for solid samples at higher Pb concentrations, atomic absorption spectroscopy (AAS), and inductively coupled plasma (ICP) atomic emission spectrometry (AES). Both XRF and ICP-AES have the advantage of being able to determine a wide range of trace elements simultaneously. The lower limit of detection (LOD) provided by energy dispersive XRF (~ 2 $\mu g/g$ for soils and sediments, and 0.5 $\mu g/g$ for plant materials) allows this method to be used for many environmental samples of interest [31]; this approach is not only inexpensive and non-destructive, but it also avoids the time and expense of sample digestion. A high sensitivity XRF spectrometer was subsequently developed which can measure Pb down to 0.1 $\mu g/g$ in biological materials [32]. Conventional methods such as graphite furnace AAS are relatively inexpensive and can be used to measure Pb in acid digests and other aqueous samples down to a few $\mu g/L$. However, inductively-coupled plasma mass spectrometry with quadrupole mass analyzer (Q-ICP-MS) has become a workhorse in many laboratories, which need to measure Pb down to the ng/L range; a wide range of other trace elements can be measured simultaneously. Total reflection X-ray spectrometry can measure Pb in the ng/L range and is also a multi-element method, but the limits of detection are strongly dependent on the concentration of matrix elements [33]. Laser excited atomic fluorescence spectroscopy (LEAFS) is a novel method for measuring Pb in the ultra-trace concentration range [34,35]. For extremely low concentrations (pg/L) of Pb in uncontaminated natural waters, including polar ice, double focusing (sector field) ICP-MS (ICP-SMS) is now the method of choice [36]. In fact, with this instrument, the main limitation is no longer detecting power, but rather eliminating all forms of possible sample contamination. As described by Boutron [37] and Nriagu et al. [38], many studies previous to the 1970s and 1980s have overestimated Pb concentrations, often by many orders of magnitude, due to inadequate contamination control. With respect to all dilute natural waters, but also with respect to many other samples pre-dating the development of metallurgy ("pre-anthropogenic samples"), there is a need for clean lab protocols to be employed at every stage of sample handling, preparation, and analyses [37,38]. Using rigorous cleaning procedures and strict clean lab techniques, detection limits as low as 50 pg/L Pb have recently been achieved in samples of polar ice [36].

In cases which require extreme accuracy and precision, isotope dilution mass spectrometry (ID-MS) may be used to measure Pb concentrations. This consists of an addition to the sample of a solution of well-known Pb concentration and isotopic composition ("spike") followed by determination of the isotopic composition of the spiked sample using mass spectrometry, Q-ICP-MS, ICP-SMS, multi-collector ICP-MS (MC-ICP-MS), or thermal ionization mass spectrometry (TIMS).

To measure Pb concentrations in small particles such as aerosols, or to study the variation in Pb concentrations within solids such as mineral grains, there is a range in available techniques such as micro-beam XRF (including synchrotron methods), proton induced X-ray emission spectrometry (PIXE), secondary ionization mass spectrometry (SIMS), and laser ablation ICP-MS (LA-ICP-MS). These methods have been used for high spatial resolution even in 3-D, as well as for rapid analyses of biological or geological structures growing incrementally [39–43] or small, specific phases [44].

3. LEAD ISOTOPES AND THEIR MEASUREMENT

3.1. Stable Isotopes

Because of the range in geological age of Pb ores, and the U and Th concentrations of the rocks from which they are derived, the isotopic composition of Pb ores is variable, and each ore deposit can be described by its characteristic Pb isotope signature [11] (Fig. 2). For example, lead from the Broken Hill mine in Australia (the predominant source of industrial Pb used for gasoline Pb additives in Europe) has a $^{206}Pb/^{207}Pb = 1.04$, compared with crustal rocks which have a $^{206}Pb/^{207}Pb = 1.20$ [45]. Uncontaminated marine sediments [46,47] as well as atmospheric soil dust dating from pre-anthropogenic times [48], both of which are derived from crustal rocks, also have a $^{206}Pb/^{207}Pb = 1.20$. Similarly, the Pb contained in coal and other fossil fuels, and in waste products, reflects the isotopic composition of the original Pb source [49,50]. As a result, there has been widespread interest in determining the isotopic composition of Pb in environmental samples, with a view to identifying predominant atmospheric emission sources [12].

While this "fingerprinting" approach is growing in popularity [51], many lead ores and coal bodies have overlapping signatures, and interpretations of predominant source areas are not always unequivocal [52]. A further caveat involves emissions of atmospheric Pb from mixtures of Pb ores. For example, secondary Pb smelters today process scrap Pb from the most diverse sources. Even during

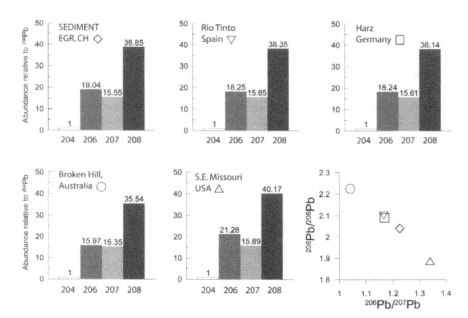

Figure 2 Isotopic composition of Pb in pre-anthropogenic aerosols [Etang de la Gruère (EGR), Switzerland (CH)], and in selected Pb ores [12].

the Medieval period, Pb ores from different mining regions were mixed to improve their metallurgical properties [53].

3.2. Measurements of Stable Lead Isotopes

Thermal ion mass spectrometry has traditionally been the method of choice to obtain precise Pb isotope determinations, with a typical RSD of 0.01% (2σ) for the $^{206}Pb/^{207}Pb$ ratio. Moreover, TIMS allows precise measurement of all four Pb isotopes, especially with the double or triple spike techniques [54], allowing the radiogenic ones (^{206}Pb, ^{207}Pb, ^{208}Pb) to be normalized to ^{204}Pb which is non-radiogenic. Also, to distinguish among anthropogenic Pb sources, it can be very instructive to make a "three isotope plot" such as $^{207}Pb/^{204}Pb$ against $^{206}Pb/^{204}Pb$ and $^{208}Pb/^{204}Pb$ vs. $^{207}Pb/^{204}Pb$, and this requires precise measurement of all four isotopes. It has been pointed out by Sangster et al. [12], however, that 86% of the discriminating power of Pb isotopes in environmental samples can be obtained without using ^{204}Pb, and that precise measurements of the three radiogenic stable Pb isotopes are more important. There is growing use of Q-ICP-MS measurements of the $^{206}Pb/^{207}Pb$ in environmental samples, but because of the poor precision (typical RSD of 0.3–0.7% (2σ) for the $^{206}Pb/^{207}Pb$ ratio), it is not considered further here.

Recently, it has been shown that Pb isotope ratios in environmental samples can be measured with acceptable accuracy and precision using ICP-SMS [55,56].

This approach is rapid (no chemical separation of Pb is required) and reasonably accurate and precise. For example, Krachler et al. [57] have reported precisions of 0.1% for the $^{206}Pb/^{207}Pb$ and $^{208}Pb/^{206}Pb$ ratios at total Pb concentrations of only 0.1 $\mu g/L$ in "real world" samples.

Equipped with an array of Faraday cups, the multiple collector-ICP-MS achieves comparable or better quality of measurements than TIMS without double spiking [58–60]. However, as for TIMS, a chemical separation before measurement is required to obtain such precision. Micro-beam techniques such as the sensitive high resolution ion microprobe (SHRIMP) [39,61] and LA-ICP-MS [58] are also useful for obtaining high resolution sampling of solid samples and simultaneous measurements of isotopic composition.

3.3. Intermediate Decay Products of U–Th Decay Series

Short-lived isotopes of Pb such as ^{210}Pb ($t_{1/2} = 22.26$ year), ^{212}Pb ($t_{1/2} = 10.6$ h), ^{214}Pb ($t_{1/2} = 26.8$ min) are useful tracers of transient signals for studying sediment fluxes in estuaries [62], deposition of atmospheric particles [63,64], scavenging of trace metals by aquatic particles [65], and oceanic circulation [66]. Lead-210 in particular, has been extensively used as a geochronological tool for studying the accumulation rates of sediments, peat, and glacial ice during the past two centuries [67,68].

4. ANCIENT AND MODERN USES OF LEAD

4.1. Ancient and Medieval Uses

Lead has been used since 6400 B.C. [69] and obtained as a by-product of silver mining for ~5000 years [3]. The Hanging Gardens of Babylon were floored with sheet lead. In classical times, Pb was used in all Mediterranean civilizations for construction, coinage, glass-making, water pipes, cosmetics and beverages [70]. Egypt had a Pb-based cosmetics industry as early as 2000 B.C. [71]. The most important Pb deposits in the Mediterranean region and in western Europe had already been prospected or/and mined in Roman times [72,73]. As a result, environmental contamination by Pb has its origins in antiquity, a fact which is clearly revealed by peat bogs, for example, in England [74a] and the harbor sediments from the Phoenician city of Sidon (Fig. 3) [74b].

Following the decline of the western Roman Empire, Pb production continued, but at a much reduced pace, as witnessed by sediments and peat bogs [75]. However, the atmospheric Pb flux did not reach background values: therefore, Pb continued to be mined and refined, especially in Asia [76]. Even at this time in Europe, Pb continued to be used for glass making and ceramic glazes [77]. Lead was needed for the cames of the stained glass windows of churches and cathedrals [53]; the abundance of these windows in Europe today is a testimony to the usefulness of metallic Pb as a construction material. Lead was also used to

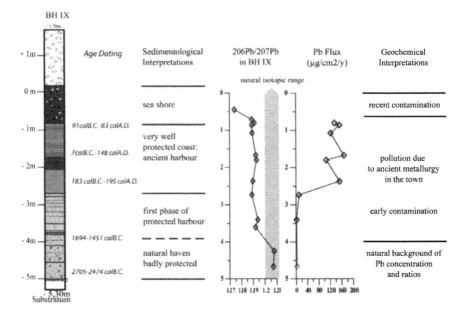

Figure 3   Pb contamination evidenced by Pb flux and Pb isotopes in sediments from the ancient harbor of the Phoenician city of Sidon [74].

make colored glass during this period, as it had been used for the previous 2500 years.

The Medieval period in Europe experienced a renaissance in Pb production, with the development of "silver mines" across the German-speaking part of central Europe, as well as the Vosges mountains of France and the Silesia region of Poland. The Harz Mountains, the Ore mountains, and the Black Forest of Germany were particularly important mining areas. Already in the 13th century, the Harz Mountains alone produced 500 ton of Pb per year. In fact, peat bogs from mining regions of Germany show that the extent and intensity of recent Pb contamination from industrialization, fossil fuel burning, and leaded gasoline use, are dwarfed by Pb contamination from the Medieval Period [78,79].

4.2.   Modern Uses

So many industrial and commercial uses have been found for Pb that it is difficult to list them all. The main uses of Pb which are most relevant to environmental contamination and human exposure are: lead-based indoor paints, lead water pipes, lead solder in cans for storing food, and leaded gasoline additives; these have either since been banned or are in the process of being banned. Action is also being taken with respect to lead glazes on pottery and ceramics, lead in ammunition and for weights in sport fishing, in certain plastics, and use of leaded crystal. In the EU, for example, Pb will be banned as a stabilizer in PVC by 2015. There are moves to replace Pb in ammunition used for hunting waterfowl; in Canada alone, several million waterfowl die annually from lead poisoning [80]. Replacements are being sought for sinkers used in sport fishing, and larger weights (used for sport fishing in deep waters) are being coated with plastic.

Since 1800, Pb use increased dramatically [81] with mining production reaching 2.8 million tonnes and total metal production of 6.7 million tonnes in 2003 [82]. With ~60% of the world's Pb obtained by recycling, it is the most re-used of all the metals [82]. Lead is mined worldwide, with the largest primary producers being Australia (19%), the US (13%), China (12%), Peru (8%), and Canada (6%). Some new Pb is also produced in Mexico, Germany, and France. The two most important Pb mining areas are Broken Hill (Australia) and southeast Missouri (USA) with cumulative outputs last century each exceeding 30 million tonnes [12].

The main use for Pb today (74%) is lead-acid batteries, especially for cars. However, in the recent past, up to 300,000 ton of Pb per year was used for gasoline additives alone [14]. Lead additives in gasoline, beginning with tetraethyl lead in 1923 and supplemented with tetramethyl lead in Europe starting in 1960, were incompatible with catalytic converters, and are gradually being phased out: maximum allowable Pb concentrations were introduced first, during the early 1970s, allowing a gradual decline in gasoline Pb concentrations, followed by the introduction of unleaded fuels, and finally by outright bans on leaded gasoline in most developed countries, starting with the main "automotive" countries: the US, Germany, and Japan. Unleaded gasoline is expected to be used exclusively in all western and nearly all eastern European countries as of 2005, according to the Aarhus Treaty signed in 1998 [83]. Many nations in Asia, South America, Africa, and the Middle East have not yet banned the use of leaded gasoline [84,85].

Lead is still used widely in building and construction (as sheet Pb for roofs, and for noise abatement and reducing vibrations, and in lead-clad steel); for lead paints used to protect steel from corrosion (red lead) and for marking highways (lead chromate); in pipes for handling corrosive gases and liquids; as a sheathing material for power cables; for casting; in decorative leaded crystal (up to 36% lead oxide), optical glassware (opthalmic and scientific applications); for glazing heat-resistant ceramics; for radiation shielding; in ammunition; as a stabilizer in plastic; in petroleum refining; as a waterproofing material; in food packaging; insecticides; for making alloys such as antimonial lead, bronze, brass, and pewter; for solders, especially in consumer and industrial electronics; and for lead weights (from the smallest used for sport fishing, through those used to balance automobile wheels to the largest used in building keels for yachts; for counterweights in lift trucks and cranes). Hair dyes containing 0.5% Pb acetate are still available, even in OECD countries. In Canada, approximately one-half of the costume jewellery sold contains Pb.

The diverse uses and applications are outlined here because many Pb-bearing consumer products end up in residential waste streams, and burning municipal solid waste (MSW) is an additional source of Pb to the environment. For example, the glass screens on TV sets and personal computers each contain several hundred grams of Pb added to protect the viewer from radiation. The US EPA estimates that 50,000 tonnes of Pb is added to the waste stream annually from consumer electronics alone.

5. EMISSIONS OF LEAD TO THE ENVIRONMENT

According to Pacyna and Pacyna [86], emissions of Pb to the global atmosphere have declined from 332,350 tonnes per year in 1983 to 119,259 tonnes per year in 1995. While this represents a tremendous reduction, the natural rates of atmospheric Pb emission estimated [87] are only 2600 ton per year, with 1200 ton from volcanic emissions and 1400 ton from soil dust. For comparison, the natural rates of atmospheric Pb deposition estimated using a peat core from a Swiss bog is only 4400 ton per year [88]. As a consequence, the emissions of Pb to the global atmosphere today are still on the order of 45 times the natural value estimated by Patterson and Settle [87] and 27 times the value estimated by Shotyk et al. [88]. Even compared to the higher estimate of natural Pb (12,000 ton/year) by Nriagu [89], the contemporary fluxes from anthropogenic sources are still an order of magnitude greater. These calculations apply to global scale emissions, so regional and local impacts must be even greater.

The single most important anthropogenic source of Pb to the global atmosphere remains vehicle emissions; this source alone exceeds the natural fluxes [87] by a factor of approximately 30 times. Other important sources of atmospheric Pb are non-ferrous metallurgy and fossil fuel combustion from stationary sources (mainly coal-burning) (Table 1). Comparing the 1995 data with 1983 data

Table 1 Predominant Anthropogenic Sources of Pb to the Global Atmosphere (tonnes per year)

Source	Emission	Natural emissions	Ratio anthropogenic/natural
Vehicles	88,739	2600	34.1
Non-ferrous metallurgy	14,815	2600	5.7
Fossil fuel combustion, stationary sources	11,690	2600	4.5
Iron and steel production	2,926	2600	1.1
Waste incineration	821	2600	0.3
Cement production	268	2600	0.1
TOTAL	119,259	2600	45.8

Source: Anthropogenic emissions for 1995 from Ref. [85], and natural emissions from Ref. [86].

shows strong declines in emissions from the production of primary and secondary Pb (6216 ton/year in 1995 vs. 21,450 ton/year in 1983), Cu (6271 ton/year in 1995 vs. 16,575 ton/year in 1983), and Zn (2123 vs. 8510 ton/year), as well as emissions from steel production (2926 vs. 7633 ton/year) and cement production (268 vs. 7129 ton/year). In contrast, emissions from stationary fossil fuel combustion actually increased over the same period, from 10,577 to 11,690 ton/year [86].

5.1. Lead in Natural vs. Anthropogenic Atmospheric Particles

While it is certainly useful to compare the magnitude of natural and anthropogenic fluxes of Pb to the atmosphere, and to quantify air Pb concentrations, there are a number of important differences in the nature of the Pb emitted. In pre-anthropogenic times, the majority of Pb in the air was supplied by atmospheric soil dust particles derived from rock weathering, and the Pb concentrations in these particles were more or less proportional to the abundance of Pb in crustal rocks [6,48]. Most of these Pb-bearing particles would have been comparatively large: on the order of 50 μm if they had been supplied by local soils, and perhaps as small as 5 μm if they had been supplied by long-range transport of Sahara dust [90]. As the Pb in these particles is hosted mainly in silicate minerals such as potassium feldspar [30,91] which has a very low solubility in aqueous solutions at atmospheric temperature and pressure, the bioavailability of this Pb is relatively low. Under natural conditions, therefore, atmospheric Pb was an insignificant source of Pb to humans [6]. In contrast, most of the anthropogenic Pb in the atmosphere is derived directly or indirectly from high temperature combustion processes such as leaded gasoline use, metallurgical processing, coal combustion, and refuse incineration. Lead emitted from these sources is released to the air in the form of sub-micron particles, with a median diameter of $\sim$0.5 μm [92]: these particles, with an average atmospheric residence time of $\sim$1 week, are not only amenable to long-range atmospheric transport (thousands of kilometers), but they are easily respirable. Moreover, the surfaces of these particles—where chemical reactions take place—are highly enriched in Pb, and this Pb has a relatively high solubility.

5.2. Atmospheric Lead from Alkyllead Fuel Additives

The chemistry of organolead compounds in the environment has been reviewed elsewhere [93–96], but a few salient aspects are summarized here. First, tetra-alkyllead compounds are volatile: Henry's Law constant of 4.7×10^4 and 6.9×10^4 (Pa m^3)/mol for tetramethyl and tetraethyllead, respectively, according to Wang et al. [97]. However, only a small fraction of the Pb leaving an automobile as exhaust is in this form, e.g., typically 0.1–10% [95]. In addition to the relatively low emission factor from leaded gasoline combustion, tetra-alkyllead compounds are rapidly decomposed by homogeneous gas phase reactions such as photolysis, reaction with ozone, triplet atomic oxygen, or hydroxyl radical [98,99] with half-lives of less than 10 h in summer and 40 h in winter. According to Harrison and Allen [94], the chemical cycle of alkyl lead compounds originating in gasoline lead additives can be summarized as

$$R_4Pb \longrightarrow R_3Pb^+ \longrightarrow R_2Pb^{2+} \longrightarrow (RPb^{3+})^* \longrightarrow Pb^{2+}$$

where ( )* indicates an unstable species. Based on these observations, it is reasonable to expect, therefore, that organolead will represent a small percentage of total Pb in atmospheric samples, and that ionic alkyllead species will dominate the inventory of organolead compounds; this is also what is observed in direct air measurements [99]. Second, washout factors (concentration ratio in rainwater compared to air) for Pb(II) is typically a factor of ten greater than that for tetraalkyllead compounds which predict a relative enrichment of alkyllead in an ageing air mass, as the alkyllead is less efficiently scavenged [94].

In the peat bog at Etang de la Gruère (EGR) at an elevation of 1005 m above sea level in Switzerland, not >0.02% of the total Pb in any one sample is in the form of alkyllead species [100]. In contrast, in snow and ice from Mont Blanc (4250 m a.s.l.), 0.1% of the total Pb is organic [96] and in Greenland snow and ice, 1% of the total lead is in the form of organolead species [101,102]. Compared to the peat bog at EGR, the total Pb concentrations in alpine snow and ice are typically 3–4 orders of magnitude lower, and in Greenland snow and ice 5–6 orders of magnitude lower. Thus, the relative importance of organolead in these archives increases with decreasing total Pb concentrations and decreasing proximity to the source; this most likely reflects the more efficient scavenging of ionic Pb species (greater washout), but other factors may be involved.

An elegant demonstration of the temporal changes in environmental fluxes of alkyllead concentrations is the study of trimethyl Pb in Châteauneuf-du-Pape wines: none was found in wine samples pre-dating the introduction of leaded gasoline, and peak concentrations were found in wine dating from 1978; since then, concentrations have declined markedly [103].

6. INPUTS AND FATE OF ANTHROPOGENIC LEAD IN THE BIOSPHERE

6.1. Lead Concentrations in Soils

Pioneering measurements of Pb concentrations in roadside plants from British Columbia, Canada [104] and Colorado, USA [105], provided some of the first indications of environmental Pb contamination, and isotopic studies by Chow [106] showed unambiguously that the anthropogenic Pb in surface soils originated from leaded gasoline. However, with respect to Pb in the soils of British Columbia, for example, it was difficult to unambiguously distinguish among natural enrichments, Pb due to geological mineralizations, contamination by lead arsenate pesticides used in agriculture, and gasoline Pb residues [104].

One approach to quantify the extent of anthropogenic enrichment in soils is to use a conservative, lithogenic element such as Zr, to take into account the degree to which Pb might have become naturally enriched within a vertical profile due to chemical weathering [107]. However, if there is no place left on Earth devoid of anthropogenic Pb [1], then the natural background concentrations of Pb in surface soil layers can only be determined using archived samples pre-dating industrialization, provided that such samples exist, or by mathematical modeling. In southern Sweden, Pb concentrations in the mor layer of forest soils are on the order of $40-100$ $\mu g/g$. However, using a combination of Pb concentrations, Pb isotope ratios, and an estimate of pre-anthropogenic rates of atmospheric Pb deposition using cores from ancient layers of peat bogs, Bindler et al. [108] have estimated that the natural concentration of Pb in the surface layers of these soils was 0.1 $\mu g/g$.

6.2. Cumulative Impact of Anthropogenic, Atmospheric Lead

In a mass balance study of Pb in soils from southern Germany, Dörr et al. [109] found that virtually all of the anthropogenic Pb could be found in the topmost 20 cm. Assuming that all of the gasoline Pb emitted in the former West Germany since 1950 ($\sim250 \times 10^3$ ton) was evenly distributed over the land surface, they estimated a total deposition of anthropogenic Pb of ~1 g/m^2. Measurements of Pb inventories in individual soil profiles ranged from 1.4 g/m^2 at a rural site to 10.8 g/m^2 at an urban site. For comparison, the cumulative mass of atmospheric, anthropogenic Pb (CAAPb) was also calculated in Switzerland, using peat cores from eight mires [110]. In Switzerland, CAAPb ranged from $1.0-10.0$ g/m^2, in good agreement with the values from southern Germany [109], and with the highest values from the south side of the Alps with direct exposure to the highly industrial region of northern Italy. In southern Sweden, Renberg et al. [111] estimated CAAPb at $2-3$ g/m^2. This range is in good agreement with our values from Stoby Mose in Denmark (3 g/m^2). A summary of our unpublished data concerning CAAPb in other European bogs is given in Table 2.

The bog at Bagno in Ukraine is west of the Carpathian Mountains, and directly exposed to the industrial regions of the former Eastern Europe, compared

Table 2 Cumulative Mass of Atmospheric, Anthropogenic Pb (CAAPb) in Different Parts of Europe

Name and location of site	CAAPb (g/m^2)
Loch Laxford, NW Scotland	0.9
Flech's Loch, Island of Foula, Shetland	1.9
Myrarnar, Faroe Islands	1.9
Staaby Mose, Denmark	3.1
Nebuga, Ukraine	0.9
Babyn Moh, Ukraine	0.9
Bagno, Ukraine	4.6

to the other bogs (Nebuga and Babyn Moh) from the same country. For comparison with the values from bogs in Europe, we have studied three peat bog profiles in southern Ontario, Canada [112]. The Sifton Bog in the city of London, Ontario, recorded 2.4 g/m^2 CAAPb, the Luther Bog in a rural area 1.6 g/m^2, and the Spruce Bog in Algonquin Park, a comparatively remote site, 1.0 g/m^2.

All of these sites show a minimum of 1 g/m^2. Extrapolating to the land area of the northern hemisphere ($\sim 100 \times 10^6$ km^2) suggests that the total cumulative burden of atmospheric, anthropogenic Pb to the continents is ~ 1 million tonnes. However, given that the annual consumption of gasoline lead exceeded 300,000 ton per year during the late 1960s and early 1970s, this estimate seems too low.

6.3. The Fate of Anthropogenic Lead in Soils

Understanding the fate of anthropogenic Pb in soils is vital to evaluating bioavailability, mobilization, and transport. In forest soils of the northeastern US, Miller and Friedland [113] suggested that Pb was being redistributed, but Wang et al. [114] concluded that Pb is efficiently retained. Dörr et al. [109] measured the distribution of ^{210}Pb, a naturally-occurring Pb isotope that is believed to be effectively independent of anthropogenic Pb emissions. Based on the distribution of this isotope, they concluded that Pb had penetrated no deeper than ~ 20 cm; they concluded, however, that the downward migration velocity of Pb was ~ 1 mm/year. Using the abundance of stable Pb isotopes (primarily ^{206}Pb, ^{207}Pb, and ^{208}Pb), Steinmann and Stille [115] concluded that Pb mobilization in soils is very limited. Studying the isotopic composition of Pb at a Medieval Pb smelting site in England, Whitehead et al. [116] estimated a Pb migration rate of ~ 8 mm/year. Bacon et al. [117] reported changes in Pb isotope ratios in deeper soil layers of some Scottish soils, but these may have been due to geological changes or lateral flow.

To help evaluate the fate of metals in soils, it can be helpful to separate total Pb into the compartments by which it is predominantly bound: exchangeable, organically-complexed, adsorbed to Fe and Mn oxides, associated with sulfide minerals, and the residual fraction which is primarily the "natural" silicate fraction [118,119]. This approach is especially useful when it is combined with the determination of the isotopic composition of Pb in these fractions. In soils from the Middle East, isotope analyses of the labile fractions were used [120] to show that there had been substantial penetration of anthropogenic Pb into deeper soil layers. However, in a comparable study of soils from central Europe, no such mobilization was found [121].

To summarize, the fate of any metal in soil is a response to a complex set of parameters including soil texture, mineralogy, pH and redox potential, hydraulic conductivity, abundance of organic matter and oxyhydroxides of Al, Fe, and Mn, in addition to climate, situation, and nature of the parent material [122–124]. As a result, it is not possible to make any general conclusions regarding the final fate of anthropogenic Pb in soils. In fact, the fate of Pb in soils will probably have to be evaluated on a soil-by-soil basis. However, the data shown in Fig. 1 clearly show that acidic soils will certainly have the greatest potential for Pb migration. The Pb isotope data from acidic soils in southern Sweden suggest that Pb is migrating and, along with Al and Fe, accumulating in the B horizon of podzols [111].

6.4.   Lead Concentrations in Solution

Because Pb has to be transferred from the solid phase to the aqueous phase before it can be taken up by plants and aquatic organisms, the concentration of Pb in natural waters can be a sensitive indicator of the potential for Pb to become biologically "available". To estimate how much of this Pb might be transferred to the aqueous phase, solutions collected from ten Swiss forest soils were measured for concentrations of total dissolved Pb (Peter Blaser, unpublished data) and found to range from <1 to >60 µg/L. Five of the profiles were acidic (pH 4–5) and these contained 20–60 µg/L Pb in the aqueous phase of the surface layers. These layers, however, are the most critical ecologically, as they represent the biologically active zone of acidic forest soils: this is also the zone which has been most impacted by anthropogenic Pb.

For comparison, with the Pb concentrations of the topmost layers, the solutions collected from the deeper soil horizons all had <1 µg/L and were below the LOD provided by Q-ICP-MS. Hirao and Patterson [125] reported 0.015 µg/L in stream runoff in the Sierra Nevada Mountains, but were only able to accomplish this because of the extreme cautions taken to avoid contaminating their samples with industrial lead, and because they used ID-MS to measure the Pb concentrations. The Sierra Nevada watershed is characterized by granitic rocks, and most of the Pb in uncontaminated streamwater is probably derived from the chemical weathering of biotite and potassium feldspar [126]. Therefore, their measured Pb concentration (15 parts per trillion) is probably not an unreasonable estimate of the natural Pb concentration in uncontaminated waters from crystalline terrains. Thus, any study of soil solution Pb has to be capable of reliably measuring Pb in this concentration range.

The problem of Pb determination at extremely low concentrations in seawater was discussed at length by Patterson and Settle [127], in polar snow and ice by Boutron [37], and in freshwaters by Nriagu [38]. Using clean lab procedures and double-focusing sector field ICP-MS, Krachler et al. [36] have shown that it is possible to now reliably measure Pb concentrations as low as 60 pg/L which is nearly three orders of magnitude less than the concentration of Pb in stream waters in [125]. Using this comparatively new approach, therefore, it is now feasible to begin detailed studies of anthropogenic Pb in soil solutions.

7.   TEMPORAL TRENDS IN ATMOSPHERIC LEAD DEPOSITION

7.1.   Lead in Sediments

Many studies have used lake sediments to document changes in the timing and intensity of Pb contamination, including cores collected in the US [128,129], the Canadian Arctic [130], Spain [131], Switzerland [132,133], Scotland [134,135], and the Middle East [136]. One of the main challenges has been to distinguish between atmospheric and non-atmospheric sources [137], but a mathematical approach to solving this problem using precise measurements of Pb isotope ratios in sediments of Lake Constance has recently been developed [138].

7.2. Lead in Bryophytes

Bryophytes such as mosses have no roots, but instead rely exclusively upon atmospheric inputs for nutrient elements. They also receive Pb from the air, and retain it efficiently, allowing moss analyses to be used as a monitoring tool for studying changes in atmospheric metal deposition. The isotopic composition of Pb in forest moss species such as *Hylocomium splendens* [139] and *Polytrichum formosum* [140] which have been collected annually during the past decades, allows a reconstruction of the predominant sources of anthropogenic Pb and their temporal variation. Lichens can be used for the same purposes [141], but because they grow on a rock substrate, more care might be needed to separate atmospheric from lithogenic inputs. Herbarium samples of *Sphagnum* moss which had been collected from peat bogs in Switzerland since 1867 [142] and Scotland since 1830 [143] have documented temporal changes in the isotopic composition of atmospheric Pb over longer time periods. The absolute chronology which herbarium samples provide is a great advantage, compared to archives such as sediments or peat, which require radiometric age dating. Herbage samples from the Rothamsted experimental station in England have been collected annually since 1856, have also provided a valuable record of atmospheric Pb deposition [144].

7.3. Lead in Tree Rings and Bark Pockets

The use of tree rings as archives of atmospheric Pb contamination has been seriously questioned [145,146], although Aberg et al. [147] reported more success. Bark pockets which become trapped in a tree as two branches grow together, show much promise for reconstructing changes in the isotopic composition of Pb in atmospheric aerosols [148,149].

7.4. Peat Bog Archives

Ombrotrophic bogs are excellent archives of atmospheric Pb deposition because they receive Pb only from the air, and because they efficiently retain this metal despite the low pH of the waters (pH 4), the abundance of natural, complex-forming organic acids and the seasonal variations in redox potential [48]. Bogs are probably the best continental archives of atmospheric Pb deposition and they are receiving increasing attention for this purpose [74,78,79,112,150–164]. With bogs commonly found in the temperate zone of both hemispheres, they offer the promise of high-resolution reconstructions of atmospheric Pb deposition worldwide.

A summary of atmospheric Pb deposition in Switzerland during the past 14,500 years is shown in Fig. 4 [165]. This graph shows that atmospheric Pb deposition in central Europe has been dominated by anthropogenic Pb continuously for 3000 years. Prior to this, Pb concentrations were proportional to those of Sc, which suggests that soil dust was the main factor regulating atmospheric Pb deposition. The background rate of atmospheric Pb deposition $(10 \ \mu g/m^2/y)$ was only found in peat samples from 6000 to 9000 years old; peats from $\sim$6000–3000 years old had elevated Pb concentrations because forest clearances and soil tillage for agriculture had already increased the dust flux. Peat samples from the early part of the Holocene ($\sim$13,000 until 9000 years ago) were more radiogenic which denotes a change in the predominant sources of atmospheric soil dust.

7.5. Relative Importance of Gasoline Lead vs. Other Sources of Industrial Lead

Using a peat core from Denmark which was cut into 1 cm slices and age dated using both ^{210}Pb and ^{14}C (atmospheric bomb pulse curve), Pb was separated into lithogenic and anthropogenic components using Ti as the reference element, and the isotopic composition of the Pb determined using TIMS [166]. The data show that the maximum concentration of anthropogenic Pb (1954) pre-dates the minimum in $^{206}Pb/^{207}Pb$ (1979) by more than two decades (Fig. 5a). In other words, the maximum impact of gasoline Pb (revealed by the isotopic composition) occurred $\sim$25 years after the peak in Pb contamination (indicated by the EF). Comparing the isotopic composition of the peat samples with gasoline leads [167] and British coals [49] (Fig. 5b) suggests that coal burning was the predominant source of this Pb contamination. Clearly, even though gasoline Pb has certainly been an important source of anthropogenic Pb to the atmosphere during the past decades, other sources of industrial Pb were even more important.

7.6. The Cumulative Input of Anthropogenic Lead

To emphasize this point, the cumulative input of anthropogenic Pb to the peat bog at EGR is shown in Fig. 6. This graphic shows that 10% of the anthropogenic Pb was already in the bog at the start of the Medieval period and >20% before the start of the Industrial Revolution. Prior to the introduction of leaded gasoline, $\sim$75% of the anthropogenic Pb inventory was already in the bog [110]. This graphic is helpful in many ways, as it indicates that reducing emissions of anthropogenic Pb by eliminating leaded gasoline is certainly not going to eliminate industrial emissions altogether. In fact, many other sources of atmospheric Pb are also important, and also have to been reduced.

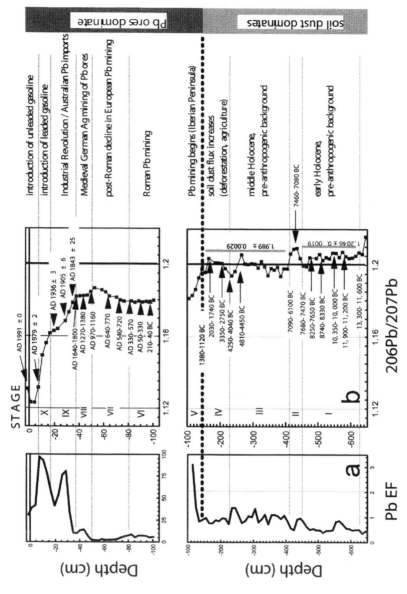

Figure 4 Pb profile in the Swiss peat bog of EGR [165]. (a) Pb enrichment factor calculated as the ratio of Pb/Sc in the peat normalized to background value. (b) Isotopic composition of Pb summarized as $^{206}Pb/^{207}Pb$.

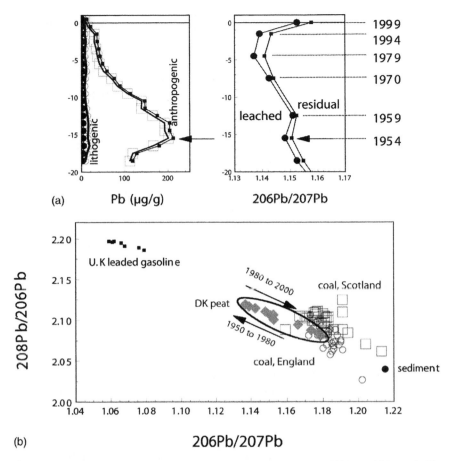

(a) Pb (µg/g)   206Pb/207Pb

(b)   206Pb/207Pb

Figure 5 Pb profile in the Danish peat bog of Storelung Mose [48]. (a) Lithogenic Pb, anthropogenic Pb vs. depth (see Ref. [48] for calculation details as well as the isotopic composition summarized by the ^{206}Pb/^{207}Pb ratio). (b) Diagram ^{206}Pb/^{207}Pb vs. ^{208}Pb/^{206}Pb for the leached fraction of the Danish peat samples, also shown UK coal [49], UK leaded gasoline in 1997 [167] and the isotopic composition of a pre-anthropogenic sediment [165].

7.7. Lead in Polar Snow and Ice

Ever since the pioneering paper by Murozumi [168] of Pb in Greenland snow and ice, these kinds of samples have provided much important information about temporal trends in atmospheric Pb deposition not only in Greenland [169–172] but also from the European Alps [173] and even Antarctica [174]. All of this work, however, was done using discrete samples, which had to be decontaminated individually: this not only introduces the risk of contamination, but is also expensive and time consuming. While these studies certainly documented the impact of gasoline additives and the subsequent declines in atmospheric Pb emissions,

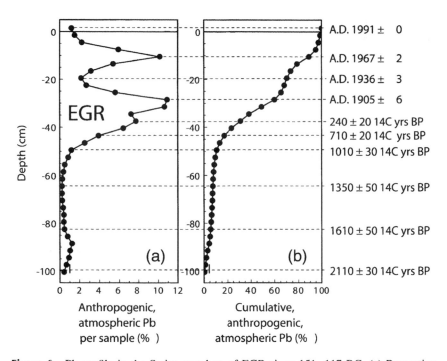

Figure 6   Pb profile in the Swiss peat bog of EGR since 151–117 BC. (a) Proportion of anthropogenic atmospheric Pb for each sample compared to the complete profile. (b) Cumulative anthropogenic atmospheric Pb.

which followed the gradual phasing out of leaded gasoline, they provided very little information about Pb emissions from the early part of the 20th century. A notable recent development has been the direct coupling of an ice melting head to an ICP-MS to allow continuous measurements of Pb on-line, offering unprecedented temporal resolution (Fig. 7). This new development is important, especially for ice which has accumulated during the industrial period, because it shows that the intensity and extent of Pb contamination prior to the introduction of leaded gasoline is comparable to that found afterward [175].

7.8.   Lead in Atmospheric Aerosols Today

Many published studies of the isotopic composition of aerosols have been summarized by Bollhöfer and Rosman [176–179] who provide precise Pb isotope data for aerosols collected at 80 sites in the northern hemisphere. While there have been dramatic declines in air Pb concentrations in many developed countries, and changes in the isotopic composition of the Pb, significant

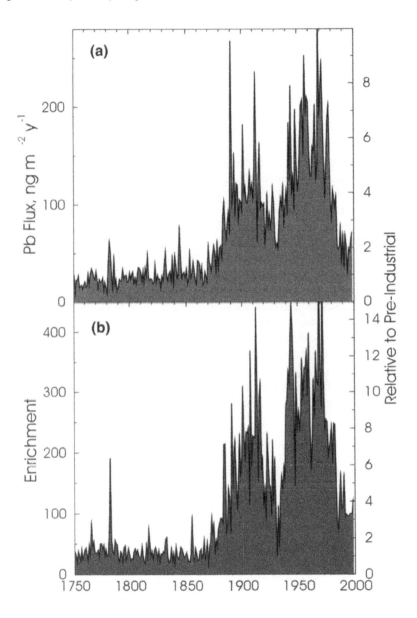

Figure 7 Annual average Pb flux (a) and crustal enrichment of Pb (b) in central Greenland. Reproduced by permission of the American Geophysical Union from Ref. [175].

atmospheric Pb contamination remains. Each of the papers cited here and describing the isotopic composition of Pb in atmospheric aerosols, as well as sediments, bryophytes, peat bogs, and polar snow and ice, shows that even the most recent samples are still contaminated by anthropogenic Pb. Aerosols today are typically enriched in Pb by 100–1000 times, relative to the abundance of Pb in crustal rocks [180]. As noted out by Bacon [181], a substantial portion of anthropogenic Pb now being deposited from the atmosphere has its origin other than in gasoline.

As an example, we show the isotopic composition of Pb in snow samples collected in the Black Forest of southern Germany using clean lab techniques and measured using ICP-SMS by Krachler et al. [57]. Compared to the isotopic composition of atmospheric dust dating from pre-anthropogenic times, the snow samples are much less radiogenic (Fig. 8) which clearly shows that industrial sources of Pb dominate the atmospheric Pb flux. For comparison, the isotopic composition of Pb from incinerators [167,182] is also shown, and these overlap with the values from the snow collected in 2003.

A second example is provided by the isotopic composition of Pb in wines [183]. Again, compared to the natural isotope signature of Pb in atmospheric dust or soils, the values for the wines are much lower which shows unambiguously that the wines are contaminated with industrial Pb. The possible sources of this Pb are many: perhaps the soil solutions were contaminated because of past use of Pb arsenate as a pesticide; the grapes may have been contaminated by Pb-rich atmospheric aerosols; processing of the grapes, fermentation of the

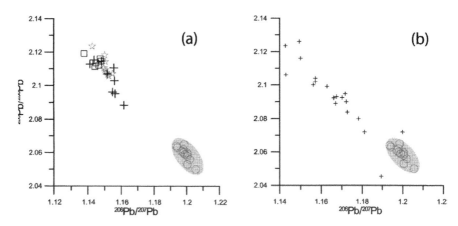

Figure 8   (a) Diagram ^{206}Pb/^{207}Pb vs. ^{208}Pb/^{206}Pb for snow samples from Black Forest (+), Germany, sampled during Winter 2002. For comparison are also shown the isotopic composition of natural dust (O) [48], isotopic composition of incinerator dust (☆) [167,182], and aerosols composition from Southern Germany (□) [179]. (b) Diagram ^{206}Pb/^{207}Pb vs. ^{208}Pb/^{206}Pb for wine samples (+) [183], also shown in the shaded zone "natural dust" (O) isotopic composition.

wines, and bottling of the final product are all possible sources of Pb contamination from contact with metal or glass surfaces.

8. ENVIRONMENTAL LEAD EXPOSURE AND HUMAN HEALTH

8.1. Blood Lead Levels (BLLs) and Their Significance

Lead is primarily absorbed via respiration and ingestion and carried throughout the body by the blood, which enables Pb to enter all tissues. For this reason, measurement of BLLs is the most common method for establishing degree of exposure in humans, and is usually reported in units of $\mu g/dL$ [184]. Environmental lead poisoning can affect both adults and children, but the greatest concern is for children as they experience symptoms at significantly lower BLL than do adults. Whereas many of the symptoms experienced by adults are reversed when exposure is ceased, children tend to develop permanent developmental and neurological problems when exposed chronically to Pb [185]. Unfortunately, once elevated BLL have been detected, it is too late to prevent the deleterious effects of Pb on the brain. Children are particularly sensitive because they absorb a greater proportion of ingested Pb from the gastrointestinal tract than do adults and retain more of that Pb. In addition, a greater proportion of systemically circulating lead gains access to the brain of children, especially those 5 years old or younger, than of adults. Finally, the developing nervous system is far more vulnerable to the toxic effects of Pb than the mature brain [184].

Low level lead exposure has serious deleterious and irreversible effects on brain function, such as lowered intelligence, and diminishes school performance, especially from exposures that occur in early life; hearing deficits and growth retardation have also been observed [186]. Prior to 1970, BLL $>60 \ \mu g/dL$ were considered critical for children. Today, subclinical lead toxicity is defined as a BLL of $10 \ \mu g/dL$ (100 parts per billion) or higher, and is estimated to affect one out of 20 children in the US [186]. The recent findings published by Canfield et al. [187], however, show that BLL, even those $<10 \ \mu g/dL$, are inversely associated with the IQ scores of children between the ages of three and five. Additional health effects in children with average BLL of $3 \ \mu g/dL$, including decreased height and delayed puberty, has been reported [188]. To summarize, D. O. Carpenter, Director of the Institute of Health and Environment, State University of New York, was recently quoted as saying that "BLL as low as $1 \ \mu g/dL$ are associated with harmful effects on children's learning and behavior. There may be no lower threshold for some of the diverse effects of lead in children" (Chem. Eng. News, April 7, 2003). Other recent developments in the area of childhood Pb neurotoxicity have been reviewed elsewhere [184].

To help put these concentration values into perspective, Patterson [6] estimated that the natural concentration of Pb in blood is $0.25 \ \mu g/dL$. A more recent estimate, based on the Pb concentrations of human bones from pre-industrial times and the relationship between bone Pb and blood Pb concentrations, is

0.016 µg/dL [189]; these values are 40 and 600 times, respectively, lower than the blood Pb concentration (10 µg/dL) currently considered critical for children.

During the past two decades, average BLLs in the US have declined by >90%. However, even today, approximately 890,000 pre-school children in the US are estimated to have BLLs exceeding 10 µg/dL. In some cities in the northeastern region of the US, more than one-third of pre-school children have BLLs >10 µg/dL [186]. To help put these values in perspective, in the Province of Ontario, Canada, in 1984, the average BLL in children was 12.0 ± 4.4 µg/dL [190]. For comparison with contemporary US values, a recent study of 421 school children in Jamaica showed an average of 9.2 µg/dL in rural areas and 16.6 µg/dL in urban areas [191]; at the site of a former lead ore processing plant the average concentration was 35 µg/dL. In the Cape Province of South Africa, >90% of the children showed BLL >10 µg/dL [192]. Today, lead poisoning is considered one of the most common pediatric health problems in the US [185].

8.2. Mechanism of Lead Poisoning

Lead enters into brain cells, both neurons and glia, by channels that under normal conditions allow the passage of Ca^{2+}. Lead enters (and damages) mitochondria via the cellular mechanism that normally functions to bring calcium into this organelle [184]. The molecular mechanisms of Pb toxicity are incompletely understood, but it is known that Pb^{2+} can replace Zn^{2+} and Ca^{2+} at their binding sites in various proteins, thereby altering their structure and function [185]. The Zn enzyme δ-aminolevulinic acid dehydratase (ALAD) has received the most attention, as this is the only one known to be inhibited by lead. The mammalian form of ALAD contains a unique catalytic zinc-binding site with three cysteine residues. Lead, being a much larger and more polarizable cation than that of Zn forms much more stable complexes with these sulfur-bearing ligands and can easily replace Zn. With respect to Ca, Pb interferes with the ability of Ca to trigger exocytosis of neurotransmitters in neuronal cells, suggesting that Pb might generally target proteins involved in Ca-mediated signal transduction. Here, Pb promotes phospholipid binding at lower concentrations than does Ca, suggesting that Pb binds to the Ca site more tightly than does Ca itself [185].

8.3. Predominant Sources of Lead Exposure

During the last decades of the 20th century, thanks largely to the pioneering investigations of Patterson and co-workers at the California Institute of Technology, it was unambiguously established that aerosols in both rural and remote locations are contaminated by industrial Pb, mainly from the extensive use of gasoline lead additives [193]. These findings were obtained after years of careful study using soil, plant, and animal samples from the rural Sierra Nevada Mountains of California [125,194–196], seawater from remote marine locations such as the mid-Atlantic [197], Bermuda [198], and the North Pacific [199]. Even snow and ice samples from glaciers in Greenland exhibit Pb concentrations which are up to 400 times higher than deeper, older ice layers dating from pre-anthropogenic times [168]. Between 1923 and 1986, 7 million ton of gasoline lead additives were consumed in the US alone. The complete history of leaded gasoline use since its introduction in 1923, has been reviewed elsewhere [14,200,201]. According to the United Nations, ~20% of the gasoline used worldwide today still contains lead additives.

Of particular interest here is the strong correlation in the temporal trend between BLLs and the concentration of lead in gasoline [202]. As leaded gasoline is replaced in developed countries by unleaded fuels, air Pb concentrations have been declining and BLLs along with it. Despite this, in the US, however, paint appears to be the single most important source of childhood lead poisoning. Children with BLLs >55 μg/dL are more likely to have paint chips that are observable in abdominal radiographs, and most pre-school children with BLLs >25 μg/dL are reported to have put paint chips in their mouths [203]. With respect to children with BLLs between 25 and 10 μg/dL, house dust contaminated with lead from deteriorating paint, and from lead-contaminated soil tracked in from outdoors, is the major source of Pb exposure [203]. More than 95% of US children who have elevated BLLs, fall within this range [186].

The lead problem of the inner cities in the US has been studied in detail by Mielke [204] who has found that >50% of the children in some areas of New Orleans and Philadelphia have BLLs above the current guideline of 10 μg/dL. As pointed out [204], soils in urban areas have become so highly contaminated with industrial Pb that they are a "giant reservoir" of Pb-rich dust particles. In the inner city areas of New Orleans, the geographic distribution of BLLs in children strongly resemble the abundance of Pb in soils [204].

There have been several major sources of Pb contamination to urban soils. Leaded gasoline consumption is the most obvious one, with many major cities having soil Pb concentrations on the order of several thousand μg/g [205]. However, as these soils were probably already contaminated to some extent with Pb from other sources even before the introduction of leaded gasoline, with industrial emissions, disposal of consumer products (e.g., paint and plumbing), coal combustion and waste incineration all contributing to Pb contamination in urban areas [204]. Referring again to the seminal work by Patterson [6], the ablated and weathered residues from applying 3.4 million ton of leaded paint in the US between 1925 and 1965 are sufficient to have increased the Pb concentrations in the top 5 cm of urban soils from the natural concentration of 15 mg/kg to 600 mg/kg. Even soils in rural areas may be contaminated with Pb, especially agricultural soils used for growing fruit, due to the past use of lead arsenate as a pesticide; at the peak of its use, several thousand tonnes was used in the US alone [6].

8.4. Other Sources of Lead Exposure

Human exposure to environmental Pb occurs from air, soil, household dust, food, drinking water, and various consumer products. In an uncontaminated environment, human exposure has been estimated as follows: ingesting food, 20 μg Pb/day; drinking water, 0.5 μg Pb/day; breathing air, 0.01 μg Pb/day [6]. During the 1960s in the US, average body burdens were approximately 100 times greater than this.

In the US, Pb was used in residential paint from 1884 to 1978; it has been banned for household use since 1978, but is still used in specialty paints. Moreover, although the use of leaded paint was banned, there was no government mandate to have lead paints removed from interior surfaces where it already existed. As a result, there may be more than 20 million homes in the US with "significant lead-based paint hazards" [184], and two million homes in the Province of Ontario, Canada, in the same category [80]. In addition, inappropriate methods of clean up such as sandblasting of painted surfaces, can exacerbate the problem [206]. In Canada, the Hazardous Substances Act has prohibited the use of lead-based paints in interior consumer paints, and paints for children's toys and furniture, since 1975 [80]. However, according to Health Canada (1999), interior paints are still allowed to contain up to 0.5% Pb. For comparison, exterior housepaints removed from houses in Toronto were found to contain up to 33% Pb.

Lead solder was used to seal canned foods in the US until 1995. At its peak, several thousand tonnes of Pb per year was consumed for this application [6]. In Canada, use of lead-solder cans has decreased by 99%, but some canned foods imported into Canada may contain lead solder.

Leaded gasoline was banned in the US in 1986, but lead additives are still used in racing fuels, as well as fuels for watercraft, light aircraft, and farm machinery [204]. In Canada, unleaded gasoline was introduced in 1975, and leaded gasoline was banned for use in motor vehicles in 1990. In Ontario, for example, ambient air Pb concentrations declined from 0.3 to 0.01 μg/m^3 between 1980 and 1990 [190]. For comparison, with these air Pb concentrations, Patterson [6] calculated natural air Pb concentration of 0.0005 μg/m^3 which is a factor of twenty lower than the lower LOD for air Pb employed by the OMEE.

Concerns remain about dietary Pb exposure from lead glazes on pottery and ceramics, and from leaded crystal. Lead exposure from surfaces used to contain food or drink (water pipes, lead solder, glazes on ceramics and pottery, and leaded crystal), is especially of concern when the foods or drinks they contain are acidic. In Canada, glazes currently may contain not >7 mg/kg Pb, but this value is being revised down to 2.5 or 0.5 mg/kg. Leaded crystal (containing up to 36% PbO) rapidly releases Pb to acidic solutions, with wine Pb concentrations increasing from 89 to 3518 μg/L after 4 months storage [207]. Even in the absence of leaded crystal glasses and decanters, the concentrations in wine are an ongoing concern. In a recent Danish study, 15 out of 50 wine bottles from EU countries tested exceeded the allowable limit for Pb in glass (250 mg/kg Pb). Rosman et al. [208] measured Pb in French wines from 1950 to 1980; the concentrations range from 78 to 227 μg/L and the Pb isotope data showed that all were contaminated by industrial Pb.

The maximum allowable concentration (MAC) of Pb in drinking water in Europe is currently 50 μg/L (which corresponds to the Pb concentration for water in contact with leaded pipes estimated by Patterson [6]), but this will be reduced to 10 μg/L by 2013. The MAC for Pb in drinking water in Canada today is 10 μg/L. The WHO recommended MAC had been 50 μg/L, but this was reduced in 1995 to 10 μg/L.

9. SUMMARY AND CONCLUSIONS

Lead has no biological function and is one of the most toxic metals. At the same time, it is one of the most useful, and perhaps no other metal has found such a wide range of industrial applications. It has been used extensively since Antiquity, which is when environmental Pb contamination began. With respect to contamination since industrialization, peat bogs and polar ice show that coal combustion and industrial emissions were as important during the first half of the 20th century as gasoline lead was during the second half.

Air Pb concentrations have generally declined since the introduction of pollution control technologies and the gradual elimination of leaded gasoline additives. However, all of the most recent published studies of the isotopic composition of Pb in aerosols and archival samples show that anthropogenic sources continue to dominate the atmospheric Pb flux by a considerable margin. The health effects of childhood Pb exposure is a growing concern, as deleterious effects are seen at BLL well below those currently believed to be safe, and safe levels are one or two orders of magnitude above the estimated natural values.

Mobilization of Pb-rich particles from highly contaminated soils in urban areas is an on-going health concern for many large cities. Even in areas far removed from industrial emission sources, Pb concentrations in the surface soil layers are far above their natural concentration range. In acidic forest soils, Pb concentrations are not only elevated in the biologically active zone, but also in their corresponding pore fluids. Accurate and precise measurements of the isotopic composition of Pb employing appropriate clean lab protocols, will continue to advance our understanding of the fate of Pb in the environment and its impact on human and ecosystem health.

ACKNOWLEDGMENTS

Thanks from W.S. to Andriy Cheburkin for building the EMMA and OLIVIA XRF spectrometers, countless Pb measurements, many years of fruitful discussions, and on-going excitement with ^{210}Pb; Bernd Kober for his insight into Pb isotope systematics and for keeping Betsy alive, and Michael Krachler for bringing new meaning to "trace" analysis of Pb in environmental samples. Special thanks to B. Haas for patience and understanding, and the usual message.

ABBREVIATIONS

AAS	atomic absorption spectrometry
ALAD	δ-aminolevulinic acid dehydratase
BC/AD	before Christ/Annum Domini (after Christ): in the text and the figures this means always calendar years
BLL	blood lead level
CAAPb	cumulative mass of atmospheric anthropogenic lead
EF	enrichment factor
EGR	Etang de la Gruère, Switzerland
Eh	standard potential
EPA	Environmental Protection Agency
EU	European Union
ICP-AES	inductively-coupled plasma-atomic emission spectrometry
ICP-SMS	inductively coupled plasma-sector field mass spectrometry
ID-MS	isotope dilution mass spectrometry
LA-ICP-MS	laser ablation inductively coupled plasma mass spectrometry
LEAFS	laser-excited atomic fluorescence spectroscopy
LOD	limit of detection
MAC	maximum allowable concentration
MC-ICP-MS	multi-collector inductively coupled plasma mass spectrometry
MSW	municipal solid waste
OECD	Organization for Economic Cooperation and Development
OMEE	Ontario Ministry of Energy and Environment
PIXE	proton induced X-ray emission spectrometry
PVC	polyvinyl chloride
Q-ICP-MS	inductively-coupled plasma mass spectrometry with quadrupole mass analyzer
RSD	relative standard deviation
SHRIMP	sensitive high resolution ion microprobe
SIMS	secondary ionization mass spectrometry
TIMS	thermal ionization mass spectrometry
WHO	World Health Organization
XRF	X-ray fluorescence spectrometry

REFERENCES

1. Jaworski JF. In: Hutchinson TC, Meema KM, eds. Lead, Mercury, Cadmium and Arsenic in the Environment. SCOPE. Chichester: John Wiley and Sons, 1987:107–117.
2. Harn OC. Lead, the Precious Metal. Johnatan Cape, 1924:323.
3. Nriagu J. J Chem Educ 1985; 62:668–674.
4. Lasdown R, Yule W. Lead Toxicity. History and Environmental Impact. Baltimore: John Hopkins University Press, 1986:286.

5. Ewers U, Schlipkter HW. In: Merian E, ed. Metals and Their Compounds in the Environment. Weinheim: VCH, 1991:971–1014.
6. Patterson CC. Arch Environ Health 1965; 11:344–360.
7. Reuer MK, Weiss DJ. Philos Trans Ser A Math Phys Eng Sci 2002; 360:2889–2904.
8. Russell RD, Farquahr RM. Lead Isotopes in Geology. New York: Interscience, 1960:243.
9. Doe RB. Lead Isotopes. Berlin: Springer Verlag, 1970:137.
10. Faure G. Principles of Isotope Geology. 2d ed. Chichester: John Wiley and Sons, 1986:589.
11. Dickin AP. Radiogenic Isotope Geology. Cambridge: Cambridge University Press, 1995:490.
12. Sangster DF, Outridge PM, Davis WJ. Environ Rev 2000; 8:115–147.
13. Robbins JA. In: Nriagu JO, ed. The Biogeochemistry of Lead in the Environment. Amsterdam: Elsevier/North Holland Biomedical Press, 1978:285–393.
14. Nriagu JO. Sci Total Environ 1990; 92:13–28.
15. Nriagu JO, ed. Biogeochemistry of Lead in the Environment, Vol. 1a. Ecological Cycles Vol. 1b. Biological Effects. Amsterdam: Elsevier/North Holland Biomedical Press, 1978.
16. Boggess WR, Wixson BG, eds. Lead in the Environment. Washington: National Science Foundation, 1979:272.
17. Committee on Lead in the Human Environment. Lead in the Human Environment. Washington DC: National Academy of Sciences, 1980:525.
18. Harrison RM, Laxen DPH. Lead Pollution, Causes and Control. New York: Chapman and Hall, 1981:168.
19. Förstner U, Wittmann GTW. Metal Pollution in the Aquatic Environment. 2d ed. Berlin: Springer Verlag, 1981:489.
20. Salomons W, Förstner U. Metals in the Hydrocycle. Berlin: Springer Verlag, 1984:346.
21. Förstner U. Contaminated Sediments. Heidelberg: Springer Verlag, 1989:157.
22. Kabata-Pendias A, Pendias HK. Trace Elements in Soil and Plants. New York: CRC Press, 1984:315.
23. Adriano DC. Trace Elements in the Terrestrial Environment: Biogeochemistry, Bioavailibility, and Risks of Metals. Vol. 2. Berlin: Springer Verlag, 2001:866.
24. Bodek I, Lyman WJ, Reehl WF, Rosenblatt DH, eds. Environmental Inorganic Chemistry. Properties, Processes and Estimations Methods. New York: Pergamon Press, 1988.
25. Lyndsay WL. Chemical Equilibria in Soils. Chichester: John Wiley and Sons, 1979:449.
26. Wedepohl KH. Geochim Cosmochim Acta 1995; 59:1217–1232.
27. Wampler JM. In: Fairbridge RW, ed. Encyclopedia of Geochemistry and Environmental Sciences. Dowden, New York: Hutchinson, and Ross, 1972:642–645.
28. Long LE. In: Fairbridge RW, ed. Encyclopedia of Geochemistry and Environmental Sciences. Dowden, NY: Hutchinson, and Ross, 1972:646–648.
29. Goldschmidt VM. Geochemistry. Oxford: Clarendon Press, 1954.
30. Wedepohl KH. In: Wedepohl KH, ed. Handbook of Geochemistry. Berlin: Springer Verlag, 1969:82C81–82.
31. Cheburkin AK, Shotyk W. Fresenius J Anal Chem 1996; 354:688–691.
32. Cheburkin A, Shotyk W. X-Ray Spectrom 1999; 28:145–148.

33. Prange A, Kramer K, Reus U. Spectrochim Acta Part B 1991; 46:1385–1393.
34. Cheam V, Lechner J, Sekerka I, Desrosiers R. J Anal At Spectrom 1994; 9:315–320.
35. Cheam V, Lechner J, Sekerka I, Desrosiers R, Nriagu J, Lawson G. Anal Chim Acta 1992; 269:129–136.
36. Krachler M, Zheng J, Fisher DA, Shotyk W. J Anal At Spectrom 2004; 5510–5517.
37. Boutron CF. Fresenius J Anal Chem 1990; 337:482–491.
38. Nriagu JO, Lawson G, Wong HK, Azcue JM. J Great Lakes Res 1993; 19:175–182.
39. Hoskin PWO, Wysoczanski RJ. J Anal At Spectrom 1998; 13:597–601.
40. Lazareth CE, Willenz P, Navez J, Keppens E, Dehairs F, André L. Geology 2000; 28:515–518.
41. Narewski U, Werner G, Schulz H, Vogt C. Fresenius J Anal Chem 2000; 366:167–170.
42. Reinhardt H, Kriews M, Miller H, Lüdke C, Hoffman E, Skole J. Anal Bioanal Chem 2003; 375:1265–1275.
43. Watmough SA, Hutchinson TC, Douglas Evans R. Environ Sci Technol 1997; 31:114–118.
44. Rauch S, Morrison GM, Motelica-Heino M, Donard OFX, Muris M. Environ Sci Technol 2000; 34:3119–3123.
45. Kramers JD, Tolstikhin ID. Chem Geol 1997; 139:75–110.
46. Hamelin B, Grousset F, Sholkovitz ER. Geochim Cosmochim Acta 1990; 54:37–47.
47. Sun SS. Philos Trans R Soc Lond A 1980; 297:409–445.
48. Shotyk W. Geochim Cosmochim Acta 2001; 65:2337–2360.
49. Farmer JG, Eades LJ, Graham MC. Environ Geochem Health 1999; 21:257–272.
50. Bacon JR. J Environ Monit 2002; 4:291–299.
51. Hurst R, Davies TE, Chinn BD. Environ Sci Technol 1996; 30:304A–307A.
52. Parkinson GS, Catchpole WM. J Inst Pet 1973; 59:59–65.
53. Wedepohl KH, Baumann A. Miner Deposita 1997; 32:292–295.
54. Thirlwall MF. Chem Geol 2000; 163:299–322.
55. Woolard D, Franks R, Smith DR. J Anal At Spectrom 1998; 13:1015–1019.
56. Prohaska T, Watkins M, Latkoczy C, Wenzel WW, Stingeder G. J Anal At Spectrom 2000; 15:365–369.
57. Krachler M, Le Roux G, Kober B, Shotyk W. J Anal At Spectrom 2004; 19:354–361.
58. Halliday AN, Lee D-C, Christiansen JN, Rehkämper M, Yi W, Luo X, Hall CM, Ballentine CJ, Pettke T, Stirling C. Geochim Cosmochim Acta 1998; 62:919–940.
59. Rehkaemper M, Mezger K. J Anal At Spectrom 2000; 15:1451–1460.
60. Weiss D, Kober B, Dolgopolova A, Gallagher K, Spiro B, Le Roux G, Mason TFD, Kylander M, Coles BJ. Int J Mass Spectrom 2004; 232:205–215.
61. Stern RA, Outridge PM, Davis WJ, Stewart REA. Environ Sci Technol 1999; 33:1771–1775.
62. Swarzenski PW, Porcelli D, Andersson PS, Smoak JM. In: Bourdon B, Henderson GM, Lundstrom CC, Turner SP, eds. Uranium-Series Geochemistry. Washington: Geochemical Society & Mineralogical Society of America, 2003:577–606.
63. Dibb JE. J Geophy Res 1990; 95:22, 407–422, 415.
64. Winkler R, Rosner G. Sci Total Environ 2000; 57–68.
65. Radakovitch O, Cherry RD, Heussner S. Deep-Sea Res Part I 1999; 46:1539–1563.
66. Henderson GM, Maier-Reimer E. Geochim Cosmochim Acta 2002; 66:257–272.

67. Appleby PG, Oldfield F. Catena 1978; 5:1–8.
68. Appleby PG, Oldfield F. In: Ivanovitch M, Harmon RS, eds. Uranium-Series Disequilibrium; Application to Earth, Marine, and Environmental Sciences. Oxford: Clarendon Press, 1992:731–778.
69. Wertime TA. Science 1973; 182:875–887.
70. Nriagu JO. Lead and Lead Poisoning in Antiquity. New York: John Wiley and Sons, 1983:434.
71. Walter P, Martinetto P. Nature 1999; 397:483–484.
72. Meier S. Blei in der Antike: Bergbau, Verhüttung, Fernhandel. PhD dissertation. Universität Zürich, Zürich, Switzerland, 1995.
73. Nriagu JO. Science 1998; 281:1622–1623.
74. (a) Le Roux G, Weiss D, Grattan JP, Givelet N, Krachler M, Cheburkin A, Rausch N, Kober B, Shotyk W. J Environ Monit 2004; 6:502–510. (b) Le Roux G, Véron A, Morhange C. Archaeology and History in Lebanon. 2003; Autumn 2003:115–119.
75. Brännvall M-L, Bindler R, Renberg I, Emteryd O, Bartnicki J, Billström K. Environ Sci Technol 1999; 33:4391–4395.
76. Nriagu JO. Science 1998; 281:1622–1623.
77. Krysko WW. Lead in History and Art. Stuttgart: Dr. Riederer Verlag, 1979:244.
78. Kempter H, Frenzel B. Sci Total Environ 1999; 241:117–128.
79. Kempter H, Frenzel B. Water Air Soil Pollut 2000; 121:93–108.
80. Hazardous Contaminants Branch, Ontario Ministry. Rationale for the Development of Soil, Drinking Water, and Air Quality Criteria for Lead. 1993:114.
81. Nriagu JO. Science 1996; 272:223–224.
82. ILZSG, Lead, Review of Trends in 2003, International Lead and Zinc Study Group (http://www.ilzsg.org/), 2004.
83. von Storch H, Costa-Cabral M, Hagner C, Feser F, Pacyna J, Pacyna E, Kolb S. Sci Total Environ 2003; 311:151–176.
84. Thomas V. Annu Rev Energy Environ 1995; 20:301–324.
85. Thomas V, Kwong A. Energy Policy 2000; 29:1133–1143.
86. Pacyna JM, Pacyna EG. Environ Rev 2001; 9:269–298.
87. Patterson CC, Settle DM. Geochim Cosmochim Acta 1987; 51:675–681.
88. Shotyk W, Krachler M, Chen B. Global Biogeochem Cycles 2004; 18:GB1016.
89. Nriagu JO. Environment 1990; 32:7–33.
90. Schütz L. In: Leinen M, Sarnthein M, eds. Paleoclimatology and Paleometeorology: Modern and Past Patterns of Global Atmospheric Transport. Dordrecht: Kluwer Academic Publishers, 1989.
91. Heinrich H, Schulz-Dobrick B, Wedepohl KH. Geochim Cosmochim Acta 1980; 44:1519–1533.
92. Rahn KA. The Chemical Composition of the Atmospheric Aerosol. Technical Report. Kingston: Graduate School of Oceanography, University of Rhode Island, 1976:265.
93. Radojevic M, Harrison RM. Enviro Sci Technol 1987; 59:157–180.
94. Harrison RM, Allen AG. Appl Organomet Chem 1989; 2:49–58.
95. Van Cleuvenbergen RJA, Adams FC. In: Hutzinger O, Wixson BG, eds. Handbook of Environmental Geochemistry. Berlin: Springer, 1990:97–153.
96. Heisterkamp M, Van de Velde K, Ferrari C, Boutron C, Adams FC. Environ Sci Technol 2000; 21:260–266.
97. Wang Y, Turnbull AB, Harrison RM. Appl Organomet Chem 1997; 11:889–901.

98. Harrison RM, Laxen DPH. Environ Sci Technol 1978; 12:1384–1392.
99. Wang Y, Allen AG. Appl Organomet Chem 1996; 10:773–778.
100. Shotyk W, Weiss D, Heisterkamp M, Cheburkin AK, Appleby PG, Adams FC. Environ Sci Technol 2002; 36:3893–3900.
101. Lobinski R, Boutron CF, Candelone JP, Hong S, Szpunar-Lobinska J, Adams FC. Environ Sci Technol 1994; 28:1459–1466.
102. Lobinski R, Boutron CF, Candelone JP, Hong S, Szpunar-Lobinska J, Adams FC. Environ Sci Technol 1994; 28:1467–1471.
103. Lobinski R, Witte C, Adams FC, Teissedre PL, Cabanis JC, Boutron C. Nature 1994; 370:24.
104. Warren HV, Delavault RE. Trans Royal Soc Can 1960; LIV:11–20.
105. Cannon HL, Bowles LC. Science 1962; 137:765–766.
106. Chow TJ. Nature 1970; 225:295–296.
107. Blaser P, Zimmermann S, Luster J, Shotyk W. Sci Total Environ 2000; 249:257–280.
108. Bindler R, Brannvall ML, Renberg I, Emteryd O, Grip H. Environ Sci Technol 1999; 33:3362–3367.
109. Dörr H, Münnich KO, Mangini A, Schmitz W. Naturwiss 1990; 77:428–430.
110. Shotyk W, Blaser P, Grunig A, Cheburkin AK. Sci Total Environ 2000; 249:281–295.
111. Renberg I, Brännvall M-L, Bindler R, Emteryd O. Ambio 2000; 29:150–156.
112. Givelet N, Roos-Barraclough F, Shotyk W. J Environ Monit 2003; 5:935–949.
113. Miller EK, Friedland AJ. Environ Sci Technol 1994; 28:662–669.
114. Wang EX, Bormann FH, Gaboury B. Environ Sci Technol 1995; 29:735–739
115. Steinmann M, Stille P. Appl Geochem 1997; 12:607–623.
116. Whitehead K, Ramsey MH, Maskall J, Thornton I, Bacon JR. Appl Geochem 1997; 12:75–81.
117. Bacon JR, Berrow ML, Shand CA. Int J Environ Anal Chem 1992; 46:71–76.
118. Jersak J, Amundson R, Brimhall G. J Environ Qual 1997; 26:511–521.
119. Lin Z, Hasbro K, Ahlgren M, Qvarfort U. Sci Total Environ 1998; 209:47–58.
120. Teutsch N, Erel Y, Halicz L, Banin A. Geochim Cosmochim Acta 2001; 65:2853–2864.
121. Emmanuel S, Erel Y. Geochim Cosmochim Acta 2002; 66:2517–2527.
122. Jenny H. Factors of Soil Formation. New York: Dover, 1941:288.
123. Bear FE. Chemistry of the Soil. 2d ed. New York: Reinhold, 1964:515.
124. Sposito G. The Chemistry of Soils. Oxford: Oxford University Press, 1989:277.
125. Hirao Y, Patterson CC. Science 1974; 184:989–992.
126. Garrels RM, MacKenzie FT. In: Gould RF, ed. Equilibrium Concepts in Natural Water Systems. Washington: Amer Chem Soc, 1967:222–242.
127. Patterson CC, Settle DM. National Bureau of Standards Special Publication, Washington 1974; 422:321–351.
128. Graney JR, Halliday A, Keeler GJ, Nriagu J, Robbins JA, Norton SA. Geochim Cosmochim Acta 1995; 59:1715–1728.
129. Callender E, Van Metre PC. Environ Sci Technol 1997; 31:424a–428a.
130. Outridge PM, Hermansson MH, Lockhart WL. Geochim Cosmochim Acta 2002; 66:3521–3531.
131. Camarero L, Masque P, Devos W, Ani-Ragolta I, Catalan J, Moor HC, Pla S, Sanchez-Cabeza JA. Water Air Soil Pollut 1998; 105:439–449.

132. Moor HC, Schaller T, Sturm M. Environ Sci Technol 1996; 30:2928–2933.
133. Von Gunten HR, Sturm M, Moser RN. Environ Sci Technol 1997; 31:2193–2197.
134. Farmer JG, Eades LJ, Mac Kenzie AB, Kirika A, Bailey-Watts TE. Environ Sci Technol 1996; 30:3080–3083.
135. Eades LJ, Farmer JG, MacKenzie AB, Kirika A, Bailey-Watts AE. Sci Total Environ 2002; 292:55–67.
136. Erel Y, Dubowski Y, Halicz L, Erez J, Kaufman A. Environ Sci Technol 2001; 35:292–299.
137. Norton SA, Dillon PJ, Evans RD, Mierle G, Kahl JS. In: Lindberg SE, Page AL, Norton SA, eds. Sources, Deposition, and Canopy Interactions. Acid Precipitation. Vol. 3. Berlin: Springer Verlag, 1990.
138. Kober B, Wessels M, Bollhöfer A, Mangini A. Geochim Cosmochim Acta 1999; 63:1293–1303.
139. Rosman KJR, Ly C, Steinnes E. Environ Sci Technol 1998; 32:2542–2546.
140. Kunert M, Friese K, Weckert V, Markert B. Environ Sci Technol 1999; 33:3502–3505.
141. Monna F, Aiuppa A, Varrica D, Dongarra G. Environ Sci Technol 1999; 33:2517–2523.
142. Weiss D, Shotyk W, Kramers JD, Gloor M. Atmos Environ 1999; 33:3751–3763.
143. Farmer JG, Lorna JE, Atkins H, Chamberlain DF. Environ Sci Technol 2002; 36:152–157.
144. Bacon JR, Jones KC, McGarth SP, Johnston AE. Environ Sci Technol 1996; 30:2511–2518.
145. Hagenmeyer J, Schäfer H. Vegetatio 1992; 101:55–63.
146. Hagenmeyer J, Schäfer H. Sci Total Environ 1995; 166:77–87.
147. Aberg G, Pacyna JM, Strayn H, Skjelkvale BL. Atmos Environ 1999; 33:3335–3344.
148. Bellis DJ, McLeod CW, Satake K. Sci Total Environ 2002; 289:169–176.
149. Bellis DJ, Satake K, Noda M, Nishimura N, McLeod CW. Sci Total Environ 2002; 295:91–100.
150. Vile MA, Novák MJV, Brizová E, Kelman Wieder R, Schell WR. Water Air Soil Pollut 1995; 79:89–106.
151. Vile MA, Wieder RK, Novak M. Biogeochemistry 1999; 45:35–52.
152. Vile MA, Wieder RK, Novak M. Environ Sci Technol 2000; 34:12–20.
153. Brännvall ML, Bindler R, Emteryd O, Nilsson M, Renberg I. Water Air Soil Pollut 1997; 100:243–252.
154. Farmer JG, MacKenzie AB, Sugden CL, Edgar PJ, Eades LJ. Water Air Soil Pollut 1997; 253–270.
155. Kempter H, Görres M, Frenzel B. Water Air Soil Pollut 1997; 100:367–377.
156. Martinez Cortiza A, Garcia-Rodeja E, Pontevedra Pombal X, Novoa Munoz JC, Weiss D, Cheburkin A. Sci Total Environ 2002; 292:33–44.
157. MacKenzie AB, Farmer JG, Sugden CL. Sci Total Environ 1997; 203:115–127.
158. MacKenzie AB, Logan EM, Cook GT, Pulford ID. Sci Total Environ 1998; 222:157–166.
159. Norton SA, Evans GC, Kahl JS. Water Air Soil Pollut 1997; 100:271–286.
160. Weiss D, Shotyk W, Appleby PG, Kramers JD, Cheburkin AK. Environ Sci Technol 1999; 1340–1352.
161. Weiss D, Shotyk W, Boyle EA, Kramers JD, Appleby PG, Cheburkin AK. Sci Total Environ 2002; 292:7–18.

162. Bell K, Kettles IM. Current Research-C3, Geological Survey of Canada 2003.
163. Nieminen TM, Ukonmaanaho L, Shotyk W. Sci Total Environ 2002; 292:81–89.
164. Klaminder J, Renberg I, Bindler R. Global Biogeochem Cycles 2003; 17.
165. Shotyk W, Weiss D, Appleby PG, Cheburkin AK, Frei R, Gloor M, Kramers J-D, Rees S, Van der Knaap WO. Science 1998; 281:1635–1640.
166. Shotyk W, Goodsite ME, Roos-Barraclough F, Frei R, Heinemeier J, Asmund G, Lohse C, Hansen TS. Geochim Cosmochim Acta 2003; 67:3991–4011.
167. Monna F, Lancelot J, Croudace IW, Cundy AB, Lewis JT. Environ Sci Technol 1997; 31:2277–2286.
168. Murozumi M, Chow JC, Patterson CC. Geochim Cosmochim Acta 1969; 33:1247–1297.
169. Boutron C, Görlach U, Candelone JP, Bolshov MA. Nature 1991; 353:153–156.
170. Hong S, Candelone JP, Patterson CC, Boutron C. Science 1994; 265:1841–1843.
171. Rosman J, Chisholm W. Anal Mag 1994; 22:M51–M53.
172. Rosman KJR, Chisholm W, Hong SM, Candelone JP, Boutron CF. Environ Sci Technol 1997; 31:3413–3416.
173. Rosman KJR, Ly C, Van de Velde K, Boutron CF. Earth Planet Sci Lett 2000; 176:413–424.
174. Vallelonga P, Van de Velde K, Candelone JP, Morgan VI, Boutron CF, Rosman KJR. Earth Planet Sci Lett 2002; 204:291–306.
175. McConnell JR, Lamorey GW, Hutterli MA. Geophy Res Lett 2002; 29.
176. Bollhöfer A, Rosman KJR. Geochim Cosmochim Acta 2002; 66:1375–1386.
177. Bollhöfer A, Rosman KJR. Phys Chem Earth (A) 2001; 26:835–838.
178. Bollhöfer A, Chisholm W, Rosman KJR. Anal Chim Acta 1999; 390:227–235.
179. Bollhöfer A, Rosman KJ. Geochim Cosmochim Acta 2001; 65:1727–1740.
180. Flament P, Bertho ML, Deboudt K, Veron A, Puskaric E. Sci Total Environ 2002; 296:35–57.
181. Bacon JR. J Environ Monit 2002; 4:291–299.
182. Hansmann W, Koppel V. Chem Geol 2000; 171:123–144.
183. Barbaste M, Halicz L, Galy A, Medina B, Emteborg H, Adams FC, Lobinsky R. Talanta 2001; 54:307–317.
184. Lidsky TI, Schneider JS. J Environ Monit 2004; 4:36N.
185. Godwin HA. Curr Opin Chem Biol 2001; 223–227.
186. Lanphear BP. Science 1998; 281:1617–1618.
187. Canfield RL, Henderson CR Jr, Cory-Slechta DA, Cox C, Jusko TA, Lanphear BP. N Engl J Med 2003; 348:1517–1526.
188. Selevan SG, Rice DC, Hogan KA, Euling SY, Pfahles-Hutchens A, Bethel J. N Engl J Med 2003; 348:1527–1536.
189. Flegal AR, Smith DR. Environ Res 1992; 58:125–133.
190. Ontario Ministry of Energy and Environment. Scientific Criteria Document for Multimedia Environmental Standards Development—Lead. Queen's Printer for Ontario, 1994.
191. Lalor GC, Rattray R, Vutchkov M, Campbell B, Lewis-Bell K. Sci Total Environ 2001; 269:171–181.
192. Nriagu J, Jinabhai C, Naidoo R, Coutsoudis A. Sci Total Environ 1996; 191:69–76.
193. Patterson CC. In: Committee on Lead in the Human Environment, ed. Lead in the Human Environment. National Academy of Sciences, 1980:525.

194. Elias R, Hinkley TK, Hirao Y, Patterson CC. Geochim Cosmochim Acta 1976; 40:523–587.
195. Shirahata H, Elias RW, Patterson CC, Koide M. Geochim Cosmochim Acta 1980; 44:149–162.
196. Elias RW, Hirao Y, Patterson CC. Geochim Cosmochim Acta 1982; 46:2561–2580.
197. Tatsumoto M, Patterson CC. Nature 1963; 199:350–352.
198. Chow JC, Patterson CC. Earth Planet Sci Lett 1966; 1:397–400.
199. Schaule BK, Patterson CC. Earth Planet Sci Lett 1981; 54:97–116.
200. Needleman HL. Environ Res 1997; 74:95–103.
201. Hagner C. Historical Review of European Gasoline Lead Content Regulations and Their Inpact on German Industrial Markets. GKSS Forschungszentrum Geesthacht GmbH, 1999:31.
202. Thomas V, Socolow RH, Fanelli JJ, Spiro TG. Environ Sci Technol 1999; 22:3942–3948.
203. Lanphear BP, Matte TD, Rogers J, Clickner RP, Dietz B, Bornschein RL, Succop P, Mahaffey KR, Dixon S, Galke W, Rabinowitz M, Farfel M, Rohde C, Schwartz J, Ashley P, Jacobs DE. Environ Res 1998; 79:51–68.
204. Mielke HW. Am Sci 1999; 87:67–73.
205. Wong MH. In: Hutchinson TC, Meema KM, eds. Lead, Mercury, Cadmium and Arsenic in the Environment. SCOPE. Chichester: John Wiley and Sons, 1987.
206. Mielke HW, Powell ET, Shah A, Gonzales CR, Mielke PW. Environ Health Perspect 2001; 109:973–978.
207. Graziano JH, Blum C. Chemical Speciation Bioavailability 1991; 3:81–85.
208. Rosman KJ, Chisholm W, Jimi S, Candelone JP, Boutron CF, Teissedre PL, Adams FC. Environ Res 1998; 78:161–167.

Subject Index

A

A.
  ambivalens, see Acidianus
  arthrocyaneus, see Arthrobacter
  cycloclastes, see Achromobacter
  faecalis, see Alcaligenes
  fulgidus, see Archaeoglobus
  johnsonii, see Acinetobacter
  vinelandii, see Azotobacter
AAS, *see* Atomic absorption
  spectroscopy
Acetaldehyde
  phosphono-, 139, 140, 146
Acetate, 36
  phosphono-, 140–142
Acetogenesis, 17, 27, 28
Acetogens, 24
Achromobacter
  cycloclastes, 88
Acidianus sp., 106, 108, 123
  ambivalens, 124, 127
Acid rain, 97
Acinetobacter ssp., 137
  johnsonii, 137
Adenosine 5'-diphosphate, *see* 5'-ADP
Adenosine monophosphate, *see* AMP
Adenosine phosphosulfate reductase, *see*
  Reductases
Adenosine 5'-triphosphate, *see* 5'-ATP
Adenylate kinase, 137

5'-ADP, 80, 81, 137, 138
Aerosols (containing) (*see also* Air
  and Atmosphere)
  cadmium, 197
  iron, 160 163, 165, 182
  lead, 256, 260, 262, 264
  phosphates, 133
  pre-anthropogenic, 246
AES, *see* Atomic emission spectrometry
Africa, 249
AFS, *see* Atomic fluorescence
  spectrometry
Agriculture
  fertilizer, *see* Fertilizer
  pesticide, *see* Pesticides
  soil, *see* Soil
Air (*see also* Atmosphere)
  carbon dioxide in, *see*
    Carbon dioxide
  dioxygen reservoir, 52
  methane in, 56, 66
Alcaligenes faecalis, 87, 88
Algae (*see also* individual names), 97
  cadmium as nutrient, 209–215
  cadmium in, 207
  green, 12, 134, 135
  iron uptake, 155
  micro-, 207–209
  nitrate reduction, 77
  nutrients, *see* Nutrients
  phosphate uptake, 134, 135

Alkaline phosphatase, 136
    replacement of zinc by cadmium,
        205, 209
Alkyllead, 251, 252
    tetra-, *see* Tetraalkyllead
Alterobactin, 156
Aluminum phosphate, 134
Amazon
    iron in, 163
    mercury in, 230
Amino acids
    synthesis, *see* Synthesis
4-Aminobutylphosphonate, 142
2-Aminoethylphosphonate, 138, 142
δ-Aminolevulinic acid dehydratase, 264
Ammonia, 2, 4, 15, 77, 81, 95
    for biosynthesis, 93
    formation, 92–94
    industrial production, 78–80
    monooxygenase,
        see Monooxygenases
    synthesis, 78–80
Ammonifex, 109
Ammonium, 77
    assimilation, *see* Assimilation
AMP, 137
Anaerobic ecosystems
    dihydrogen cycling, 17–32
    organic decomposition, 17–19
Anaerobic organisms, *see*
        individual names
Anammox process, 76, 78, 94
Andesite, 68
Anglesite, 244
Anhydrite, 61, 62
Animals (*see also* individual species)
    digestive systems, 32
    methanogenesis, 12
Anodic stripping voltammetry, 202
Antarctica, 178, 182
    correlation between cadmium and
        phosphate concentrations, 196
    dust deposition, 181
    East, 181
    lead deposition, 259
    mercury in atmosphere, 224
    Vostoc ice core, 181, 182
Anthopleura elegantissima, 138

Anthropogenic (emission of)
    (*see also* Environment)
    cadmium sources, 197–199
    iron, 160, 162
    lead, 241, 242, 246, 250–267
    mercury, 224
    pre-, 245, 246
Antibiotics (*see also* individual
        names), 139
Apatites, 132
    calcium, 134
Arabian Sea
    iron in, 161
    nutrients in, 170
Archaea (*see also* individual names),
        10, 25, 54, 57, 77
    acidophilic, 108, 124
    anaerobic, 106
    chemolithoautotrophic, 124
    hyperthermophilic, 90, 123, 124, 127
    nitrate-reducing, 85
    sulfur reducers, 123, 124
    sulfur respiration, 123, 124, 127
    thermophilic, 107, 127
Archaean Era, 32, 146
Archaeoglobus fulgidus, 85
Arctic
    lakes, 227
    lead deposition, 256
    mercury in atmosphere, 224
Arsenic(III)
    as electron donor, 85
Arthrobacter arthrocyaneus, 140
Asia, 249
    lead mining, 247
Assimilation (of)
    ammonium, 96
    hypophosphite, 143, 144
    nitrate, 77, 85, 96
    pathways, 95, 96
    phosphite, 143, 144
Astrobiology, 16
Atlantic Ocean
    cadmium in, 201
    correlation between cadmium and
        phosphate concentrations, 196, 201
    equatorial, 170
    iron deposition, 161, 163, 166

[Atlantic Ocean]
  lead in, 264
  North, 160, 166, 170, 177, 201
  Rim, 166
  Saharan dust, 161
  subtropical, 167
  tropical, 167
Atmosphere (*see also* Aerosols,
    Air, Stratosphere, *and*
    Troposphere), 32
  cadmium in, 197, 199
  carbon dioxide in, 180–184
  dihydrogen in, 66
  dioxygen content, 15, 49–71
  early Earth, 3, 78
  iron in, 160–162, 179
  lead in, 240–242, 247,
    250–263
  mercury in, 222–227
  methane in, 66
  modern, 10, 11
  nitrogen fixation, 155
  oxygen-poor, 69
  prebiotic, 10, 14
  redox chemistry, 39, 40
  reducing, 15
Atomic absorption spectroscopy, 244
  graphite furnace, *see*
    Graphite furnace atomic
    absorption spectroscopy
Atomic emission spectrometry
  inductively coupled plasma, *see*
    Inductively coupled
    plasma-atomic emission
    spectrometry
Atomic fluorescence spectrometry
  laser excited, 244
5'-ATP, 137, 138, 207
  generation, 27, 38
  hydrolysis, 81, 96
  Mg-, 80, 82
  number per dinitrogen fixed, 38
  synthesis, *see* Synthesis
Australia
  Broken Hill mine, 245, 246, 249
  lead production, 249
Azotobacter sp., 79
  vinelandii, 80

B

B.
  acetoformans, see Bacillus
  azotoformans, see Bacillus
  caldolyticus, see Bacillus
  cereus, see Bacillus
  subtilis, see Bacillus
Babylon
  Hanging Gardens, 247
Bacillus sp., 136
  acetoformans, 6
  azotoformans, 78, 90
  caldolyticus, 143
  cereus, 140
  subtilis, 106, 135, 136
Bacteria(l) (*see also* individual names),
    54, 57, 77, 106
  acetogenic, 17, 27, 28, 90
  anaerobic, 17, 90
  cadmium export systems, 207
  cleavage of C–P linkages, 140–143
  cyano-, *see* Cyanobacteria
  denitrifying, 85, 87
  entero-, 93
  eu-, *see* Eubacteria
  gram-negative, 90, 207
  gram-positive, 6, 90, 137, 207
  heterotrophic, 135, 137, 155, 165, 207
  hydrogenases, *see* Hydrogenases
  lithoautotrophic, 94
  marine, 156
  mesophilic, 107, 108, 127
  nitrate reduction, 77, 137, 145
  nitrifying, 86, 94
  nitrite-ammonifying, 92, 93
  nitrogen fixation, 79, 81
  phosphate uptake, 134–137
  phosphate-accumulating, 137
  phosphate-solubilizing, 136
  photosynthetic, 32, 208
  phototrophic, 32–36, 136, 155
  polysulfide sulfur respiration, 108, 109
  proteo-, 92, 109
  purple, 12
  sulfate-reducing, 12, 24–26, 78, 92,
    144, 145, 228, 243
  sulfur-reducing, 108, 109

Baltic Sea
  iron in, 170
Basalt, 13, 16, 52, 65, 69
  arc, 53
  iron in, 163
  lead in, 243
  mid-oceanic ridge, 53, 59, 61,
    62, 163
  oxidation, 61, 62
BASTA herbicide, 139
Batteries
  lead-acid, 249
  nickel-cadmium, 197
Bermuda, 264
Bialaphos, 139, 140
Bioavailability (of)
  iron, 158, 162, 175
  lead, 251, 254, 255
  mercury, 228
Biocides (*see also* individual names), 147
Biogeochemistry and biogeochemical
    cycles (of), 1–6
  cadmium, 195–217
  consequences of dihydrogen as
    energy source, 41
  dihydrogen, 9–41
  Earth, *see* Earth
  iron, 153–184
  mercury, 221–234
  microbial mats, 37
Biomass, 132, 172, 174, 200, 201,
    210, 211
  production, 134, 162
  sludge, 137
Biosphere
  "deep", 16, 40
  rock-hosted, 16, 17, 40
Biosynthesis (*see also* Synthesis)
  phosphonates, 138–140
Biota, 51, 53, 54, 57
  marine, 68
  (methyl)mercury in, 229
Biotite (*see also* Mica), 255
Black Forest
  lead in snow, 262
  lead mining, 248
Black Sea
  iron in, 170

Blood
  lead concentrations, 242, 263–265
  mercury in, 222
Bone (containing) (*see also* Skeleton)
  cadmium, 205
  lead, 263
  phosphorus, 123
Boulangerite, 244
Bournonite, 244
British Columbia, 252
Bryophytes (*see also* individual names)
  lead in, 256, 262

C

C.
  bovis, see Corynebacterium
  debilis, see Chaetoceras
  glutamicum, see Corynebacterium
  pasteurianum, see Clostridium
  reinhardtii, see Chlamydomonas
Cadmium (II) (in)
  aerosols, *see* Aerosols
  anthropogenic sources, *see*
    Anthropogenic
  as algal nutrient, 209–215
  as carcinogen, 205
  as nutrient for phytoplankton, 197, 199,
    200, 205, 207, 209–215
  as paleotracer, 215, 216
  atmosphere, *see* Atmosphere
  ATP-dependent transmembrane
    transporter, 207
  biogeochemistry, 195–217
  biological effects, 204–209
  bone, *see* Bone
  chemical speciation, *see* Speciation
  chloride complex, 201
  correlation with phosphate
    concentrations, 196, 200, 201,
    215, 216
  detoxification, *see* Detoxification
  diatoms, *see* Diatoms
  dissolved, 199–201, 203, 209–211
  drinking water, *see* Drinking water
  environment, *see* Environment
  export systems in bacteria, 207

[Cadmium (II) (in)]
  global cycle, 199
  metabolism in diatoms, 212
  -nickel batteries, *see* Batteries
  ocean, *see* Ocean
  particulate, 200, 201, 204,
     209–211, 213
  remineralization, 212–215
  resistance, 207
  soil, *see* Soil
  substitution of other metal ions, 204,
     205, 207, 208
  toxicity, *see* Toxicity
Calcite, 134
Calcium(II) (in), 204
  apatite, 134
  mediated signal transduction, 264
  regulation, 205
  substitution by cadmium, 204, 205
California, 169
  cadmium in coastal water, 214
  iron in coastal water, 164, 177
  Sierra Nevada mountains, 264
Calvin cycle, 209
Canada, 198, 222, 226, 230, 249,
     252, 256
  allowed lead concentration in
     drinking water, 267
  lead exposure of children, 263, 264
  peat bogs, 254
Carbon
  biological partition, 62, 63
  burial, 66–68
  cycle, 3, 55, 153–184
  fixation, 4
  isotopes, 56, 60, 64, 65
  organic, *see* Organic carbon
  -phosphorus bond cleavage, 140–143,
     146, 147
  -phosphorus linkage formation, 139
  reservoir, 53, 60
  residence time, 66
Carbonate(s), 52
  burial, 67
  cycle, 59–61, 66
  in sediments, *see* Sediments
  marine, 60
  reduction, 70

[Carbonate(s)]
  reservoir, 70
  residence time, 66
  weathering, 66
Carbon dioxide (in), 2–4, 11, 17
  as oxygen reservoir, 63
  atmosphere, *see* Atmosphere
  ocean, *see* Ocean
  pre-industrial concentration in air, 52
  reduction, 144
  role of iron in regulation, 180–184
  seawater, *see* Seawater
  sequestering of anthropogenic carbon
     dioxide, 180
Carbonic anhydrase
  cadmium in, 209, 210, 212, 213
  zinc in, 209
Carbon monoxide, 4
Cell
  iron function in, 154, 155
Cement production, 250, 251
Cerussite, 244
Chaetoceras debilis, 174
Châteauneuf-du-Pape wine
  lead in, 252
Chemoautotrophs, 11, 16, 17, 54
  microbes, *see* Microbes
Children
  exposure to lead, 242, 263–265
China
  lead mining, 249
Chlamydomonas reinhardtii, 135
Chloride
  cadmium complex, 201
Citrate, 83, 84
  cycle, 4
  homo-, *see* Homocitrate
Clay (*see also* Silicates), 134
  alumino-silicate, 158
  marine, 163, 166
Cleavage of C-P linkages, 140–143,
     146, 147
Climate
  and iron, 182, 183
  changes, 233
  cold, 180
  glacial-interglacial changes, 181, 184
  global, 182–184, 216

Clostridium pasteurianum, 80
Clusters
　Fe_4S_3, 85
　Fe_4S_4, 81, 111, 114, 116
　FeS, 86
　P-, 81, 84, 85
Coal
　British, 257, 259
　combustion, 251, 257, 265
　lead in, 243, 245, 257
Cobalamin
　methyl-, 140
Cobalt (different oxidation states) (in)
　phytoplankton, *see* Phytoplankton
　substitution of other metals, 207
Coccolithophores (*see also* individual
　　names)
　cadmium in, 207, 208
　iron uptake, 155
Colorado, 252
Congo river
　iron in, 163
Coordination sphere of nitrogenase, 82
Corals
　aragonite skeleton, 216
　cadmium/calcium ratio, 216
Corynebacterium
　bovis, 135
　glutamicum, 136
Crystal structure of
　formate dehydrogenase-N, 116
　nitrite reductase, 87
　nitrous oxide reductase, 91
　polysulfide sulfur transferase, 125, 126
Cyanobacteria (*see also* individual
　　names), 3, 33, 35–39, 54, 95, 134,
　　155, 179
　cadmium in, 205
　cadmium toxicity, 208
　dihydrogen in the metabolism, 32–34
　in sulfide-rich seawater, 58
　mats, 35
　phosphate uptake, 135–137
Cyclotella nana, 135
Cytochrome *c*, 93–95
　oxidase, *see* Oxidases
Cytochrome cd_1
　nitrite reductase, 86, 88

D

D.
　desulfuricans, *see* Desulfovibrio
　phosphitoxidans, *see* Desulfotignum
　vulgaris, *see* Desulfovibrio
Danube river
　iron in, 163
Data bases of proteins
　PDB, *see* Protein Data Bank
Dehydratases
　δ-aminolevulinic acid, 264
Dehydrogenases, 120, 121
　D-hydroxyacid, 144
　electron transfer, 119
　formate, 86, 110–113, 115–120, 123
　glucose, 136
　NAD^+-dependent phosphite, 144,
　　145, 147
Denitrification, 6, 76–78, 85–92
　ocean, 78
Denmark, 253, 257
Desulfotignum phosphitoxidans, 144, 145
Desulfovibrio sp., 92, 117
　desulfuricans, 90
　vulgaris, 25
Desulfurobacterium, 109
Desulfuromonas sp., 106, 109
Detergents
　additives, 140
　laundry, 137, 147
　polyphosphate in, 137, 147
Detoxification (of)
　cadmium, 205–207
　nitrite, 93
　reactive oxygen species, 154
Devonian dragonfly, 55
Diamond, 64
Diatoms (*see also* individual names)
　bloom, 164, 165, 173–175, 177, 179
　cadmium export, 207
　cadmium in, 209, 211, 212
　coastal, 157
　growth, 207, 208
　iron uptake, 155, 157, 165, 174
　iron-stressed, 158
　nitrate uptake, 172
　oceanic, 157, 171

[Diatoms (*see also* individual names)]
  phosphate uptake, 135
  silicate availability, 178, 183
Diazotrophs, 79, 80
  iron uptake, 165, 179
Dihydrogen (*see also* Hydrogen)
  abiotic production, 13, 14, 16, 17, 40
  as energy source, 41
  biogeochemistry, 9–41
  consuming processes, 27, 32–34, 37
  cycling in anaerobic ecosystems,
    17–32
  cycling in photosynthetic microbial
    mats, 36–39
  cycling in phototrophic ecosystems,
    32–39
  environmental concentrations, 26
  escape to space, 65, 66, 68, 69
  factors controlling concentration,
    26–32
  from the planetary matrix, 13–17
  generation, 13, 16, 34–37, 61
  in photosynthesis, 57
  interspecies transfer, 19, 21–24, 26
  lithotrophic production, 17
  microbial metabolism, 17–21
  photoproduction, 34
  photosynthetic metabolism, 32, 33
  producers, 32, 33
  reversible biological reaction, 23–25
  temperature dependence of activity, 29
  volcanic emission, 58
  volcanic gases, 14
Dimethylmercury, 222
Dimethylsulfoxide reductase, *see*
    Reductases
Dinitrogen (*see also* Nitrogen)
  fixation, *see* Nitrogen fixation
  production, 78
  reduction, 81, 82, 84
Dinitrogenases (*see also*
    Nitrogenases), 81
  reductase, *see* Reductases
Dinitrogen monoxide (or nitrous oxide),
    6, 77
  formation, 89, 180
  redox potential, *see* Redox potential
  reductase, *see* Reductases

[Dinitrogen monoxide (or nitrous oxide)]
  reduction, 90
Dinoflagellates
  iron uptake, 157, 174
Dioxygen (*see also* Oxygen)
  cycle, 2, 50
  formation, 35
  in atmosphere, *see* Atmosphere
  over geological time, 49–71
  reduction, 34, 88
Dissolved gaseous mercury, 226, 227
  photooxidation, 227
  photoproduction, 230
  photoreduction, 227
Dissolved organic matter, 227, 228
Disulfide
  molybdenum(IV), 4
Dolomite, 52
Dragonfly
  Devonian, 55
  modern, 55
Drinking water (containing) (*see also*
    Water)
  cadmium, 204, 205
  lead, 266, 267
  nitrate, 77
Dust (containing)
  deposition during ice ages,
    181, 182
  from pre-anthropogenic time, 245
  iron, 160–162, 166
  lead, 245, 250, 251, 257, 265
  phosphates, 133
  Sahara, 257

                    E

E.
  coli, see Escherichia
  herbicola, see Erwinia
  huxleyi, see Emiliana
Earth
  a solar planet, 50
  age, 51
  ancient ecosystems, 56–59
  atmosphere, *see* Atmosphere
  biogeochemistry, 11, 12

[Earth]
 crust, *see* Earth crust
 deep, rock-hosted biosphere, 40
 early, 3, 11, 13, 32, 39, 40, 50, 55,
  57, 62, 66, 68, 78, 146
 evolution, *see* Evolution
 history, 12, 13, 32, 39, 40
 interior, 68
 iron formations, 57
 life on, 54
 mantle, *see* Earth mantle
 modern, 39, 41, 58
 surface, 3, 13, 16, 55, 58, 59, 61,
  161, 241
Earth crust (containing) (*see also*
  Minerals, Ores, Rocks,
  and Soil), 13, 16
 continental, 51–53, 63
 early, 15, 39
 iron, 158
 lead, 243
 modern geochemistry, 13
 oceanic, 51, 53, 61–64, 67
 oxygen reservoir, 51–53, 68, 69
 phosphorus, 132
 ultramafic, 15, 39
Earth mantle, 15, 51, 65
 carbon cycle, 59–61
 early, 14
 oxidation, 62, 63
 oxygen cycle, 59–66, 68
 oxygen reservoir, 63
 sulfur cycle, 59–61
Ecosystems (*see also* Environment)
 anaerobic, *see* Anaerobic ecosystems
 ancient, 56–59
 anoxic, 11, 32
 dihydrogen cycling, *see* Dihydrogen
 hydrothermal, 5
 methanogenic, 30
 photosynthesis-independent, 16, 17
 phototrophic, 32–39
 sulfate-reducing, 30
Ectothiorhodospira sp., 85
Egypt
 ancient, 247
Electron transfer (in)
 dehydrogenase, 119

[Electron transfer (in)]
 intermolecular, 81, 82
 photosynthesis, 33
ELISA, *see* Enzyme-linked
  immunosorbent assay
Emiliana huxleyi, 207, 208
England, 247, 254, 256
Environment(al)
 anoxic, 143
 cadmium in, 196, 197
 geochemistry of lead, 239–267
 mercury in, 221–234
 mercury speciation, 222, 223
Environmental Protection Agency of
  the United States
 cadmium in drinking water, 204
 lead distribution, 250
 mercury toxicity, 222
Environment Canada
 mercury toxicity, 222
Enzyme-linked immunosorbent assay
 formate dehydrogenase, 118
Erosion, 65, 67, 69
Erwinia herbicola, 136
Escherichia coli, 86, 92, 93, 143, 144, 146
 formate dehydrogenase-N, 116–119
 phosphate uptake, 135, 136
Estuaries
 cadmium in, 198, 199
 Gironde, 198, 203, 204
 iron in, 163
Ethanol, 36
 methanogenic oxidation, 22
Ethanolamine
 phosphate, 138
 phosphono-, 139, 140
Ethylmercury, 222
Eubacteria (*see also* individual
  names), 108
Eucaria, 77
Eukaryotes (or eukaryotic) (*see also*
  individual names)
 anaerobic, 12
 cadmium in, 205
 iron uptake, 155, 165
 nitrogen cycle, 77
 phytoplankton, 155, 208, 211
Europa (planet), 16, 17

Europe
  allowed lead concentration in
    drinking water, 267
  Alps, 259
  Central, 254, 257
  lead deposition, 254, 257
  lead mining, 247, 248
  silver mines, 248
Eutrophication of
  lakes, *see* Lakes
  surface waters, *see* Surface water
Evolution (of)
  driving forces of, 6
  Earth, 2
  life, 2, 10, 14, 15

Γ

Fayalite, 13
Feldspars
  potassium, 243, 244, 251, 255
FeMo cofactor (FeMoco), 79, 81, 82–84
Fe-protein, 80, 81
  reduction, 82
Fermentation, 19, 31, 35–37
  anaerobic, 33, 36
  glucose, 36, 38
  organic matter, 21
  propionate, 20, 21
  saccharide, 35
Ferredoxins (*see also* individual
    names), 80, 82
Fertilizer (containing)
  cadmium, 199, 204
  iron in ocean, *see* Iron enrichment
    experiments in oceans
  mineral, 76, 77
  phosphate, 133, 147
  production, 96
Fish
  mercury in, 222, 223, 229
Flavin mononucleotide, 90, 158
Flavodoxins, 82, 175
  and iron-stressed microorganisms, 158
Florida, 179
  Everglades, 228
FMN, *see* Flavin mononucleotide

Food chain
  mercury in, 223
Foraminifera
  cadmium/calcium ratio, 215, 216
  fossil, 196, 215
Formate (or formic acid), 110, 115
  formation, 36
  hydrogen lyase, 24
  -nitrite oxidoreductase, *see*
    Oxidoreductases
Fossils, 55, 56, 196, 215
Fossil fuels
  combustion, 197, 250
  lead in, 243, 245, 248–251
Fragilariopsis kerguelenis, 174
France, 198, 204
  lead mining, 248, 249
  rivers, *see* Rivers
Freshwater (containing)
  cadmium, 201
  lakes, *see* Lakes
  lead, 255
  rivers, *see* Rivers
  speciation of cadmium, 203
Fumarate, 115
  reductase, *see* Reductases
  reduction, 34
  respiration, *see* Respiration
Fungi (*see also* individual names)
  nitrate reduction, 77
  phosphate uptake, 136
  symbiosis with plants, 136

G

G.
  metallireducens, *see Geobacter*
Galapagos Islands, 164, 171, 173
Galena, 240, 244
Gasoline (*see also* Fossil fuels)
  leaded, 242, 245, 248, 249, 251–254,
    257–260, 264, 266, 267
Geobacter sp, 109
  metallireducens, 93
Geothermal
  emanations, 10, 11, 13
  gradient, 6

Germany, 253, 262
  lead mining, 246, 248, 249
GFAAS, *see* Graphite furnace
    absorption spectroscopy
Glacial ice
  lead in, 247, 252
  mercury deposition, 223
  pre-anthropogenic, 265
Glass
  lead in, 247–249, 266
Global (cycle of)
  cadmium, 199
  climate, *see* Climate
  ecology, 55, 57, 71
  mercury, 227
  methane, 12
  phosphorus, 133
Glucose
  degradation, 23
  dehydrogenase, *see* Dehydrogenases
  fermentation, *see* Fermentation
  phosphorylation, *see* Phosphorylation
Glutamine synthetase, *see* Synth(et)ases
Glycine
  N-(phosphonomethyl)-, *see*
    Glyphosate
Glycogen
  stores, 36
Glyphosate
  decomposition, 140, 143
Gneiss
  graphitic, 53
Granite, 13, 14, 16, 69
  lead in, 243, 255
  red, 52
Graphite furnace atomic absorption
    spectroscopy, 244
"Grasshopper effect" of mercury,
    224, 225
Great Lakes
  mercury volatilization, 226
Greenland
  cadmium accumulation, 198
  lead accumulation, 252, 259–261, 264
  snow, 198, 259, 264
Greenhouse
  effect, 77, 96
  gas, 66, 70, 77, 182

Groundwater (containing) (*see also*
    Water)
  mercury, 231–233
  nitrate, 77

H

Haber-Bosch process, 78, 79
Harz mountain, 246, 248
Hawaii, 56
Health Canada
  mercury toxicity, 222
Hematite, 57
Herbicides (*see also* individual names)
  degradation, 140, 143
  phosphorus containing, 139, 143, 147
High nutrient low chlorophyll regions,
    see Ocean
Himalaya
  iron in rivers, 163
Holocene
  carbon dioxide concentration, 183
  peat samples, 257
Homocitrate
  in FeMoco, 82–84
Human
  exposure to cadmium, 204
  exposure to lead, 242, 263–265
Humic acids, 134, 158, 162, 229
Hungary, 145
Hydride transfer, 144
Hydrogen (*see also* Dihydrogen)
  fuel cells, 41
  production, 5
Hydrogenases, 110–113, 117–119, 123
  bacterial, 10, 26, 33, 34
  de-, *see* Dehydrogenases
  NiFe, 124, 127
  redox potential, *see* Redox potential
Hydrogen peroxide, 154, 159, 226, 227
Hydrogen sulfide (*see also* Sulfide),
    2–4, 33, 170
  photolysis, *see* Photolysis
Hydrolases (*see also* individual
    names), 431
  phosphonoacetaldehyde, 142
  phosphonoacetate, 140–142
  phosphonopyruvate, 140, 141

Hydrothermal vents
  dihydrogen flux, 58
  ferrous sulfides in, 13
  iron in, 160, 163
  organisms in, 53, 58, 109, 123
Hydroxamate siderophores, *see*
  Siderophores
Hydroxide
  iron, 158–160, 163, 170, 243, 254
  manganese, 243, 254
Hydroxylamine, 88, 94
  oxidoreductase, *see* Oxidoreductases
Hylocomium splendens, 256
Hypophosphite, 146
  assimilation, *see* Assimilation

I

Ice age, 180
  dust input in oceans, 181
ICP-AES, *see* Inductively coupled
  plasma-atomic emission
  spectrometry
ICP-MS, *see* Inductively coupled
  plasma-mass spectrometry
ID-MS, *see* Isotope dilution-mass
  spectrometry
Indian Ocean
  cadmium in, 201
  correlation between cadmium and
    phosphate concentrations, 196
  iron deposition, 161
  nutrients in, 170
Inductively coupled plasma-atomic
  emission spectrometry, 244
Inductively coupled plasma-mass
  spectrometry, 244, 260
  double-focusing sector field,
    244–246, 255, 262
  laser ablation, 245, 247
  multiple collector, 245, 247
  with quadrupole mass analyzer,
    244–246
Industry (or industrial)
  battery, *see* Batteries
  cadmium in, 197, 198
  cosmetics, 247

[Industry (or industrial)]
  metal plating, 197
  mining, *see* Mining
  pigments, *see* Pigments
  plastics, 197, 249
  uses of lead, 247–250
Iron (different oxidation states) (in)
  aerosols, *see* Aerosols
  as essential nutrient, 154–158
  as limiting nutrient, 2, 170–180
  atmospheric fluxes, 160–162
  bioavailability, *see* Bioavailability
  biogeochemical models of cycling, 160
  biological cycling, 165, 166
  bottle incubation experiments, 171,
    172, 177
  carrier-mediated uptake, 157
  chemical forms, 158
  chemistry in seawater, 158–160
  colloids, 157, 158, 161
  control of ecosystems, 170–180
  cycle, 153–184
  cycling in the ocean, 160–170
  dissolved, 158, 159, 163, 164,
    167–169, 175, 177, 182
  distribution in oceans, 166–170
  dust, 160–162, 166
  enrichment experiments, *see* Iron
    enrichment experiments in oceans
  function in cells, 154, 155
  hydrothermal sources, *see*
    Hydrothermal vents
  hydroxides, *see* Hydroxides
  hypothesis, 180–184
  interaction with other limiting
    factors, 177–179
  marine organisms, 154–158
  ocean, *see* Ocean
  particulate, 158, 162, 169
  phosphate, *see* Phosphate
  photosynthesis, *see* Photosynthesis
  -protein, *see* Fe-protein
  redox chemistry, 159
  regulation of atmospheric carbon
    dioxide, 180–183
  rivers, *see* River
  rocks, *see* Rocks
  sediments, *see* Sediments

[Iron (different oxidation states) (in)]
  speciation, *see* Speciation
  stress, 157, 158, 177–179
  sulfide, *see* Sulfides
  uptake mechanisms, 155, 156, 165, 174
  uptake rates, 156, 157
Iron(III)
  dissolved, 2, 157, 159, 160, 167
  hydrous oxide, 202
  hydroxide, *see* Hydroxide
  ligands, 2
  reduction, 2, 155, 159
Iron enrichment experiments in
      oceans, 172–181
  and carbon export, 180
  EisenEx, 173–175
  IRONEX I, 173–175, 178, 179
  IRONEX II, 173–175, 179
  SEEDS, 173–175
  SERIES, 173, 175, 179–181
  SoFex, 173–176, 179–181
  SOIREE, 173–175, 183
Iron-sulfur proteins, *see* individual names
Isotope dilution-mass spectrometry,
      245, 255
Itai-itai disease, 196, 205
Italy, 253

J

Jamaica
  lead blood levels of children, 264
Jamesonite, 244
Japan
  itai-itai disease, 205
  Minamata Bay disease, 196

K

Kerguelen Islands, 167
Kidney damage, 205
Kinase
  adenylate, 137
Klebsiella
  aerogenes, 144
  pneumoniae, 80
Kluyveromyces fragilis, 142

L

Lactate, 36
Lakes (*see also* Water)
  Adirondack, 229
  arctic, 227
  Big Dam West, 230–234
  Constance, 256
  ELA, 229
  Etang de la Gruère, 246, 252, 257,
      258, 260
  eutrophic, 97, 134, 135
  freshwater, 134, 222, 223, 226, 227
  Gardsjon, 229
  Kejimkujik, 229
  mercury (cycling) in, 222, 223,
      226–233
  mesotrophic, 134
  oligotrophic, 134
  Onondaga, 231, 233
  phosphate in, 134
  Tantare, 198
Laughing gas, *see* Dinitrogen oxide
Lead (different oxidation states) (in)
  ^{210}Pb as geochronological tool, 247
  abundance, 243, 244
  -acid batteries, *see* Batteries
  anthropogenic emission, *see*
      Anthropogenic
  atmosphere, *see* Atmosphere
  bioavailability, *see* Bioavailability
  biogeochemistry, 239–267
  blood, *see* Blood
  bryophytes, *see* Bryophytes
  chemistry, 242–245
  cycling, 239–267
  deleterious brain effects, 263
  dietary exposure, 266
  environment, *see* Environment
  environmental exposure, 263–267
  human health, 263–267
  isotope ratios, 241, 245–247, 254, 258,
      259, 262
  isotopes, 56, 58
  measurement of concentrations,
      244, 245
  mechanism of poisoning, 264
  natural water, *see* Water

[Lead (different oxidation states) (in)]
ores, *see* Ores
organolead compounds, *see*
    individual names
personal computers, 250
pesticides, *see* Pesticides
sources of exposure, 264, 265
stable isotopes, 241, 245–247
substitution of other elements, 243, 264
sulfide, *see* Sulfides
TV sets, 250
unstable isotopes, 241, 247
uses, 247–250
Legume roots, 79
Lichens
lead in, 256
Life
origin of, *see* Origin of life
thermodynamics, 54, 55
Limestone, 52
lead in, 243
Lipid bilayer, 155
Liposomes, 120
proteo-, *see* Proteoliposomes
Loess deposits, 181
Long Island Sound
mercury volatilization, 226
Loon
mercury in, 222, 230
Lyases (*see also* individual names)
C–P, 140–143, 147
formate hydrogen, 24
Lyngbya sp., 38

M

M.
barkeri, *see Methanosarcina*
edulis, *see Mytilis*
omelianskii, *see Methanobacillus*
thermoacetica, *see Moorella*
MAC, *see* Maximum allowable
    concentration
Magma, 53
Magnesium(II), 65, 80, 82
Magnetite, 13, 61, 64, 65
Manganese hydroxide, 243, 254

Mariana Trench
bottom, 28
Mars, 16, 17, 50, 57, 64
mantle, 63
Mass spectrometry
inductively coupled plasma, *see*
    Inductively coupled plasma-mass
    spectrometry
inductively coupled plasma-atomic
    emission, *see* Inductively coupled
    plasma-atomic emission mass
    spectrometry
isotope dilution, *see* Isotope
    dilution-mass spectrometry
secondary ion, *see* Secondary ion
    mass spectrometry
thermal ionization, *see* Thermal
    ionization-mass spectrometry
Maximum allowable concentration of
lead in drinking water, 267
Mediterranian Sea
iron in, 169
Membrane
bilayer, 15
transporter, 156, 157
Menaquinones, 111, 113, 114, 121
8-methyl-, 110, 111, 113–115,
    119–123, 127
redox potential, *see* Redox potential
reduction, 117, 118, 123
Mercury (different oxidation states) (in)
anthropogenic input, *see*
    Anthropogenic
aquatic transport, 226–228
atmosphere, *see* Atmosphere
bioaccumulation, 222, 223, 228–230
biogeochemistry, 221–234
blood, *see* Blood
demethylation, 228
dissolved gaseous, *see* Dissolved
    gaseous mercury
environment, *see* Environment
ethyl-, *see* Ethylmercury
fish, *see* Fish
global emission, 227
"grasshopper effect", 224, 225
halogenated species, 226
lakes, *see* Lakes

[Mercury (different oxidation states) (in)]
  methyl-, *see* Methylmercury
  methylation, 228
  natural sources, 224
  nitrate, *see* Nitrate
  particulate, 224, 226
  phenyl-, 222
  reactions with sulfite, 225, 226
  reactive species in atmosphere, *see*
      Reactive gaseous mercury
  redox potental, *see* Redox potential
  redox reactions, 226–228
  speciation, *see* Speciation
  toxicity, *see* Toxicity
  transport, 229, 231
  volatilization, 224–227, 231–234
Mercury(0), 222, 224, 226, 228
Mercury(I), 227
  Hg_2^{2+}, 222
Mercury(II), 222, 224, 226
Metabolism (of)
  cadmium in diatoms, 212
  cyanobacteria, 32–34
  dihydrogen, 18–21, 32–34
  microbial, 18–21
Metallothioneins, 162
  cadmium, 205, 206
Metazoa, 71
Meteorites, 15, 63, 146
Methane, 2, 4, 15, 54, 68–70, 95
  anaerobic oxidation, 24, 25
  global emission, 12
  in air, 56, 66
  in marine sediments, 24
  in swamps, *see* Swamps
  monooxygenase, *see*
      Monooxygenases
  production, 25, 26
  self-ignition, 146
Methanobacillus omelianskii, 22
Methanogenesis, 12, 17, 21, 27, 28, 30,
    31, 38, 58
  carbon dioxide reduction, 20
Methanogens (*see also* individual
    names), 12, 24–26, 53, 228
  symbionts, 12
Methanosarcina barkeri, 25, 26
Methylcobalamin, 140

Methyllead
  tetra-, *see* Tetramethyllead
8-Methylmenaquinone, 110, 111,
    113–115, 119–123, 127
Methylmercury, 222, 226, 228
  bioavailability, *see* Bioavailability
  degradation, 233
  di-, 22
  in wetlands, *see* Wetlands
Mica, 243, 244
Microbes (or microbial) (*see also*
    individual names), 77
  anaerobic, 12
  and dihydrogen, *see* Dihydrogen
  chemoautotrophic, 58
  lithoautotrophic, 58
  lithotrophic, 16
  mats, 12, 36–39
  metabolism, *see* Metabolism
  methylating, 228
  oxygen-producing, 56
  regulators of energy metabolism, 21
Microcoleus sp., 36
Microorganisms (*see also* individual
    names *and* species)
  anaerobic, 20
  dihydrogen metabolism, 32–36
  iron-stressed, 158
  photosynthetic, 37
  phototrophic, 32–36
  release of iron-binding ligands, 160
Middle East, 254
Minamata Bay disease, 196
Minerals (*see also* Ores *and* individual
    names)
  alumino-silicate, 160
  iron oxide, 159
  lead, 241, 244, 251
  phosphorus, 131–133
  silicate, 13, 14, 243, 251, 254
  sulfide, 164, 254
  ultramafic, 13
Mining
  cadmium pollution, 197–199
  lead, 244, 247–249
  phosphorus, 133
  silver, 240, 241, 247, 248
  zinc, 197, 205

Mississippi river
  iron in, 163
Missouri
  lead mining, 246, 249
Moco, *see* Molybdenum cofactor *and*
  Molybdopterin
MoFe-protein, 80, 81, 82, 84
Molybdenum(III)
  in dinitrogen conversion, 82
Molybdenum(IV), 4
  in nitrogenases, 82
Molybdenum cofactor, 86
Molybdenum-iron protein, *see*
  MoFe-protein
Molybdopterin, 85
  guanine dinucleotide, 111, 112,
    114, 116
Monooxygenases
  ammonia, 95
  methane, 95
Mont Blanc, 252
Moon, 64
Moorella thermoacetica, 90
Moss
  lead in, 256
Mutagenesis
  site-directed, 111, 112
Mutases
  phosphoenolpyruvate, 142, 146
  phosphonopyruvate, 140, 141
Mycorrhiza, 136
Mytilis edulis, 139

N

N.
  europaea, see Nitrosomonas
  gonorrhoeae, see Neisseria
NAD^+, 93, 144, 145
  -dependent phosphite dehydrogenase,
    144, 145, 147
NADH, 4
  nitrite oxidoreductase, *see*
    Oxidoreductases
Neisseria gonorrhoeae, 88
Netherlands
  waste water treatment, 138

Neurotoxicity of
  lead, 263
Neurotransmitters (*see also*
    individual names)
  exocytosis, 264
New York, 226
Nickel-cadmium batteries,
    see Batteries
Nicotinamide adenine dinucleotide,
    see NAD^+
Nicotinamide adenine dinucleotide
    (reduced), *see* NADH
Nitrate(s) (in), 2, 4, 6
  ammonification, 76, 77
  as electron acceptor, 30
  as nutrient, *see* Nutrients
  assimilation, *see* Assimilation
  assimilatory reduction, 76, 77
  dissimilation, 85
  drinking water, *see* Drinking water
  groundwater, *see* Groundwater
  ocean, *see* Ocean
  pollution, 96
  reaction with mercury, 226
  redox potential, *see*
    Redox potential
  reductase, *see* Reductases
  reduction, 6, 28, 33, 34, 77, 85–93,
    96, 145, 154, 183
  residence time in ocean, 182
  respiration, 78, 85
Nitrate reductase, 171, 154
  assimilatory, 86
  dissimilatory, 86
  molybdenum-dependent, 85
Nitric oxide, 6, 77, 88
  aircraft exhaust, 97
  formation, 87, 88
  redox potential, *see* Redox potential
  reductase, *see* Reductases
  reduction, 90
Nitrification, 77, 97
Nitrite, 6
  ammonification, 92–94
  detoxification, *see* Detoxification
  oxidoreductase, *see* Oxidoreductases
  redox potential, *see* Redox potential
  reductase, *see* Reductases

[Nitrite]
　reduction, 33, 34, 77, 85, 88,
　　92–94, 154
Nitrobacteraceae, 94
Nitrogen (*see also* Dinitrogen)
　biological cycle, 75–98
　fixation, *see* Nitrogen fixation
Nitrogenases, 81, 34, 154
　active site, 82
　FeMo cofactor, *see* FeMo
　　cofactor
　iron-only, 84
　mechanistic aspects, 81–84
　molybdenum, 84
　P cluster, *see* Clusters
　sequences, 80
　structures, 80, 81
　vanadium, 84
Nitrogen fixation, 1, 33–35, 37, 38,
　　76–85, 92, 96, 134, 154,
　　179, 182
　biological aspects, 79, 80
　chemical aspects, 78, 79
　from the atmosphere, 155
Nitrogen monoxide, *see*
　　Nitric oxide
Nitrosomonas europaea, 94, 95
Nitrous oxide, *see* Dinitrogen
　　monoxide
North America(n), 222
　Shelf, 201, 202
Northern Hemisphere
　dust, 181
　iron in, 160
Nova Scotia
　Kejimkujik National Park, 222,
　　230–234
Nutrients (in ocean)
　cadmium, 197, 199, 200, 201
　cobalt, 208
　copper, 208
　depletion, 177, 178
　essential, 154–158
　iron, 2, 154–158
　micro-, 2, 208
　nitrate, 173, 175, 178
　phosphate, 173, 175
　silicate, 173, 175

O

O.
　terebriformis, see Oscillatoria
Ocean (*see also* Seawater *and*
　　individual names)
　cadmium in, 196, 199–201
　carbon dioxide reservoir, 66, 179
　coastal, 169
　correlation between cadmium and
　　phosphate concentrations, 196,
　　200, 201, 215, 216
　crust, *see* Earth crust
　deep, 133, 163, 165, 167,
　　168, 184
　denitrification, 78
　dioxygen in, 52
　dissolved iron, 167–169
　external iron sources, 160–164
　fertilization, *see* Iron
　　enrichment experiments
　　in oceans
　high nutrient low chlorophyll
　　region, 157, 160, 165, 166,
　　170–180
　iron enrichment, *see* Iron
　　enrichment experiments
　　in oceans
　iron in, 2, 160–170
　lead in, 247
　mercury cycle, 224, 227
　modern, 11
　nitrate depletion, 175
　nitrate in, 168, 170–172
　phosphate in, 133, 170
　redox chemistry, 39
　silicate depletion, 175
　sulfate in, *see* Sulfate
　sulfur reservoir, 60
　surface (*see also* Surface water), 28,
　　79, 171, 213, 227
　zinc in, 202
OECD countries, 249
Olivine, 13
Ontario
　Experimental Lakes Area, 229
　lead in paint, 266
　peat bogs, 254

Ores (*see also* Minerals *and*
    individual names)
  lead, 240, 245, 246, 264
  sulfide, 240
  zinc, 197
Oregon, 177
Organic carbon
  burial, 53, 67, 68, 70
  cycle, 66, 67
  dissolved, 229, 230
  in sediments, *see* Sediments
  particulate, 175, 176, 180
  reservoir, 57
  subduction flux, 64, 65, 68
  weathering, 66, 67
Organic matter, 11, 16, 17, 35, 40, 50,
    54, 56, 60, 64
  burial, 3, 51, 53, 57
  decomposition, 26, 37, 166
  dissolved, *see* Dissolved organic
    matter
  fermentation, *see* Fermentation
  oxidation, 59
  particulate, 213
  reaction with uranium, 57
Organization for Economic Coorperation
  and Development, *see* OECD
Origin of life, 40, 53, 64, 78
  theories, 10, 11, 13–15
Orinoco river
  iron in, 163
Oscillatoria terebriformis, 36
Oxidases (*see also* individual names)
  cytochrome *c*, 5, 78, 89, 90
  multicopper, 87
  quinol, 93
Oxidoreductases, 85
  formate-nitrite, 93
  hydroxylamine, 94, 95
  molybdo-, 114
  NADH:NO, 90, 93
  nitrite, 86
  quinol-nitrate, 86
Oxygen (*see also* Dioxygen)
  cycle, 51–53, 59–66
  fixed, 51
  geological record, 55
  long term crustal reservoir, 51–53

[Oxygen (*see also* Dioxygen)]
  modern cycle, 55
  origin in air, 50
  residence time, 51
Ozone
  depletion, 233
  layer, 96
  reaction with mercury, 226, 233
  reaction with tetraalkyllead, 251
  stratospheric, 77

P

P.
  abyssi, see Pyrodictium
  aerophilum, see Pyrobaculum
  aeruginosa, see Pseudomonas
  cepacia, see Pseudomonas
  denitrificans, see Paracoccus
  fluorescens, see Pseudomonas
  furiosus, see Pyrococcus
  nautica, see Pseudomonas
  oxalicum, see Penicillium
  pantotrophus, see Paracoccus
  putida, see Pseudomonas
  stutzeri, see Pseudomonas
  turgidula, see Pseudonitzschia
Pacific Ocean
  cadmium in, 199, 201, 213, 215
  correlation between cadmium and
    phosphate concentrations, 196,
    200, 215, 216
  equatorial, 164, 165, 167, 171,
    173–175, 181, 183
  high nutrient low chlorophyll region,
    see Ocean
  iron deposition, 160, 166
  iron enrichment experiments,
    see Iron enrichment experiments
    in oceans
  iron stress, 158
  North, 159, 166, 202, 203, 210, 211,
    215, 264
  Northwest, 160
  South, 160, 177
  subarctic, 158, 160, 167, 168, 171,
    173–175, 179–181, 183

Paracoccus
  denitrificans, 88, 89
  pantotrophus, 88, 89
Patagonia, 181
Peat bogs
  Canada, 253, 254
  Denmark, 253, 257, 259
  lead in, 241, 247, 248, 250, 252–254,
    256, 257, 262
  (methyl)mercury in, 223, 229
  Scotland, 253, 256
  Switzerland, 246, 252, 257–260
  Ukraine, 253, 254
Pelobacter sp., 109
Penicillium oxalicum, 142
Pentasulfide, 108
Peru
  lead mining, 249
Pesticides (*see also* individual names)
  lead-arsenate, 252, 262, 265
  organophosphonates, 138–143
Phanerozoic, 60, 61
Phenylmercury, 222
Phizobium ssp., 136
Phosphate
  aluminum, 134
  as nutrient, *see* Nutrients
  correlation with cadmium
    concentration, 196, 200, 201,
    215, 216
  in aerosols, *see* Aerosols
  in biology, 133–138
  in growth media, 136
  iron, 134
  lead, 243
  limitation, 183
  poly-, *see* Polyphosphate
  reduction to phosphine, 145
  transport systems, 136
  uptake kinetics, 135
Phosphine, 132, 133
  formation, 145–147
Phosphinotricin, 140
  biosynthesis, 139
  L-alanyl-L-alanyl-, 139
Phosphite, 146
  assimilation, *see* Assimilation
  dehydrogenase, *see* Dehydrogenases

[Phosphite]
  dissimilation, 144–147
  hypo-, 143
  oxidation to phosphate, 144–146
  radical, 146
Phosphoenolpyruvate, 139, 142, 146
  mutase, *see* Mutases
3-Phosphoglycerate, 209
Phospholipids, 114, 120, 121, 264
Phosphonatase, 140–143
Phosphonate(s), 138–143
  2-aminoethyl, 138, 142
  4-aminobutyl-, 142
  biosynthesis, 138–140
  degradation, 140–143
  naturally occurring, 139
Phosphonoalanine
  degradation, 142
Phosphonolipids, 138
Phosphonomycin, 139
  degradation, 140
Phosphonopyruvate, 139, 140–142, 146
  mutase, *see* Mutases
Phosphoric acid
  anhydride linkages, 134
Phosphorite, 132
Phosphorus
  and organic carbon burial, 67, 68
  biological cycling, 131–148
  chemistry of minerals, 131–133
  compounds with C–P linkages,
    138–143
  C–P linkage formation, 139
  global cycle, 133
  in bone, 132
  redox potential, *see* Redox potential
Phosphorylation
  glucose, 137
  oxidative, 4, 12, 106, 107
  substrate, 107
Photolysis, 15, 56
  hydrogen sulfide, 34
  water, 3, 33, 34
Photosynthesis (or photosynthetic), 1,
    3, 16, 17, 32, 35, 50–71
  anoxygenic, 12, 32–36, 39, 40, 54,
    58, 59
  bacterial, *see* Bacteria

[Photosynthesis (or photosynthetic)]
  dark reaction, 209
  electron transfer in, *see* Electron
    transfer
  hydrogen-based, 57, 58, 71
  iron requirement, 155
  iron-based, 58, 65, 69, 71
  limiting nutrient, 134
  oxygenic, 12, 32–36, 39, 40, 54, 56,
    58, 59, 71, 77
  phytoplankton, 157
  sulfide-based, 58, 59, 70, 71
Phylogenetic domains of life, 10
Phytochelatins
  and cadmium, 206, 207, 212
  synthase, *see* Synth(et)ases
Phytoplankton
  bloom, 165, 166, 172–174,
    176–179
  cadmium as nutrient, 197, 199, 200,
    205, 207, 209–215
  cadmium in, 205–209, 211, 212, 215
  cadmium/phosphorus ratios, 214
  cobalt in, 207, 215
  eukaryotic, *see* Eukaryotes
  growth, 153–184, 207, 212, 213
  iron source, 159
  iron-stressed, 157, 158
  pelagic, 154
  zinc in, 207, 210, 215
  zinc/phosphorus ratios, 214
Pigments
  cadmium in, 197
Pike
  (methyl)mercury in, 230
  northern, 230
PIXE, *see* Proton-induced X-ray
    emission
Planet(s) (*see also* individual names)
  extrasolar, 50
  habitability, 50
Planctomycetales, 78
Plankton, 2
  marine, 50
  microzoo-, 174
  phyto-, *see* Phytoplankton
  pico-, 174
  zoo-, *see* Zooplankton

Plant (*see also* individual names *and*
    species)
  cadmium in, 205, 206
  nitrate reduction, 77
  nutrition, 136
  phosphate uptake, 134, 136
  photosynthesis, 50, 54
  symbiosis with fungi, 136
Poland
  lead mining, 248
Polar regions (*see also* Antarctic *and*
    Arctic)
  ice, 241, 244, 245, 255, 259–262
  lead in, 241, 255, 262
  mercury in, 224–226
Polyphosphate
  degradation, 137, 138
  synthesis, *see* Synthesis
Polysaccharide(s), 36
  matrix, 35
Polytrichium formosum, 256
Plastocyanin, 88
Polysulfide
  cadmium compound, 201
  reductase, *see* Reductases
Polysulfide sulfur
  electron transport in respiration,
    109–123
  mechanism of respiration, 122
  proton translocation in respiration,
    119–123
  redox potential, *see* Redox potential
  reduction, 107, 113, 115, 118, 120,
    124, 125
  respiration of bacteria, 107–127
  transferase, 112, 124–127
Primary production, 158, 165
  cadmium as nutrient, 197
  iron limitation, 170–180
Prodictium, 106
Prokaryotes (*see also* individual names)
  anaerobic, 106
  cadmium in, 205
  iron uptake, 155, 165
  nitrogen cycle, 77
  nitrogen fixation, 79
Propionate, 36
  fermentation, *see* Fermentation

Prorocentrum micans, 157
Protein(s) (*see also* individual names)
  Fe-, *see* Fe-protein
  MoFe-, *see* MoFe-protein
Protein Data Bank (files of protein
    structures)
  nitrite reductase, 87
  nitrous oxide reductase, 91
Proteoliposomes, 110, 113, 120, 121,
    123, 125
Proterozoic Era, 50
Proton-induced X-ray emission, 245
Protoporphyrin IX, 95
Protozoa(n), 138
  iron uptake, 155
  release of iron-binding ligands, 160
Pseudomonas sp., 140
  aeruginosa, 88, 89, 135, 140
  aureofaciens, 88
  cepacia, 136
  fluorescens, 141
  neutica, 91
  putida, 91, 142
  stutzeri, 89–91, 144, 145, 147
Pseudonitzschia turgidula, 174
Pterins, 86
  molybdo- *see* Molybdopterin
Pyrite, 3, 4, 13, 14, 56, 60
Pyrobaculum sp., 123
  aerophilum, 90
Pyrococcus furiosus, 4, 5
Pyrodictium sp., 123
  abyssi, 124, 127
Pyroxene, 13
Pyruvate, 12
  phosphoenol-, 139, 142, 146
  phosphono-, 139–142, 146

Q

Quartz (*see also* Silica), 13
Quinol, 86
  mena-, 90
  oxidase, *see* Oxidases
Quinones (*see also* individual names),
    85, 124
  mena-, *see* Menaquinones

[Quinones (*see also* individual names)]
  naphtho-, 113, 114
  reduction, 116–119
  structures, 111

R

R.
  albus, see Ruminococcus
  huakuii, see Rhizobium
Radicals (*see also* individual names)
  hydroxyl, 142, 228, 251
  in C–P cleaving enzymes, 142, 143
  phosphite, 146
  phosphonyl, 142
Radiolysis
  dihydrogen generation, 16
Radon
  ^{222}Rn, 241
Rate constants
  for nitrite reduction, 89
Reactive gaseous mercury, 224, 226
Reactive oxygen species
  detoxification, *see* Detoxification
Redox cycles
  biological, 53–59
Redox potential (of)
  Hg(I)/Hg(0), 227
  hydrogenase, 117, 118
  menaquinone, 119, 122
  NAD^+/NADH, 144
  nitrate reduction, 145
  nitrate, 85
  nitric oxide, 85, 87
  nitrite, 85, 92
  nitrous oxide, 85, 89, 90
  phosphite oxidation, 144
  polysulfide sulfur, 108
  substrates, 106, 107
  sulfate reduction, 145
Reductases
  adenosine phosphosulfate, 145
  dimethylsulfoxide, 85, 114, 124
  dinitrogenase, 81
  fumarate, 113
  molybdooxido-, 114
  nitrate, *see* Nitrate reductase

[Reductases]
  nitric oxide, 6, 78, 89, 90
  nitrite, 86–89, 92, 154
  nitrous oxide, 6, 78, 90–92
  oxido-, *see* Oxidoreductases
  polysulfide, 85, 110–115, 119–121,
    124, 125, 127
Reduction potential, *see* Redox
  potential
Respiration, 1, 3, 85–95
  anaerobic, 106, 109, 115, 118
  fumarate, 109, 110, 113, 115,
    117–119, 121
  iron requirement, 155
  polysulfide sulfur, *see* Polysulfide
    sulfur
  sulfate, 57
  sulfur, 106, 107, 123, 124
Rhizobium sp., 79, 142
  huakuii, 140
Rhizosphere, 136
Riboflavin
  5'-monophosphate, *see* Flavin
    mononucleotide
Ribulose-1,5-bisphosphate
  carboxylase/oxygenase, 209
Rice
  cadmium contaminated, 205
River(s) (*see also* Water *and* individual
    names)
  cadmium in, 198, 199, 201, 205
  Dordogne, 203
  Garonne, 198, 203
  iron in, 160, 162, 163
  Jinzu, 196, 205
  Lot, 198
  mercury in, 227, 231
  phosphate in, 134
  Rio Tinto, 246
Rocks (containing) (*see also* Minerals
    and individual names)
  arc, 64
  Canadian shield, 14
  crustal, 14
  East European platform, 14
  Fennoscandian shield, 14
  hard, 51, 52, 63, 68, 69
  igneous, 53, 60, 63, 69

[Rocks (containing) (*see also* Minerals
    and individual names)]
  iron, 13, 16, 53, 161
  lead in, 241, 243, 251, 262
  sedimentary, *see* Sediments
  soft, 52
  thorium, 14, 245
  ultramafic, 17
  uranium, 14, 245
  volcanic, 62
  Witwatersrand Basin, 14
Roman times, 247
Ruminococcus albus, 22

S

S.
  deleyianum, see Sulfurospirillum
  fradiae, see Streptomyces
  marcescens, see Serratia
  typhimurium, see Salmonella
  viridochromogenes, see
    Streptomyces
Saccharides
  fermentation, *see* Fermentation
  poly-, *see* Polysaccharides
Sahel region, 161
Salmonella typhimurium, 140
Sandstone
  lead in, 243
Sargasso Sea
  cadmium in 201, 202
Scotland, 254
  peat bogs, 253, 256
Sea anemone (*see also*
    individual names), 138
Seawater (containing) (*see also* Ocean
    and Water)
  cadmium, 199–204, 209–211, 216
  carbon dioxide, 209, 210
  iron, 155, 156, 158–162
  lead, 255, 264
  phosphate, 134
  speciation of cadmium, 203
  sulfate as oxidant, 61, 62
  sulfide-rich, 58
Secondary ion mass spectrometry, 245

Sediment(s) (containing), 3, 51, 52, 133
  ancient, 55, 56
  anoxic, 20, 109, 145, 228, 243
  cadmium, 197, 199, 201
  carbon reservoirs, 53, 60, 62,
    64, 65
  carbonate, 64–66, 70
  deep sea, 16
  freshwater, 133
  from Cape Lookout Bight (NC), 21
  in estuaries, 247
  iron, 159, 160
  lake, 197
  lead, 241–248, 255, 256,
    259, 262
  marine, 24, 132, 133, 163, 181,
    183, 245
  mercury, 223, 228, 231–233
  methanogenic, 28, 30
  mobilization of iron, 163, 164
  organic carbon, 66, 68
  phosphate, 134
  pore water, 26, 163, 164
  sulfate, 52
  sulfate-reducing, 28, 31
  sulfur reservoir, 60
Sensitive high resolution ion
    microprobe, 247
Serpentinites, 65, 69
Serpentinization, 13–15, 17, 39, 68
Serratia marcescens, 136
Sewage
  sludge, 146, 199
  treatment, 145, 147
Shale
  lead in, 243
SHRIMP, *see* Sensitive high resolution
    ion microprobe
Siderophores (*see also*
    individual names)
  degradation, 155
  hydroxamate, 156
  iron(III), 2
  transport systems, 155
Sidon
  Phoenician city, 247, 248
Silesia region
  lead mining, 248

Silicate(s), 251, 254
  alumino-, 158
  as nutrient, *see* Nutrients
  lead in, 243, 244
  limitation in ocean, 179, 183
  magnesium, 65
  minerals, 13, 14, 243
  weathering, 70
Silver mining, *see* Mining
SIMS, *see* Secondary ion mass
    spectrometry
Smelting
  cadmium emission, 197, 199
  lead, 240, 254
Snow
  cadmium in, 197, 198
  Greenland, *see* Greenland
  lead in, 262
  mercury deposition, 224
Soil (containing)
  aerobic, 79
  agricultural, 137, 204, 243, 265
  cadmium, 197–199, 204
  iron, 161
  lead, 241–243, 252–255, 265
  over-fertilized, 137
  phosphate, 136
South Africa
  Cape Province, 264
  lead exposure of children, 264
  waste water treatment, 138
  Witwatersrand Basin, 14
South America, 177, 181, 249
Southern hemisphere
  dust, 181
Southern Ocean
  dust input during glacial periods,
    181, 182
  high nutrient low chlorophyll
    regions, *see* Ocean
  iron enrichment experiments, *see* Iron
    enrichment experiments in oceans
  iron in, 161, 164–167, 171, 172, 183
Spain, 246, 256
Speciation (of)
  cadmium, 201–204
  carbon systems, 15
  iron, 159

[Speciation (of)]
mercury, 222, 223, 233
nitrogen systems, 15
sulfur systems, 15
Sphagnum, 256
Sphalerite, 197
Steel production, 250, 251
Stratosphere
ozone in, 77, 97
Streptomyces
fradiae, 139
thermoautotrophicus, 84
viridochromogenes, 139, 140
Stromatolites, 12, 39
Stygioglobus, 108, 123
Sulfate, 2, 4, 24, 59
(bacterial) reduction, 3, 26–31, 33, 34, 38, 61, 62, 68, 69, 92, 144, 145, 154
as oxidant in seawater, 61, 67
in ocean, 52, 60, 61, 68, 69, 183
in sediments, 52
in the biogeochemical cycle of sulfur, 107, 127
redox potential, *see* Redox potential
reducers, *see* Bacteria
reservoir, 60, 69
respiration, *see* Respiration
thio-, *see* Thiosulfate
Sulfide
cadmium, 201
iron, 4, 13, 57, 170
lead, 243
mercury, 228
mineral, *see* Minerals
photosynthesis, *see* Photosynthesis
tetra-, 108
Sulfite
in the biogeochemical cycle of sulfur, 107, 127
reaction with mercury, 225, 226
reduction, 34
Sulfur (different oxidation states)
chemistry, 107, 108
cycle, 3, 59–61, 105–127
in biology, 106, 107
isotopes, 56, 60
reduction, 109, 124
respiration, *see* Respiration

Sulfurospirillum sp., 106, 109
deleyianum, 92, 93, 95
Superoxide, 154
Superoxide dismutase
replacement of zinc by cadmium, 209
Surface water (containing) (*see also* Water)
entrophic, 133, 137
cadmium, 201, 209–211
iron, 165, 169, 170, 182
phosphate, 135
Swamps
methane in, 58, 146
Sweden
lead concentrations in soil, 253, 255
methylmercury in wetlands, 229
Switzerland, 253, 255–258
peat bogs, *see* Peat bogs
Symbiosis
methanogens, 12
Synechococcus sp., 36, 136, 174, 208
Synthesis
amino acid, 96
ammonia, 78–80
ATP, 4, 54, 85, 94, 106, 107, 109, 110, 123, 144
polyphosphate, 137, 138
Synth(et)ases
ATP, 4, 127
glutamine, 139
phytochelatin, 206

T

T.
cruzi, see Trypanosoma
fluviatilis, see Thalassiosira
oceanica, see Thalassiosira
pseudonana, see Thalassiosira
pyriformis, see Tetrahymena
weisflogii, see Thalassiosira
Terrestrial Planet Finder, 50
Tetraalkyllead, 249, 251
Tetraethyllead, 249, 251
Tetramethyllead, 249, 251
Tetrahymena pyriformis, 139
Tetraphenylphosphonium, 109, 111, 113

Tetrasulfide, 108
Thalassiosira
 fluviatilis, 135
 oceanica, 157
 pseudonana, 157, 209
 weisflogii, 205, 206, 208–210
Thiosulfate
 reduction, 33, 34
Thermal ionization-mass spectrometry,
 245–247, 257
Thermoplasma, 108
Thermoproteus, 106, 123
Thorium (different oxidation states)
 ^{232}Th radioactive decay, 243
 decay series, 247
 in rocks, *see* Rocks
 isotopes, 56
TIMS, *see* Thermal ionization mass
 spectrometry
Tobacco smoke, 204
Total reflection X-ray spectrometry, 244
Toxicity (of)
 cadmium, 196, 197, 204, 205, 208
 lead, 241
 mercury, 222
 neuro-, *see* Neurotoxicity
Transferases (*see also* individual names)
 polysulfide sulfur, 112, 124–127
Trees
 fire-resistant, 56
 lead in bark pockets, 256
 Ohia, 56
Trichodesmium sp., 79, 179
Troposphere, 65
 dinitrogen monoxide in, 77
Trypanosoma cruzi, 139

U

Ukraine
 peat bogs, 253, 254
United States (*see also* individual states)
 body burden of lead, 266
 Environmental Protection Agency,
 204, 222, 250
 lead exposure of children, 263–265
 mercury in, 224

Uranium (different oxidation states)
 ^{235}U radioactive decay, 243
 ^{238}U decay, 241, 243
 decay series, 247
 in rocks, *see* Rocks
 isotopes, 56
 U/Th ratio, 57

V

Variovorax, 142
Venezuela
 Cariaco Basin, 216
Vesta, 64
Vitamin K, 119
Vivianite, 132
Volcanoes (or volcanic)
 arc, 63, 65
 dihydrogen flux, 58
 emissions, 39, 146, 197
 gases, 14, 39, 59, 62–64
 hot springs, 109, 123
 lead emission, 250
 mercury release, 224
Voltammetry
 anodic stripping, 202
Vosges mountains
 lead mining, 248
Vostoc ice core, 181, 182

W

W.
 succinogenes, *see Wolinella*
Waste
 from a zinc mine, 205
 incineration, 197, 250, 251, 265
 lead in, 245, 250
 municipal, 250
Wastewater
 removal of ammonia, 78
 treatment, 137, 138
Water
 drinking, *see* Drinking water
 ground, *see* Groundwater
 lead in, 255
 photolysis, 3

[Water]
  reduction, 13
  sea-, *see* Seawater
  surface, *see* Surface water
  waste-, *see* Wastewater
Wavellite, 132
Wetlands
  (methyl)mercury in, 229, 232, 233
  Allequash Creek, 229
WHO, *see* World Health Organization
Wine
  lead in, 252, 262, 266
Wisconsin
  methylmercury in wetlands, 229
Wolinella succinogenes, 23, 90, 93–95,
    106–110, 115–125, 127
  from bovine rumen, 109–123
World Health Organization
  recommendation for lead
    concentrations, 267
Wurtzite, 197

X

X-ray fluorescence spectrometry,
  244, 255

X-ray spectroscopy
  near-edge, 209

Y

Yeast (*see also* individual names)
  cadmium in, 207

Z

Zinc(II) (in), 204
  dissolved, 214, 215
  mining, *see* Mining
  ocean, *see* Ocean
  ores, *see* Ores
  phytoplankton, *see* Phytoplankton
  regulation, 205
  substitution by cadmium, 204,
    205, 207
Zooplankton, 165, 170
  iron in, 159
  iron recycling, 167
  methylmercury in, 230
  micro-, 171, 177, 178
  release of iron-binding ligands, 160